Applications of the Surface Renewal Model of Mass Transfer

Applications of the Surface Renewal Model of Mass Transfer

Hengshuo Huang
Assistant Research Fellow
School of Materials Science and Engineering
Peking University, Beijing, China

Siddharth G. Chatterjee
Associate Professor Emeritus, Department of Chemical Engineering
SUNY College of Environmental Science and Forestry
Syracuse, New York, USA

WILEY

Published by John Wiley & Sons, Inc., Hoboken, New Jersey.

Library of Congress Cataloging-in-Publication Data

Names: Huang, Hengshuo author | Chatterjee, Siddharth G. author
Title: Applications of the surface renewal model of mass transfer /
Hengshuo Huang, Siddharth G. Chatterjee.
Description: Hoboken, New Jersey : Wiley, [2026] | Includes index.
Identifiers: LCCN 2025030991 (print) | LCCN 2025030992 (ebook) | ISBN 9781394312825 cloth | ISBN 9781394312849 adobe pdf | ISBN 9781394312832 epub
Subjects: LCSH: Mass transfer | Gases–Absorption and adsorption | Membrane separation
Classification: LCC TP156.M3 H86 2026 (print) | LCC TP156.M3 (ebook)
LC record available at https://lccn.loc.gov/2025030991
LC ebook record available at https://lccn.loc.gov/2025030992

Cover Design: Wiley
Cover Image: © filo/Getty Images

Set in 9.5/12.5pt STIXTwoText by Straive, Chennai, India

To our parents for their unwavering support and boundless love:

Dianli Huang and Jing Tie

Gourishankar Chatterjee and Jamuna Chatterjee

Contents

Preface

The work presented in this book is a result of contemplating on the surface renewal model of mass transfer for over three decades. This model was originally published by P. V. Danckwerts in 1951 for describing gas–liquid mass transfer at the surface of a turbulent liquid and has had many extensions and refinements since then. In contrast to the film model proposed by W. K. Lewis and W. G. Whitman in 1924, which pictures stagnant gas and liquid films at the interface with the dissolved gas crossing the liquid film by steady-state molecular diffusion, the surface renewal model visualizes the interface to be replenished by fresh liquid elements arriving continuously from the bulk liquid. During the residence time of a liquid element at the gas–liquid interface, which can vary from element to element, mass transfer of dissolved gas is assumed to occur in it by the mechanism of unsteady-state molecular diffusion.

Although the surface renewal model is believed to be closer to physical reality, as evidenced by many empirical studies, the film model, owing to its simpler mathematical structure, is the current standard for the design of gas–liquid contacting equipment. This book is an attempt to establish a parallel but alternative standard that is based on the surface renewal model.

After introducing the reader to the surface renewal model in Chapter 1 of the book, Chapters 2 to 7 present a systematic and rigorous exposition and application of the model to physical and chemical gas absorption whose underlying aim is the analysis and design of gas–liquid stirred tank absorbers/reactors and packed towers.

The use of the surface renewal concept to theoretically model crossflow ultrafiltration was pioneered by A. Kołtuniewicz and coworkers in the 1990s. In contrast to the film and boundary layer models of membrane filtration, the surface renewal model has the potential to describe the transfer of dissolved/suspended solids due to random hydrodynamic impulses generated at the membrane surface owing to membrane roughness or by the use of spacers or turbulence promoters, or by the deliberate introduction of controlled instabilities into the main flow, which by inducing back migration of accumulated solute molecules or particulates away from the membrane surface, can enhance permeation rates significantly. Chapters 8 to 10 present surface renewal models of crossflow ultrafiltration and microfiltration and consider both constant pressure and constant flux operation.

Some of the material in the book can be incorporated into existing courses on unit operations and mass transfer or the book could be used in a specialized one-semester course on surface renewal theory. In the latter situation, the following are suggested: (a) for advanced or senior-level undergraduate students, Chapters 1, 2, 3, 6, and 8 to 10 and (b) for graduate students, Chapters 1 to 10.

A knowledge of advanced calculus and differential equations is presumed of the reader. Some familiarity with partial differential equations would be helpful but is not mandatory. Appendix A provides some basic mathematical information that will be useful for navigating through the contents of the book.

Each chapter, except the first, has homework problems, which are of two types: qualitative and quantitative. In the former category, there belong problems where the student is asked to derive some of the key mathematical results in the book. The latter category consists of numerical problems.

This book is only a first step in the application of the surface renewal concept to the analysis and design of gas–liquid contactors and membrane filters. The field is wide open, for example, the consideration of other types of gas–liquid reactions like second order, reversible, parallel, and consecutive, taking a closer look at the exchange of dissolved/suspended solids between the bulk liquid and membrane wall because of the surface renewal mechanism, and accounting for the variation of permeate flux and cake buildup in the main flow direction of a membrane filter.

The cosmic philosophical principle underlying the surface renewal model is that of emergence, existence, and dissolution, which can be captured succinctly by the term "impermanence."

We thank Satvinder Kaur Sandhu and Summers Scholl of Wiley for their constant help and encouragement during the preparation of this book.

October 2025

Hengshuo Huang
School of Materials Science and Engineering
Peking University, Beijing, China

Siddharth G. Chatterjee
Department of Chemical Engineering
SUNY College of Environmental Science and Forestry
Syracuse, New York, USA

About the Companion Website

This book is accompanied by a companion website:

www.wiley.com/go/huang/surfacerenewalmodel

The website includes Appendices (Online) and Homework Solutions (Online).

1

The Surface Renewal Model of Mass Transfer*

1.1 Introduction

The surface of a turbulent liquid is characterized by bursting and chaotic movements of eddies that emerge from below the surface and by the presence of turbulent sweeps, upwellings, downwellings, and vortices that profoundly affect the interfacial mass-transfer process (Komori *et al.*, 1989; Banerjee, 2007; Turney and Banerjee, 2008). In 1951, Danckwerts published his classic paper on gas absorption in a turbulent liquid, in which he presented the surface renewal model of mass transfer (Danckwerts, 1951). This model visualizes the gas–liquid interface, where the absorption occurs, to be replenished by fresh liquid elements that emerge continuously from the bulk liquid. During the exposure time of an element at the gas–liquid interface, mass transfer of dissolved gas is assumed to occur in it by the mechanism of unsteady-state molecular diffusion. In contrast to the penetration model of Higbie (1935), which had assumed that all surface elements had the same residence time at the gas–liquid interface (i.e., uniform age distribution), Danckwerts derived an exponential age distribution by using the postulate that all surface elements, irrespective of their individual ages, had an equal probability of being replaced by fresh elements arriving from the bulk liquid. The only parameter that appears in this distribution is the frequency or rate of surface renewal S, which is the mean rate of production of fresh surface according to Danckwerts (1951) and which depends upon the prevailing hydrodynamic conditions. This well-known age distribution, whose experimental confirmation has been provided by Lamb *et al.* (1969) in the case of a stirred liquid and Lesage *et al.* (2002) in the case of a single liquid phase flowing through a packed-bed reactor, has been widely used in chemical engineering, and a variety of applications and extensions of the surface renewal model have appeared in the literature over the years.

*The material in this chapter has been adapted from the works of Mondal and Chatterjee (2017), Horvath and Chatterjee (2018), Huang (2023), Hasan *et al.* (2013), and Idan and Chatterjee (2015).

Applications of the Surface Renewal Model of Mass Transfer, First Edition.
Hengshuo Huang and Siddharth G. Chatterjee.

Companion website: www.wiley.com/go/huang/surfacerenewalmodel

In mass-transfer studies, the surface renewal model is believed to be a more realistic representation of interfacial mass transfer than the film model[1] since it predicts that the liquid-side mass-transfer coefficient k_L is proportional to the square root of the diffusion coefficient D of the dissolved gas in the liquid – a result that has often been confirmed experimentally. For example, Kuthan and Brož (1989) obtained experimental values of k_L for the absorption of helium, nitrogen, and propane in a liquid film of aqueous ethylene glycol flowing over a smooth wetted wall and an expanded metal sheet. For the case of the wetted wall, they found $k_L \propto D^{0.5}$, which agrees with the prediction of the surface renewal model. For the expanded metal sheet, $k_L \propto D^{0.64}$, and the film-penetration model (Dobbins, 1956; Toor and Marchello, 1958) was found to be more appropriate. According to Astarita (1967), for a flowing liquid in contact with a solid or a more viscous liquid phase, $k_L \propto D^{2/3}$. Richter and Jähne (2010) measured the transfer velocity (i.e., k_L) of five sparingly soluble gas tracers as a function of wind speed (1–10 m s^{-1}) in the Heidelberg Aeolotron and a small circular wind-wave facility. Their experiments revealed that the exponent of D varied from 2/3 to 1/2 as the water surface transitioned from smooth to rough or wavy with increasing wind speed, a finding later confirmed by the extensive experimental measurements of Krall (2013). Empirical correlations for k_L in towers filled with random packings indicate that $k_L \propto D^{1/2}$ (Onda *et al.*, 1968a; 1968b; Richardson *et al.*, 2003). All of this evidence is in sharp contrast to the prediction of the film model, according to which $k_L \propto D$.

1.2 Literature Review

Since the publication of the original surface renewal model by Danckwerts (1951), many studies have been conducted on heat- and mass-transfer phenomena near the surface of a turbulent liquid, including measurements of gas absorption rates in wetted-wall columns, stirred tanks, and packed towers. We present only a brief review of the literature, mostly drawn from chemical engineering, on the surface renewal model, although a considerable body of work has developed on it in fields like oceanography and meteorology – see, for instance, the paper by Zorzetto *et al.* (2021) and the references cited therein.

Johnson and Huang (1956) measured the dissolution rates of five different organic solids from a stationary surface into a turbulent vessel with and without baffles and proposed a dimensionless correlation for the Sherwood number in

1 The film model, proposed by Lewis and Whitman (1924), pictures stagnant gas and liquid films at the interface with the dissolved gas crossing the liquid film, whose thickness depends upon the prevailing hydrodynamics, by steady-state molecular diffusion.

which the Schmidt number exponent was set equal to 0.5 in accordance with the Danckwerts model and obtained satisfactory results. Further, they found good consistency between the experimental and theoretical values of the Schmidt number exponent, which indicated that the surface renewal model could also be used to describe mass transport from a plane solid surface into a turbulent liquid. Perlmutter (1961) likened the Higbie and Danckwerts age distributions to residence time distributions in plug-flow and well-mixed vessels and derived a variety of surface renewal models based on this analogy. The models included multiple capacitances, dead time, and interfacial nonequilibrium effects; however, no estimates of the model parameters drawn from model comparisons with experimental data were presented.

An extension of the surface renewal model to describe mass transfer from a turbulent fluid to a solid wall was made by Harriott (1962), who assumed randomly arriving eddies at arbitrary intervals to approach random distances from the wall, thereby removing accumulated solute in its vicinity. In the intervals between the arrival of eddies, mass transfer of the solute was assumed to occur by molecular diffusion alone. The gamma distribution was chosen to fit the random sequences of distances and times, and it was found that the random distance model was consistent with available data for mass transfer in pipes. Angelo *et al.* (1966) extended Higbie's penetration model to account for surface stretch (i.e., the creation of new surface) to describe mass transfer to oscillating and forming drops. In the case of the former, their model was in reasonable agreement with experimental data on the extraction efficiency for some liquid–liquid systems.

Analytical solutions for the hydrodynamic behavior of very small scales of turbulent motion in the vicinity of the surface of a turbulent liquid were obtained by Lamont and Scott (1970), who related the energy dissipation rate to the turbulent energy spectrum and the collective outcome of eddies with diverse sizes. For turbulent bubbly flow in a 5/16-in. tube, the predicted mass-transfer coefficient agreed quite well with its measured value at a Reynolds number of 10,000. However, the exponent of the Reynolds number in the theoretical equation for the mass-transfer coefficient (0.69) was higher than the experimental exponent (0.52). Chung *et al.* (1971) presented a general transient age distribution from which different age-distribution functions could be obtained by substituting different steady-state age distributions into the general equation. Using these, they developed penetration and surface renewal models and derived four different expressions for the instantaneous rate of mass transfer for different boundary conditions (see Table 1 in their paper). These models were expected to more accurately reflect physical reality compared to steady-state models, especially during the start-up period of interphase transport processes. Also, the transient models had better agreement with experimental data for large flow rates than for small ones. Babu and Narsimhan (1980) derived a transient form of the generalized gamma age-distribution function for interfacial transport phenomena in free or fixed configurations. By fitting their model to published experimental measurements of the Nusselt number in the thermal entrance region of a pipe and response measurements from a continuous tubular flow reactor, the shape factor of the distribution was determined to be five, and an empirical linear correlation was established between the fractional surface renewal rate and Reynolds number lying within the range of 20,000–55,000.

Sada *et al.* (1979) measured the rates of absorption of CO_2 into water, NaOH solution (0.2 mol L^{-1}), and glycerol solution (40 weight percent) in a stirred vessel. They found that the experimental data of the enhancement factor agreed well with predicted values obtained from the Danckwerts surface renewal model for the first two solutions. However, the data for the glycerol solution were lower than the calculated values, probably due to a decrease in the average surface renewal rate in a system with high viscosity. Their findings also suggested that both bulk turbulence and interfacial turbulence have a similar impact on the rate of gas absorption in the presence or absence of a chemical reaction. Luk and Lee (1986) proposed a pseudo-steady, two-dimensional single-eddy model in which deterministic analyses were applied to velocity or concentration fluctuation data to estimate the eddy exposure time distribution and the contribution of each eddy to the overall mass transfer. The mass-transfer coefficients for both velocity and concentration fluctuation data were measured in the proximity of the air–water interface in a free-surface unbaffled stirred tank and exhibited close alignment with experimental findings obtained from the local concentration profile. The distribution of eddy exposure time, measured at a depth of 90 μm from the air–water surface, was skewed significantly toward lower exposure times in contrast to the classical Danckwerts age-distribution function, which gives a much higher value for an exposure time of zero. Seo and Lee (1988), using the single-eddy model, obtained exposure-time distributions from velocity fluctuations measured with a hot-film anemometer in the immediate vicinity of the interface and estimated the contribution of an individual eddy to the overall mass transfer. They drew congruent conclusions with those of Luk and Lee (1986), that is, the experimental eddy exposure times were skewed toward lower values. They found that a gamma-type distribution containing an adjustable parameter enabled a more precise description of the surface age distribution compared to the Danckwerts model.

Asher and Pankow (1991) measured aqueous-phase carbon dioxide concentration fluctuation time scales at carbon dioxide–water interfaces under several turbulence conditions and used them in the surface renewal model to calculate the liquid-phase mass-transfer coefficient. The results showed a good match between experimental and estimated values for clean interfaces and high turbulence intensity. However, the model was not able to accurately capture the mass-flux values for film-covered water surfaces. Fan *et al.* (1993) developed a rigorous self-consistent surface renewal model that incorporated Markovian stochastic processes. They derived expressions for three kinds of contact-time distribution of liquid elements and the rate of transfer of molecules through the interface, demonstrating mathematically that the rate of initial transient mass transfer was greater than or equal to the rate of mass transfer at steady state. Maucci *et al.* (2001) proposed an alternative surface renewal model containing three parameters for the prediction of the instantaneous liquid–solid mass-transfer coefficient from the instantaneous velocity in liquid systems or liquid–solid fixed beds and validated the model using shape and response-time profiles. Lesage *et al.* (2002) compared theoretical predictions of the surface renewal model with experimental measurements of the velocity gradient in a fixed-bed reactor with microelectrodes inserted into the packing. They resorted to the variable-interval time-averaging method, in which two extra parameters were the local interval time-averaging length and threshold level. They found that the age-distribution function was an exponential curve, and the surface renewal model was more appropriate for turbulent-like flow. Jajuee *et al.* (2006) presented a surface renewal stretch model based on the equation of continuity, including both turbulence and convective mass transfer at liquid–liquid and gas–liquid interfaces. A simplified expression for the time-averaged overall mass-transfer coefficient was derived, and a more accurate correlation for it (based on dimensionless groups) was developed and compared against published experimental data.

Mondal and Chatterjee (2017) provided a deeper theoretical understanding of the Danckwerts surface renewal model by invoking the concept of a frequency quantum, which is analogous to the energy quantum postulate used by Planck to derive his famous formula for the energy density of black-body radiation. By assuming that surface elements of a particular age have renewal frequencies that are integral multiples of this fundamental frequency quantum and by further assuming a Boltzmann-type distribution for the renewal frequency, they showed that a population balance for these elements led to the Danckwerts surface age distribution. The basic frequency is what has been conventionally called the rate of surface renewal, S. The Higbie surface age distribution resulted if the renewal frequency distribution of such elements was assumed to be continuous.

Some studies by the physical oceanography community have reported that the age distribution of surface elements in air–sea heat and gas exchange, as measured both directly and indirectly, does not follow the (conventional) exponential age distribution of Danckwerts but rather the logarithmic normal (LN) or Chi distributions (Garbe *et al.*, 2002; Garbe *et al.*, 2004; Veron *et al.*, 2011). Assuming steady-state conditions, Mondal and Chatterjee (2017) generalized the conventional one-parameter Danckwerts age distribution to a two-parameter age distribution, which they called the generalized Danckwerts (GD) age distribution.[2] Like the two-parameter LN distribution, the GD distribution was also able to capture the bell-shaped nature of the experimentally measured surface age distribution observed in air–sea gas and heat exchange.[3] Mondal and Chatterjee (2017) used the LN and GD distributions to calculate values of the steady-state liquid-side mass-transfer coefficient for the absorption of hydrogen and oxygen in water at three different wind speeds that were used by Garbe *et al.* (2002) in their experiments on air–water heat exchange on the Heidelberg Aeolotron. They found that, for both distributions, estimates of the mass-transfer coefficient were very close to one another and comparable to the experimental values reported by Hutchinson and Sherwood (1937) who investigated the absorption of eight different pure gases at 25°C in a stirred flask containing water whose surface was exposed to the gas. Subsequently, the GD age distribution was extended to unsteady-state conditions by Horvath and Chatterjee (2018).

There is a vital aspect of the surface renewal model that has received sparse attention in the vast literature on gas absorption in liquids, where the focus has been chiefly on the rate of interphase transfer, that is, the rate of absorption at the gas–liquid interface. Thus, theoretical modeling has often concentrated on obtaining analytical and numerical approximations for the enhancement factor for different types of gas–liquid reactions in which the reaction regime in the bulk liquid is assumed a priori. However, the rate of absorption is only half of the story since, due to the surface renewal mechanism, there is a perennial exchange of liquid elements between the bulk liquid and gas–liquid interface with a consequent net convective transport of dissolved gas from the interface to the bulk liquid, a phenomenon which has no parallel in the film model. In the case of physical absorption, except at steady state, these two rates are not the same because of the accumulation of absorbed gas in the surface elements (Mondal and Chatterjee, 2017; Horvath and Chatterjee, 2018; Huang and Chatterjee, 2021), whereas for absorption with chemical reaction, they are not equal even under steady-state conditions, as theoretically shown by Chatterjee (2010a) and Huang and Chatterjee (2023) for the case of a first-order reaction between the dissolved gas and a liquid-phase reagent. This is due to the consumption of dissolved gas by the chemical reaction occurring in the surface elements. We note that it is the rate of transfer of dissolved gas to the bulk liquid and not the rate of absorption at the gas–liquid interface that forms a part of the mass balance for dissolved gas in the bulk liquid, and without such consideration, rational analysis and design of gas–liquid contacting equipment are not possible.

---

2 This distribution is essentially the same as the gamma-type distribution proposed earlier by Bullin and Dukler (1972), who, however, did not provide its theoretical basis unlike Mondal and Chatterjee (2017).

3 The multiple-capacitance model presented by Perlmutter (1961) is also able to capture this bell-shaped character; see eq. 15 and Figure 1 in his work.

As noted by Chatterjee (2010a), the conceptual error of conflating the rate of absorption with the rate of dissolved-gas transfer has been made by several investigators and even by Danckwerts himself in Chapter 8 of his well-known book on gas–liquid reactions (Danckwerts, 1970), as can be seen in his general mass-balance equations for the dissolved gas and a liquid-phase reactant in a packed column, although he presented an extensive discussion of this problem in the context of the film model in Chapter 6 of this book. To our knowledge, Loffler and Merchuk (1972) and Merchuk (1985), who used the film-penetration model of mass transfer, were the first to provide the general form of the equation for the rate of dissolved-gas transfer to the bulk liquid for the case of absorption with a first-order reaction in a continuous-flow stirred-tank reactor, assuming no gas-phase resistance to mass transfer. The case of a finite gas-phase mass-transfer resistance was considered by Chatterjee (2010a) who, using the surface renewal and film-penetration models, derived analytical expressions for the dissolved-gas transfer rate for the case of absorption with a first-order reaction under steady-state conditions. This development made possible an analysis with the surface renewal model of a gas–liquid reactor in which the dissolved gas is consumed by a first-order reaction (Chatterjee, 2010a; Huang and Chatterjee, 2023), which analysis has been conventionally performed with the film model (Danckwerts, 1970; Westerterp *et al.*, 1984).[4]

The surface renewal concept has also been used in a number of publications to theoretically model crossflow constant pressure/flux ultrafiltration and microfiltration. Examples are the works of Kołtuniewicz (1992), Koltuniewicz and Noworyta (1994), Koltuniewicz and Noworyta (1995), Constenla and Lozano (1996), Arnot *et al.* (2000), Chatterjee (2010b), Sarkar *et al.* (2011), Hasan *et al.* (2013), Jiang and Chatterjee (2015), Zhang and Chatterjee (2015), Idan and Chatterjee (2015), and Dąbkowska-Susfał *et al.* (2024). In contrast to the film (e.g., gel polarization) and boundary layer models of membrane filtration, the surface renewal model has the potential to describe the transfer of dissolved/suspended solids due to random hydrodynamic impulses generated at the membrane surface owing to membrane roughness or by the use of spacers or turbulence promoters. Koltuniewicz and Noworyta (1994) list three types of forces that influence mass transport in the region near the membrane wall: (1) inertial forces caused by membrane roughness, (2) drag forces that can stimulate chaotic movements of fluid elements in both laminar and turbulent flow, and (3) lifting forces caused by the "lateral migration" of colloidal particles and fluid elements that lead to the phenomenon of "flux paradox," that is, high permeate flux. Further, flow

4 For the film model, the rate of dissolved-gas transfer to the bulk liquid can be obtained in a straightforward fashion by evaluating the concentration gradient of the dissolved gas at the inner edge of the liquid film either analytically or numerically. However, it is not elementary to obtain this rate in the case of the more sophisticated surface renewal model as will be seen later.

instabilities, when deliberately introduced into the main flow (e.g., by means of controlled centrifugal instabilities called Dean vortices), induce back migration of accumulated solute molecules or particulates away from the membrane surface, thereby significantly enhancing permeation rates. For example, flux improvements of up to 43% were reported by Mallubhotla and Belfort (1997) in the crossflow microfiltration of particulate suspensions, and Gehlert *et al.* (1998) observed enhancements of 30–120% in the ultrafiltration and microfiltration of relatively low-concentrated baker's yeast suspensions and 30–90% in the microfiltration of relatively concentrated beer yeast. The flux enhancement was found to increase with the feed flow rate and transmembrane pressure drop and decrease with increasing feed concentration of suspended matter (Mallubhotla and Belfort, 1997). It is not possible to explain such a phenomenon by the film or boundary layer models of crossflow membrane filtration.

Finally, the parameter S of the surface renewal model has been found to increase with the velocity of the main flow (Kołtuniewicz, 1992; Koltuniewicz and Noworyta, 1994; Koltuniewicz and Noworyta, 1995) and can be thought of as a "scouring" term that represents the removal of deposited material from the membrane wall (Arnot *et al.*, 2000), which depends upon the level of flow instability.

References

Angelo, J.B., Lightfoot, E.N. and Howard, D.W. (1966). "Generalization of the penetration theory for surface stretch: Application to forming and oscillating drops," *AIChE Journal*, 12(4), pp. 751–760. Available at: http://dx.doi.org/10.1002/aic.690120423.

Arnot, T.C., Field, R.W. and Koltuniewicz, A.B. (2000). "Cross-flow and dead-end microfiltration of oily-water emulsions," *Journal of Membrane Science*, 169(1), pp. 1–15. Available at: http://dx.doi.org/10.1016/S0376-7388(99)00321-X.

Asher, W.E. and Pankow, J.F. (1991). "Prediction of gas/water mass transport coefficients by a surface renewal model," *Environmental Science & Technology*, 25(7), pp. 1294–1300. Available at: http://dx.doi.org/10.1021/es00019a011.

Astarita, G. (1967). *Mass transfer with chemical reaction*. Amsterdam: Elsevier.

Babu, D.R. and Narsimhan, G. (1980). "A surface renewal model for unsteady-state transfer processes," *The Chemical Engineering Journal*, 20(3), pp. 169–175. Available at: http://dx.doi.org/10.1016/0300-9467(80)80001-3.

Banerjee, S. (2007). "Modeling of interphase turbulent transport processes," *Industrial & Engineering Chemistry Research*, 46(10), pp. 3063–3068. Available at: http://dx.doi.org/10.1021/ie061097m.

Bullin, J.A. and Dukler, A.E. (1972). "Random eddy models for surface renewal: formulation as a stochastic process," *Chemical Engineering Science*, 27(2), pp. 439–442. Available at: http://dx.doi.org/10.1016/0009-2509%5b72%5d85081-4.

Chatterjee, S.G. (2010a). "A theoretical analysis of dissolved-gas transfer to the bulk liquid in gas absorption with a first-order reaction," *Indian Chemical Engineer*, 51(3), pp. 177–193. Available at: http://dx.doi.org/10.1080/00194500903315476.

Chatterjee, S.G. (2010b). "On the use of the surface-renewal concept to describe cross-flow ultrafiltration," *Indian Chemical Engineer*, 52(3), pp. 179–193. Available at: http://dx.doi.org/10.1080/00194506.2010.547927.

Chung, B.T.F., Fan, L.T. and Hwang, C.L. (1971). "Surface renewal and penetration models in the transient state," *AIChE Journal*, 17(1), pp. 154–160. Available at: http://dx.doi.org/10.1002/aic.690170131.

Constenla, D.T. and Lozano, J.E. (1996). "Predicting stationary permeate flux in the ultrafiltration of apple juice," *Lebensmittel-Wissenschaft & Technologie*, 29, pp. 587–592.

Dąbkowska-Susfał, K., Półgrabska, A., Sobieszuk, P. and Kołtuniewicz, A.B. (2024). "Mathematical modeling of the enzymatic hydrolysis of lignocellulosic waste in membrane bioreactor considering transport phenomena," *Chemical Engineering Research and Design*, 208, pp. 656–665. Available at: http://dx.doi.org/10.1016/j.cherd.2024.07.034.

Danckwerts, P.V. (1951). "Significance of liquid-film coefficients in gas absorption," *Industrial & Engineering Chemistry*, 43(6), pp. 1460–1467. Available at: http://dx.doi.org/10.1021/ie50498a055.

Danckwerts, P.V. (1970). *Gas-liquid reactions*. New York: McGraw-Hill.

Dobbins, W.E. (1956). *Nature of the oxygen transfer coefficient in aeration systems in biological treatment of sewage and industrial wastes*. New York: Reinhold (as cited by Metzger and Dobbins, 1967).

Fan, L.T., Shen, B.C. and Chou, S.T. (1993). "The surface-renewal theory of interphase transport: a stochastic treatment," *Chemical Engineering Science*, 48(23), pp. 3971–3982. Available at: http://dx.doi.org/10.1016/0009-2509(93)80376-2.

Garbe, C.S., Jähne, B. and Haußecker, H. (2002). "Measuring the sea surface heat flux and probability distribution of surface renewal events," in *Gas transfer at water surfaces* (eds. Donelan, M.A., Drennan, W.M., Saltzman, E.S. and Anninkhof, R.W.). Washington, DC: American Geophysical Union, pp. 109–114.

Garbe, C.S., Schimpf, U. and Jähne, B. (2004). "A surface renewal model to analyze infrared image sequences of the ocean surface for the study of air-sea heat and gas exchange," *Journal of Geophysical Research: Oceans*, 109(C8). Available at: http://dx.doi.org/10.1029/2003JC001802.

Gehlert, G., Luque, S. and Belfort, G. (1998). "Comparison of ultra- and microfiltration in the presence and absence of secondary flow with polysaccharides, proteins, and yeast suspensions," *Biotechnology Progress*, 14(6), pp. 931–942. Available at: http://dx.doi.org/10.1021/bp980101p.

Harriott, P. (1962). "A random eddy modification of the penetration theory," *Chemical Engineering Science*, 17(3), pp. 149–154. Available at: http://dx.doi.org/10.1016/0009-2509(62)80026-8.

Hasan, A., Peluso, C.R., Hull, T.S., Fieschko, J. and Chatterjee, S.G. (2013). "A surface-renewal model of cross-flow microfiltration," *Brazilian Journal of Chemical Engineering*, 30(1), pp. 167–186. Available at: http://dx.doi.org/10.1590/S0104-66322013000100019.

Higbie, R. (1935). "The rate of absorption of a pure gas into a still liquid during short periods of exposure," *Transactions of the American Institute of Chemical Engineers*, 35, pp. 365–389.

Horvath, I.R. and Chatterjee, S.G. (2018). "A surface renewal model for unsteady-state mass transfer using the generalized Danckwerts age distribution function," *Royal Society Open Science*, 5(5), 172423. Available at: http://dx.doi.org/10.1098/rsos.172423.

Huang, H. (2023). *An integral theoretical view of physical and chemical gas absorption*, PhD dissertation. Syracuse, New York: SUNY College of Environmental Science and Forestry.

Huang, H. and Chatterjee, S.G. (2023). "A rigorous surface renewal model for transient gas absorption with first-order chemical reaction in a stirred liquid," *Chemical Engineering Science*, 268, 118393. Available at: http://dx.doi.org/10.1016/j.ces.2022.118393.

Huang, H. and Chatterjee, S.G. (2021). "Transient physical gas absorption in a stirred liquid using the surface renewal model of mass transfer," *Chemical Engineering Science*, 234, 116449. Available at: http://dx.doi.org/10.1016/j.ces.2021.116449.

Hutchinson, M.H. and Sherwood, T.K. (1937). "Liquid film in gas absorption," *Industrial & Engineering Chemistry*, 29(7), pp. 836–840. Available at: http://dx.doi.org/10.1021/ie50331a023.

Idan, G. and Chatterjee, S.G. (2015). "Application of a surface-renewal model to permeate-flux data of constant-pressure cross-flow microfiltration with Dean vortices," *Brazilian Journal of Chemical Engineering*, 32(2), pp. 609–627. Available at: http://dx.doi.org/10.1590/0104-6632.20150322s00003417.

Jajuee, B., Margaritis, A., Karamanev, D. and Bergougnou, M.A. (2006). "Application of surface-renewal-stretch model for interface mass transfer," *Chemical Engineering Science*, 61(12), pp. 3917–3929. Available at: http://dx.doi.org/10.1016/j.ces.2006.01.026.

Jiang, S. and Chatterjee, S.G. (2015). "A surface-renewal model for constant flux cross-flow microfiltration," *Journal of Applied Polymer Science*, 132(21). Available at: http://dx.doi.org/10.1002/app.41778.

Johnson, A.I. and Huang, C. (1956). "Mass transfer studies in an agitated vessel," *AIChE Journal*, 2(3), pp. 412–419. Available at: http://dx.doi.org/10.1002/aic.690020322.

Kołtuniewicz, A. (1992). "Predicting permeate flux in ultrafiltration on the basis of surface renewal concept," *Journal of Membrane Science*, 68(1–2), pp. 107–118. Available at: http://dx.doi.org/10.1016/0376-7388(92)80153-B.

Koltuniewicz, A. and Noworyta, A. (1994). "Dynamic properties of ultrafiltration systems in light of the surface renewal theory," *Industrial & Engineering Chemistry Research*, 33(7), pp. 1771–1779. Available at: http://dx.doi.org/10.1021/ie00031a016

Koltuniewicz, A. and Noworyta, A. (1995). "Method of yield evaluation for pressure-driven membrane processes," *Chemical Engineering Journal*, 58, pp. 175–182.

Komori, S., Murakami, Y. and Ueda, H. (1989). "The relationship between surface-renewal and bursting motions in an open-channel flow," *Journal of Fluid Mechanics*, 203, pp. 103–123. Available at: http://dx.doi.org/10.1017/S0022112089001394.

Krall, K. (2013). *Laboratory investigations of air-sea gas transfer under a wide range of water surface conditions, doctoral dissertation.* Heidelberg: University of Heidelberg. Available at: http: //www.ub.uni-heidelberg.de/archiv/14392.

Kuthan, K. and Brož, Z. (1989). "Mass transfer in liquid films during absorption Part III. Dependence of the liquid-side mass transfer coefficient on the molecular diffusivity of gases at high values of the Schmidt number," *Chemical Engineering and Processing: Process Intensification*, 25(2), pp. 75–84. Available at: http://dx.doi.org/10.1016/0255-2701(89)80033-9.

Lamb, W.B., Springer, T.G. and Pigford, R.L. (1969). "An interface impedance bridge," *Industrial & Engineering Chemistry Fundamentals*, 8(4), pp. 823–827. Available at: http://dx.doi.org/10.1021/i160032a039.

Lamont, J.C. and Scott, D.S. (1970). "An eddy cell model of mass transfer into the surface of a turbulent liquid," *AIChE Journal*, 16(4), pp. 513–519. Available at: http://dx.doi.org/10.1002/aic.690160403.

Lesage, F., Midoux, N. and Latifi, A.M. (2002). "Momentum transfer in a fixed-bed reactor described by the surface renewal model," *Chemical Engineering Science*, 57(24), pp. 5115–5122. Available at: http://dx.doi.org/10.1016/S0009-2509(02)00426-8.

Lewis, W.K. and Whitman, W.G. (1924). "Principles of gas absorption," *Industrial & Engineering Chemistry*, 16(12), pp. 1215–1220. Available at: http://dx.doi.org/10.1021/ie50180a002.

Loffler, D.G. and Merchuk, J.C. (1972). "Steady and unsteady state absorption with first order chemical reaction in a CFSTR, evaluated with film-penetration model," *Latin American Journal of Chemical Engineering and Applied Chemistry*, 2, pp. 175–183.

Luk, S. and Lee, Y.H. (1986). "Mass transfer in eddies close to air-water interface," *AIChE Journal*, 32(9), pp. 1546–1554. Available at: http://dx.doi.org/10.1002/aic.690320915.

Mallubhotla, H. and Belfort, G. (1997). "Flux enhancement during Dean vortex microfiltration. 8. Further diagnostics," *Journal of Membrane Science*, 125(1), pp. 75–91. Available at: http://dx.doi.org/10.1016/S0376-7388(96)00125-1.

Maucci, E., Briens, C.L., Martinuzzi, R.J. and Wild, G. (2001). "Modeling of transient particle–liquid mass transfer in liquid and liquid–solid systems," *Chemical Engineering Science*, 56(15), pp. 4555–4570. Available at: http://dx.doi.org/10.1016/S0009-2509(00)00461-9.

Merchuk, J.C. (1985). "On the use of the film model for analysis of gas-liquid reactions in CSTR," *Chemical Engineering Communications*, 39(1–6), pp. 389–390. Available at: http://dx.doi.org/10.1080/00986448508911685.

Metzger, I. and Dobbins, W.E. (1967). "Role of fluid properties in gas transfer," *Environmental Science & Technology*, 1(1), pp. 57–65. Available at: http://dx.doi.org/10.1021/es60001a006.

Mondal, C. and Chatterjee, S.G. (2017). "A frequency quantum interpretation of the surface renewal model of mass transfer," *Royal Society Open Science*, 4(7), 170103. Available at: http://dx.doi.org/10.1098/rsos.170103.

Onda, K., Sada, E. and Takeuchi, H. (1968a). "Gas absorption with chemical reaction in packed columns," *Journal of Chemical Engineering of Japan*, 1(1), pp. 62–66. Available at: http://dx.doi.org/10.1252/jcej.1.62.

Onda, K., Takeuchi, H. and Okumoto, Y. (1968b). "Mass transfer coefficients between gas and liquid phases in packed columns," *Journal of Chemical Engineering of Japan*, 1(1), pp. 56–62. Available at: http://dx.doi.org/10.1252/jcej.1.56.

Perlmutter, D.D. (1961). "Surface-renewal models in mass transfer," *Chemical Engineering Science*, 16(3–4), pp. 287–296. Available at: http://dx.doi.org/10.1016/0009-2509(61)80039-0.

Richter, K. and Jähne, B. (2010). "A laboratory study of the Schmidt number dependency of air-water gas transfer," in *Gas transfer at water surfaces* (eds. Komori, S., McGillis, W. and Kurose, R.). Kyoto: Kyoto University Press, pp.322–332.

Richardson, J.F., Harker, J.H. and Backhurst, J.R. (2003). *Coulson and Richardson's chemical engineering*. Fifth edition. New Delhi: Elsevier.

Sada, E., Katoh, S., Yoshii, H. and Ban, Y. (1979). "Rates of gas absorption with interfacial turbulence caused by micro-stirrers," *The Canadian Journal of Chemical Engineering*, 57(6), pp. 704–706. Available at: http://dx.doi.org/10.1002/cjce.5450570609.

Sarkar, D., Datta, D., Sen, D. and Bhattacharjee, C. (2011). "Simulation of continuous stirred rotating disk-membrane module: An approach based on surface renewal theory," *Chemical Engineering Science*, 66(12), pp. 2554–2567. Available at: http://dx.doi.org/10.1016/j.ces.2011.02.056.

Seo, Y.G. and Lee, W.K. (1988). "Single-eddy model for random surface renewal," *Chemical Engineering Science*, 43(6), pp. 1395–1402. Available at: http://dx.doi.org/10.1016/0009-2509(88)85112-1.

Toor, H.L. and Marchello, J.M. (1958). "Film-penetration model for mass and heat transfer," *AIChE Journal*, 4(1), pp. 97–101. Available at: http://dx.doi.org/10.1002/aic.690040118.

Turney, D. and Banerjee, S. (2008). "Transport phenomena at interfaces between turbulent fluids," *AIChE Journal*, 54(2), pp. 344–349. Available at: http://dx.doi.org/10.1002/aic.11427.

Veron, F., Melville, W.K. and Lenain, L. (2011). "The effects of small-scale turbulence on air–sea heat flux," *Journal of Physical Oceanography*, 41(1), pp. 205–220. Available at: http://dx.doi.org/10.1175/2010JPO4491.1.

Westerterp, K.R., van Swaaij, W.P.M. and Beenackers, A.A.C.M. (1984). *Chemical reactor design and operation*. New York: John Wiley & Sons.

Zhang, W. and Chatterjee, S.G. (2015). "Influence of residence time distribution on a surface-renewal model of constant-pressure cross-flow microfiltration," *Brazilian Journal of Chemical Engineering*, 32(1), pp. 139–154. Available at: http://dx.doi.org/10.1590/0104-6632.20150321s00003129.

Zorzetto, E., Peltola, O., Grönholm, T. and Katul, G.G. (2021). "Intermittent surface renewals and methane hotspots in natural peatlands," *Boundary-Layer Meteorology*, 180(3), pp. 407–433. Available at: https://doi.org/10.1007/s10546-021-00637-x.

2

Age-Distribution Function of the Surface Renewal Model

2.1 Introduction

A variety of age-distribution functions for the surface renewal model have been proposed in the literature over time. Our chief focus, however, in this chapter will be on those of the Danckwerts type, whose original derivation (assuming steady-state conditions) was provided by Danckwerts (1951).

2.2 Danckwerts Age Distribution

The derivation in this section follows the lines of reasoning given by Mondal and Chatterjee (2017).

The surface, assumed to be of unit area, of a turbulent liquid is visualized as being populated by a mosaic of chaotic eddies or liquid elements. At any time t since the start of the process (e.g., absorption of a gas in a well-mixed liquid and crossflow membrane filtration), such elements will have different surface ages ranging from 0 to t. Let $F(\tau, t)$ be the cumulative fraction of surface elements that have ages between 0 and τ at process (or "clock") time t, and let $\overline{R}(\tau)$ be the average growth or renewal rate of this fraction. Performing a population balance at process time t for elements having ages lying within the time interval τ to $\tau + \mathrm{d}\tau$ gives

$$\frac{\partial F(\tau, t)}{\partial \tau} = F(\tau, t)\overline{R}(\tau) \tag{2.1}$$

which upon integration yields

$$F(\tau, t) = \int_0^{\tau} F(\lambda, t)\overline{R}(\lambda)\,\mathrm{d}\lambda \tag{2.2}$$

$\overline{R}(\tau)$ is taken to be a weighted average of the individual renewal frequencies R of the surface elements. This quantity, which reflects the energy content of individual

Applications of the Surface Renewal Model of Mass Transfer, First Edition.
Hengshuo Huang and Siddharth G. Chatterjee.

Companion website: www.wiley.com/go/huang/surfacerenewalmodel

or groups of surface elements, is assumed to come in discrete packets (due to the random, discontinuous motions of the elements), which are integral multiples of a basic or fundamental frequency. This basic frequency is taken to be equal to what has been conventionally called the rate of surface renewal S. Thus, $R = 0, S, 2S, 3S, \ldots$ etc. If we now assume a Boltzmann distribution for the individual renewal frequency R, the average renewal frequency $\overline{R}(\tau)$ of elements with an age of τ will be given by

$$\overline{R}(\tau) = \frac{\sum_{N=0}^{\infty} Re^{-R\tau}}{\sum_{N=0}^{\infty} e^{-R\tau}} = \frac{\sum_{N=0}^{\infty} NSe^{-NS\tau}}{\sum_{N=0}^{\infty} e^{-NS\tau}} = \frac{Se^{-S\tau}/(1-e^{-S\tau})^2}{1/(1-e^{-S\tau})} = \frac{Se^{-S\tau}}{1-e^{-S\tau}} \tag{2.3}$$

According to eq. 2.3, $\overline{R}(\tau \to 0) = 1/\tau$ and $\overline{R}(\tau \to \infty) = 0$, that is, younger surface elements have a higher average renewal rate.

Substituting eq. 2.3 into eq. 2.2 yields

$$F(\tau, t) = \int_0^{\tau} F(\lambda, t) \frac{Se^{-S\lambda}}{1-e^{-S\lambda}} d\lambda \tag{2.4}$$

whose solution is

$$F(\tau, t) = A(1 - e^{-S\tau}) \tag{2.5}$$

Here, A is a parameter whose value depends on the initial state of the surface when the process was started at $t = 0$.

Differentiating eq. 2.5 with respect to τ yields the (density) age distribution $f(\tau, t)$, which satisfies the equation

$$\frac{\partial f(\tau, t)}{\partial \tau} = -Sf(\tau, t) \tag{2.6}$$

We will now consider two ideal cases.

2.2.1 Case 1

The surface is instantaneously and completely formed at $t = 0$ with liquid elements flowing into it from the bulk liquid and departing from it to the bulk liquid at a constant rate for $t \geq 0$. Since

$$F(t, t) = 1 \tag{2.7}$$

the value of A in eq. 2.5 is given by

$$A = \frac{1}{1 - e^{-St}} \tag{2.8}$$

which transforms eq. 2.5 into

$$F(\tau, t) = \frac{1 - e^{-S\tau}}{1 - e^{-St}} \quad \text{for} \quad 0 \leq \tau \leq t \tag{2.9}$$

Differentiating eq. 2.9 with respect to τ yields the density age distribution

$$f(\tau,t) = \frac{\partial F(\tau,t)}{\partial \tau} = \frac{Se^{-S\tau}}{1-e^{-St}} \quad \text{for} \quad 0 \le \tau \le t \tag{2.10}$$

which has been used in the modeling of crossflow ultrafiltration and microfiltration (Koltuniewicz and Noworyta, 1994; Koltuniewicz and Noworyta, 1995; Chatterjee, 2010; Hasan *et al.*, 2013; Jiang and Chatterjee, 2015) and unsteady-state physical and chemical gas absorption in a stirred cell (Huang and Chatterjee, 2021; Huang and Chatterjee, 2023).

Equation 2.10 will be used extensively in this work because of its simple form. As $t \to \infty$, it reduces to the well-known steady-state age distribution $Se^{-S\tau}$, which was first proposed by Danckwerts (1951).

2.2.2 Case 2

In this case, it is assumed that there are liquid elements, visualized as being "blue," that are already present on the surface when the process starts at $t = 0$, and "red" elements begin displacing the blue elements by the mechanism of surface renewal. At any time t, the surface will therefore consist of a mixture of both red and blue elements, the population of the latter decreasing as t increases. The red elements will have ages lying within $0 \le \tau < t$, while the blue elements will have ages exactly equal to t. From this mental picture, and eqs. 2.5 and 2.7, it follows that

$$A = 1 \tag{2.11}$$

and

$$F(\tau,t) = \begin{cases} 1 - e^{-S\tau} & \text{for} \quad 0 \le \tau < t \\ 1 & \text{at} \quad \tau = t \end{cases} \tag{2.12}$$

Equation 2.12 has a discontinuity at $\tau = t$ that becomes vanishingly small as $t \to \infty$. This discontinuity is due to the fraction of blue elements present on the surface at process time t, which is equal to e^{-St}. Equation 2.12 can be compactly expressed as

$$F(\tau,t) = 1 - e^{-S\tau} + u(\tau - t)e^{-S\tau} \quad \text{for} \quad 0 \le \tau \le t \tag{2.13}$$

where $u(t)$ is the unit step function. Differentiating eq. 2.13 with respect to τ yields

$$f(\tau,t) = Se^{-S\tau}[1 - u(\tau - t)] + \delta(\tau - t)e^{-S\tau} \quad \text{for} \quad 0 \le \tau \le t \tag{2.14}$$

where $\delta(t)$ is the delta function.

Chung *et al.* (1971) and Sada *et al.* (1979) have previously presented eq. 2.14 whereas a derivation of this equation, based on a stochastic population balance of interfacial fluid elements, has been given by Fan *et al.* (1993). Zhang and

Chatterjee (2015) used this age distribution in the derivation of analytical expressions for the permeate flux decline and cake buildup in constant-pressure crossflow microfiltration. It is once again observed that as $t \to \infty$, eq. 2.14 reduces to the conventional Danckwerts age distribution.

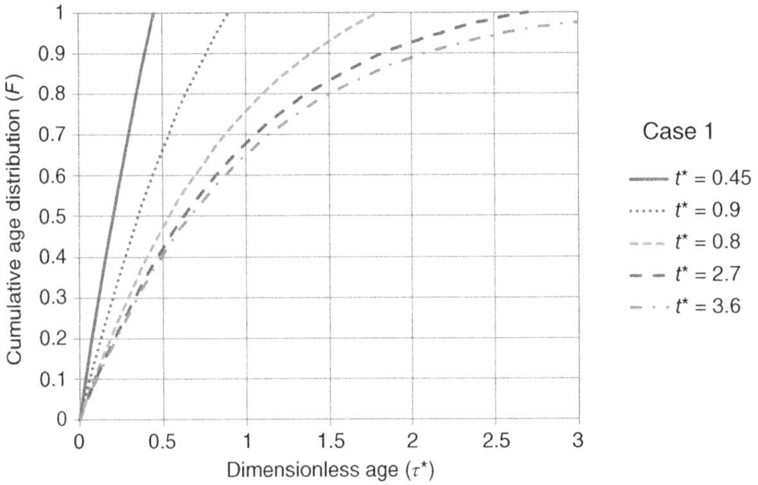

Figure 2.1 Cumulative Danckwerts age distribution for Case 1 (eq. 2.9). Note that $\tau^* = S\tau$ and $t^* = St$. Source: Zhang and Chatterjee, 2015 / Springer Nature.

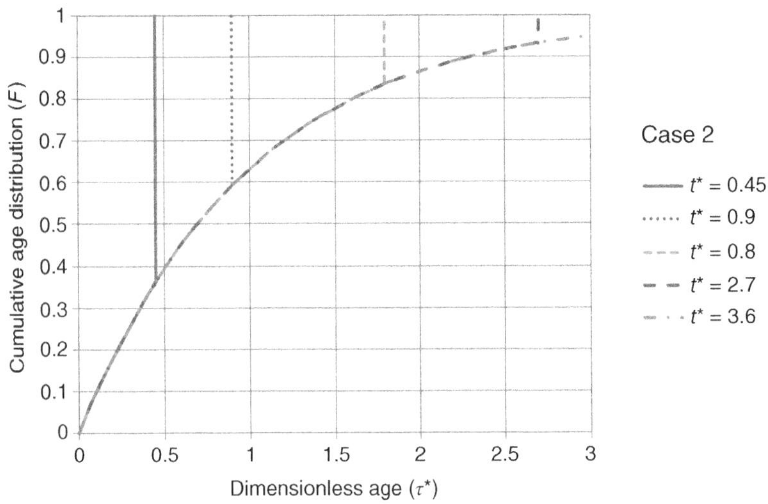

Figure 2.2 Cumulative Danckwerts age distribution for Case 2 (eq. 2.12). Note that $\tau^* = S\tau$ and $t^* = St$. Source: Zhang and Chatterjee, 2015 / Springer Nature.

Figures 2.1 and 2.2 show the cumulative Danckwerts age distribution F for the two cases as a function of the dimensionless age $\tau^* = S\tau$ and dimensionless process time $t^* = St$. As t^* increases, the initial state of the interface becomes unimportant, and F for both cases approaches the cumulative, steady-state age distribution:

$$F(\tau^*, t^* \to \infty) = 1 - e^{-\tau^*} \tag{2.15}$$

2.3 Generalized Danckwerts Age Distribution

It was indicated in Chapter 1 that the experimentally measured density surface age distributions reported in a number of investigations do not follow the (conventional) exponential Danckwerts age distribution but rather have a bell-shaped character. The generalized Danckwerts (GD) age distribution (Mondal and Chatterjee, 2017; Horvath and Chatterjee, 2018) is able to capture this behavior.

The basic differential equation that describes the unsteady-state GD age-distribution function $f(\tau, t)$ is given by (Horvath and Chatterjee, 2018) as

$$\frac{\partial f}{\partial \tau} + Sf\left(2a + 1 - \frac{a}{S\tau}\right) = 0 \tag{2.16}$$

The two parameters of the distribution are the surface renewal rate S and a, which is related to the width of the distribution, as will be discussed later.[1] The solution of this equation is

$$f(\tau, t) = K\tau^a e^{-(2a+1)S\tau} \tag{2.17}$$

where K is a parameter. Substitution of eq. 2.17 into

$$\int_0^t f(\tau, t)\, \mathrm{d}\tau = 1 \tag{2.18}$$

yields

$$K = \frac{[(2a + 1)S]^{a+1}}{\Gamma(a + 1) - \Gamma(a + 1, (2a + 1)St)} \tag{2.19}$$

where $\Gamma(z, y)$ is the incomplete gamma function defined by

$$\Gamma(z, y) = \int_y^\infty \lambda^{z-1} e^{-\lambda}\, \mathrm{d}\lambda \tag{2.20}$$

Substituting eq. 2.19 in eq. 2.17 yields the unsteady-state GD age-distribution function, that is,

$$f(\tau, t) = \frac{S(2a + 1)^{a+1}(S\tau)^a e^{-(2a+1)S\tau}}{\Gamma(a + 1) - \Gamma(a + 1, (2a + 1)St)} \tag{2.21}$$

---

1 Note that for $a = 0$, eq. 2.16 reduces to eq. 2.6.

Letting $a = 0$ and using the relations $\Gamma(1) = 1$ and $\Gamma(1, St) = e^{-St}$ reduce eq. 2.21 to eq. 2.10, which is the (exponential) unsteady-state Danckwerts age distribution of Case 1.

As $t \to \infty$, eq. 2.21 simplifies to

$$f(\tau, t \to \infty) = S\frac{(2a+1)^{a+1}}{\Gamma(a+1)}(S\tau)^a e^{-(2a+1)S\tau} \tag{2.22}$$

which is the steady-state GD age distribution derived by Mondal and Chatterjee (2017). For $a = 0$, eq. 2.22 reduces to $f(\tau, t \to \infty) = Se^{-S\tau}$, which is the conventional steady-state Danckwerts age distribution.

The mean eddy renewal or burst time t_{ren} can be obtained from

$$t_{\text{ren}}(t) = \int_0^t \tau f(\tau, t)\,\mathrm{d}\tau \tag{2.23}$$

Substituting eq. 2.21 into eq. 2.23 gives

$$t_{\text{ren}}(t) = \frac{1}{S(2a+1)}\left[\frac{\Gamma(a+2) - \Gamma(a+2, (2a+1)St)}{\Gamma(a+1) - \Gamma(a+1, (2a+1)St)}\right] \tag{2.24}$$

As $t \to 0$, eq. 2.24 reduces to

$$t_{\text{ren}}(t \to 0) = \left(\frac{a+1}{a+2}\right)t \tag{2.25}$$

that is, during the initial moments, the mean eddy renewal time is independent of the frequency of surface renewal S and increases linearly with process time t. For $a = 0$, eq. 2.25 predicts that $t_{\text{ren}}(t \to 0) = 0.5t$. As $t \to \infty$, that is, as steady state is approached, eq. 2.24 simplifies to

$$t_{\text{ren}}(t \to \infty) = \frac{1}{S}\left(\frac{a+1}{2a+1}\right) \tag{2.26}$$

which result was previously given by Mondal and Chatterjee (2017). For $a = 0$ (i.e., the conventional Danckwerts age distribution), eq. 2.26 gives $t_{\text{ren}}(t \to \infty) = 1/S$, that is, the average eddy renewal time t_{ren} (at steady state) is equal to the reciprocal of the surface renewal rate S, which is a well-known result. However, this is not true for other values of a.

As mentioned in Chapter 1, Garbe *et al.* (2002) experimentally measured the age distribution of surface renewal events on the Heidelberg Aeolotron in air–water heat exchange. They found a good fit of the logarithmic normal (LN) distribution to these data and reported values of the two parameters of this distribution at wind speeds of 2, 4.2, and 8 m s^{-1}. Mondal and Chatterjee (2017) showed that the two-parameter (steady-state) GD age distribution, given by eq. 2.22, was equivalent to the LN age distribution, which Garbe *et al.* (2002) had used to represent their experimental data, that is, it could also capture the bell-shaped nature of the distribution of surface ages observed empirically by these researchers. The values of the parameters a and S of the GD distribution obtained

Table 2.1 Values of the parameters of the generalized Danckwerts age-distribution function reported by Mondal and Chatterjee (2017) for the experimental data of Garbe *et al.* (2002).

Wind speed m s^{-1}	a	S (s^{-1})
2.0	0.5	0.036
4.2	2.5	0.204
8.0	4.2	0.417

by Mondal and Chatterjee (2017) for the three different experimental wind speeds of Garbe *et al.* (2002) are shown in Table 2.1. It is observed that both a and S increase with wind speed, that is, with the turbulence intensity at the gas–liquid interface.

Figure 2.3 shows unsteady-state GD age distributions determined from eq. 2.21 using the values of a and S reported in Table 2.1 for wind speeds of 2, 4.2, and 8 m s^{-1}. It is clearly seen that at each wind speed, the distribution has a sharp peak during the initial moments. However, with the progress of process time, it flattens out and acquires a bell-shaped character as steady state is approached. The time taken to reach a steady state is a strong and inverse function of the wind speed. It takes practically 150, 9, and 4 s to reach steady state at wind speeds of 2, 4.2, and 8 m s^{-1}, respectively. Thus, as the wind speed increases by a factor of 4, the time taken to reach steady state decreases by a factor of 25. Further, as the wind speed rises, the age distribution at steady state narrows significantly, its skewness decreases, and its maximum peak becomes larger. This is because of an increase in the value of the parameter a with wind speed (Table 2.1). It can therefore be postulated that the age distribution at steady state will acquire a delta-function-like character at extremely high wind speeds (i.e., at high turbulence intensities).

In Figure 2.4, the mean eddy renewal or burst time t_{ren} is plotted as a function of process time t (see eq. 2.24) at the three aforementioned wind speeds. As anticipated from eq. 2.25, t_{ren} increases in a linear fashion with t initially and then approaches a steady-state value, which is inversely proportional to the surface renewal rate S as shown by eq. 2.26. This steady-state value is practically 20.83, 2.86, and 1.33 s for the wind speeds of 2, 4.2, and 8 m s^{-1}, respectively. Thus, a four-fold increase in wind speed decreases the steady-state value of t_{ren} by a factor of 16, which shows the great effect of turbulence on the hydrodynamics at the gas–liquid interface. This is clear evidence that the surface renewal rate S is a hydrodynamic parameter although it is often measured by mass-transfer experiments. It is this

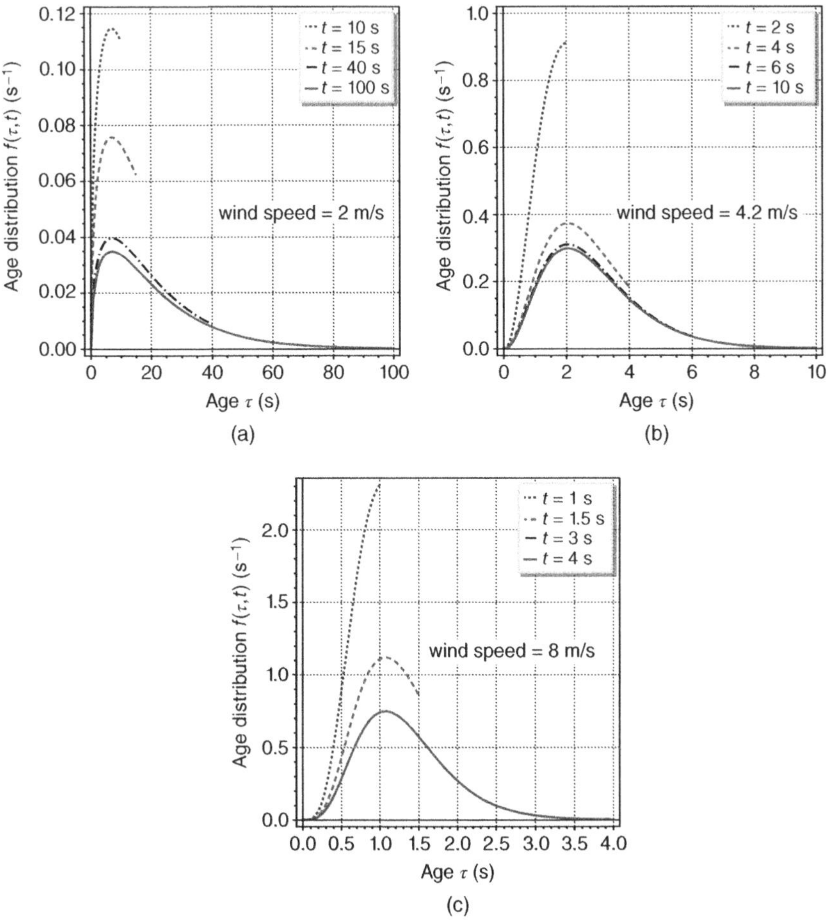

Figure 2.3 Unsteady-state generalized Danckwerts density age distributions (eq. 2.21) at wind speeds of (a) 2, (b) 4.2, and (c) 8 m s^{-1}. The values of the parameters a and S are shown in Table 2.1. Source: Horvath and Chatterjee, 2018 / The Royal Society / CC BY 4.0.

rate that governs the exchange of liquid elements between the gas–liquid interface and the bulk liquid, which determines the convective net rate of transport of dissolved gas from the former to the latter, as shown by Chatterjee and co-workers (Mondal and Chatterjee, 2017; Horvath and Chatterjee, 2018; Huang and Chatterjee, 2021, 2023), and, as mentioned in Chapter 1, has no parallel in the film model, according to which the transfer of dissolved gas from the interfacial region to the bulk liquid occurs solely by molecular diffusion across the hypothetical liquid film.

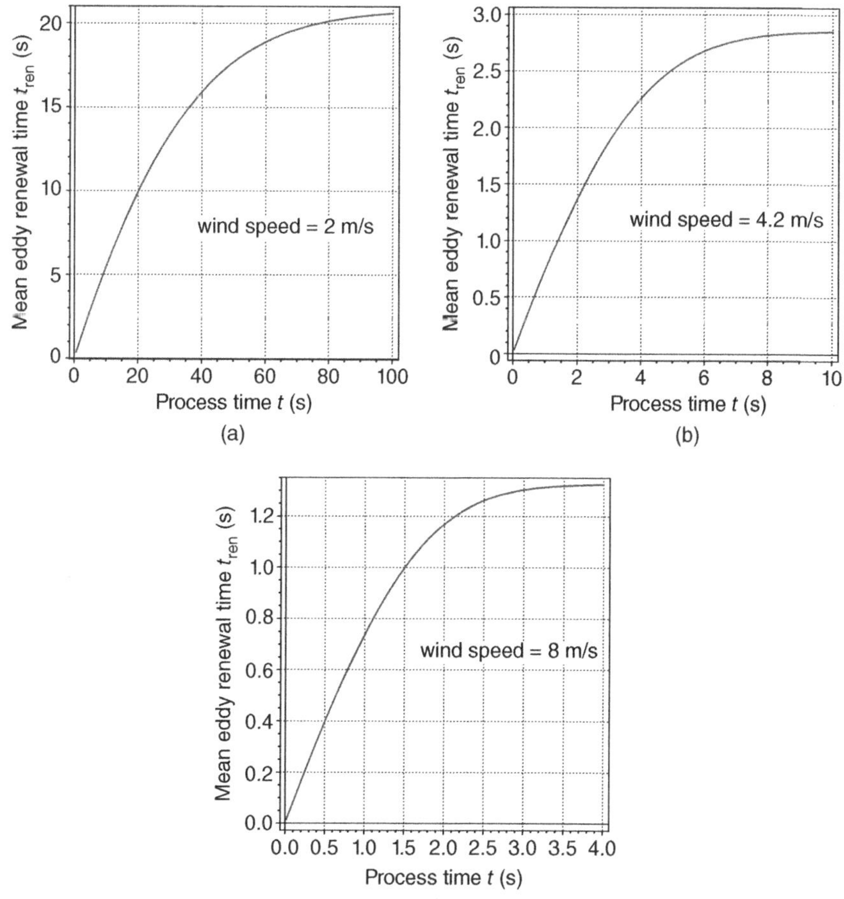

Figure 2.4 Behavior of the mean eddy renewal or burst time with process time (eq. 2.24) at wind speeds of (a) 2, (b) 4.2, and (c) 8 m s^{-1}. The values of the parameters a and S are given in Table 2.1. Source: Horvath and Chatterjee, 2018 / The Royal Society / CC BY 4.0.

Notation

a	parameter of the generalized Danckwerts age distribution
A	parameter in eq. 2.5
$f(\tau, t)$	density age distribution of surface elements at process time t (s^{-1})
$F(\tau, t)$	cumulative age distribution of surface elements at process time t
K	parameter in eq. 2.17
N	$0, 1, 2, 3, \ldots$

R	individual renewal frequency of surface elements (Hz or s^{-1})
$\overline{R}(\tau)$	average renewal frequency of surface elements with an age of τ (Hz or s^{-1})
S	fundamental renewal frequency of surface elements (Hz or s^{-1})
t	process time (s)
t^*	dimensionless process time; equal to St
t_{ren}	mean eddy renewal or burst time (s)
$u(t)$	unit step function
y	parameter of $\Gamma(z, y)$
z	parameter of $\Gamma(z, y)$

Greek Letters

$\Gamma(z, y)$	incomplete gamma function; defined by eq. 2.20
$\delta(t)$	delta function (s^{-1})
λ	variable of integration
τ	age of a surface element (s)
τ^*	dimensionless age of a surface element; equal to $S\tau$

Homework Exercises

i) Show that eq. 2.6 follows from eq. 2.5.
ii) Derive the cumulative age distribution $F(\tau, t)$ given by eq. 2.5 by directly integrating eq. 2.6.
iii) Demonstrate that as $t \to \infty$ in eq. 2.21, eq. 2.22 results.
iv) Verify the limiting forms for t_{ren} given by eqs. 2.25 and 2.26 as $t \to 0$ and $t \to \infty$, respectively, in eq. 2.24.
v) Show that as $a \to \infty$ (i.e., extremely high wind speeds), eq. 2.22 for $f(\tau, t \to \infty)$ approaches $S\delta(S\tau - 0.5)$.

References

Chatterjee, S.G. (2010). "On the use of the surface-renewal concept to describe cross-flow ultrafiltration," *Indian Chemical Engineer*, 52(3), pp. 179–193. Available at: http://dx.doi.org/10.1080/00194506.2010.547927.

Chung, B.T.F., Fan, L.T. and Hwang, C.L. (1971). "Surface renewal and penetration models in the transient state," *AIChE Journal*, 17(1), pp. 154–160. Available at: http://dx.doi.org/10.1002/aic.690170131.

Danckwerts, P.V. (1951). "Significance of liquid-film coefficients in gas absorption," *Industrial & Engineering Chemistry*, 43(6), pp. 1460–1467. Available at: http://dx.doi.org/10.1021/ie50498a055.

Fan, L.T., Shen, B.C. and Chou, S.T. (1993). "The surface-renewal theory of interphase transport: a stochastic treatment," *Chemical Engineering Science*, 48(23), pp. 3971–3982. Available at: http://dx.doi.org/10.1016/0009-2509(93)80376-2.

Garbe, C.S., Jähne, B. and Haußecker, H. (2002). "Measuring the sea surface heat flux and probability distribution of surface renewal events," in *Gas transfer at water surfaces* (eds. Donelan, M.A., Drennan, W.M., Saltzman, E.S. and Anninkhof, R.W.). Washington, DC: American Geophysical Union, pp.109–114.

Hasan, A., Peluso, C.R., Hull, T.S., Fieschko, J. and Chatterjee, S.G. (2013). "A surface-renewal model of cross-flow microfiltration," *Brazilian Journal of Chemical Engineering*, 30(1), pp. 167–186. Available at: http://dx.doi.org/10.1590/S0104-66322013000100019.

Horvath, I.R. and Chatterjee, S.G. (2018). "A surface renewal model for unsteady-state mass transfer using the generalized Danckwerts age distribution function," *Royal Society Open Science*, 5(5), 172423. Available at: http://dx.doi.org/10.1098/rsos.172423.

Huang, H. and Chatterjee, S.G. (2021). "Transient physical gas absorption in a stirred liquid using the surface renewal model of mass transfer," *Chemical Engineering Science*, 234, 116449. Available at: http://dx.doi.org/10.1016/j.ces.2021.116449.

Huang, H. and Chatterjee, S.G. (2023). "A rigorous surface renewal model for transient gas absorption with first-order chemical reaction in a stirred liquid," *Chemical Engineering Science*, 268, 118393. Available at: http://dx.doi.org/10.1016/j.ces.2022.118393.

Jiang, S. and Chatterjee, S.G. (2015). "A surface-renewal model for constant flux cross-flow microfiltration," *Journal of Applied Polymer Science*, 132(21). Available at: http://dx.doi.org/10.1002/app.41778.

Koltuniewicz, A. and Noworyta, A. (1994). "Dynamic properties of ultrafiltration systems in light of the surface renewal theory," *Industrial & Engineering Chemistry Research*, 33(7), pp. 1771–1779. Available at: http://dx.doi.org/10.1021/ie00031a016.

Koltuniewicz, A. and Noworyta, A. (1995). "Method of yield evaluation for pressure-driven membrane processes," *Chemical Engineering Journal*, 58, pp. 175–182.

Mondal, C. and Chatterjee, S.G. (2017). "A frequency quantum interpretation of the surface renewal model of mass transfer," *Royal Society Open Science*, 4(7), p. 170103. Available at: http://dx.doi.org/10.1098/rsos.170103.

Sada, E., Katoh, S., Yoshii, H. and Ban, Y. (1979). "Rates of gas absorption with interfacial turbulence caused by micro-stirrers," *The Canadian Journal of Chemical Engineering*, 57(6), pp. 704–706. Available at: http://dx.doi.org/10.1002/cjce.5450570609.

Zhang, W. and Chatterjee, S.G. (2015). "Influence of residence time distribution on a surface-renewal model of constant-pressure cross-flow microfiltration," *Brazilian Journal of Chemical Engineering*, 32(1), pp. 139–154. Available at: http://dx.doi.org/10.1590/0104-6632.20150321s00003129.

3

Physical Gas Absorption in a Large Volume of Liquid*

3.1 Introduction

In this chapter, we will use the generalized Danckwerts (GD) age distribution derived in Chapter 2 to study the physical absorption of a gas in a large body of liquid like a pond or lake.

3.2 Rates of Absorption and Dissolved-Gas Transfer

Consider a large body of turbulent liquid whose surface is exposed to a gas that is being absorbed into it. Since the volume of liquid is large, the concentration of dissolved gas in the bulk of the liquid will remain approximately constant over a short period of time. Our goal is to derive expressions for the rates of absorption and dissolved-gas transfer to the bulk liquid using the surface renewal model. If the gas-side mass-transfer resistance is assumed to be negligible, the concentration profile and the instantaneous rate of absorption $R_{\text{inst}}(\tau)$ of the gas per unit area in a surface element, assumed to be of infinite depth and having an age of τ, are given by (Astarita, 1967; Danckwerts, 1970)

$$\frac{c(x,\tau)-c_{\text{b}}}{c_{\text{s}}-c_{\text{b}}}=\text{erfc}\left(\frac{x}{2\sqrt{D\tau}}\right) \tag{3.1}$$

and

$$R_{\text{inst}}(\tau)=\left(c_{\text{s}}-c_{\text{b}}\right)\sqrt{\frac{D}{\pi\tau}} \tag{3.2}$$

Here, $c(x, \tau)$ is the dissolved-gas concentration in the element at a depth of x (measured from the gas–liquid interface) and time τ, c_{s} and c_{b} are the dissolved-gas

*The material in this chapter is adapted from Horvath and Chatterjee (2018).

Applications of the Surface Renewal Model of Mass Transfer, First Edition.
Hengshuo Huang and Siddharth G. Chatterjee.

Companion website: www.wiley.com/go/huang/surfacerenewalmodel

surface and bulk-liquid concentrations (assumed to be constant), respectively, and D is the molecular diffusion coefficient of the dissolved gas in the liquid.

The average rate of absorption [$R_{abs}(t)$] of the gas at time t since the start of the absorption operation can be obtained from

$$R_{abs}(t) = \int_0^t R_{inst}(\tau) f(\tau, t)\, d\tau \tag{3.3}$$

where $f(\tau, t)$ is the unsteady-state surface age-distribution function, which is assumed to be the GD age distribution given by eq. 2.21 of Chapter 2.

The transient net rate of transfer $R_{trans}(t)$ of the dissolved gas to the bulk liquid for surface elements of infinite depth is given by (Mondal and Chatterjee, 2017)

$$R_{trans}(t) = S \int_0^t \int_0^\infty \left[c(x, \tau) - c_b\right] f(\tau, t)\, dx\, d\tau \tag{3.4}$$

where the first and second terms on the right-hand side of eq. 3.4 represent the two opposite directions of convective mass transfer due to the surface renewal mechanism. These are, first, the convective transfer of dissolved gas to the bulk liquid from the gas–liquid interface, and second, that from the bulk liquid to the interface.

Upon substitution of eq. 2.21 and eq. 3.2 into eq. 3.3, it may be shown that

$$R_{abs}(t) = k_L^{abs}(t)\left[c_s - c_b\right] \tag{3.5}$$

where $k_L^{abs}(t)$ is the dynamic liquid-side gas-absorption coefficient given by

$$k_L^{abs}(t) = \sqrt{\frac{DS(2a+1)}{\pi}} \left[\frac{\Gamma(a+1/2) - \Gamma(a+1/2, (2a+1)St)}{\Gamma(a+1) - \Gamma(a+1, (2a+1)St)}\right] \tag{3.6}$$

Substituting $a = 0$ and using the relations $\Gamma(1/2) = \sqrt{\pi}$ and $\Gamma(1/2, St) = \sqrt{\pi}\left[1 - \text{erf}\left(\sqrt{St}\right)\right]$ reduce eq. 3.6 to

$$k_L^{abs}(t) = \sqrt{DS}\frac{\text{erf}\left(\sqrt{St}\right)}{1 - e^{-St}} \tag{3.6a}$$

which was obtained previously by Mondal and Chatterjee (2017) – see eqs. 3.6, 3.7, and 3.12 of Case 1 in their paper. As $t \to 0$, eq. 3.6 simplifies to

$$k_L^{abs}(t \to 0) = 2\left(\frac{a+1}{2a+1}\right)\sqrt{\frac{D}{\pi t}} \tag{3.7}$$

demonstrating that during the initial moments of absorption, the gas-absorption coefficient k_L^{abs} is independent of the surface renewal rate S.

Similarly, by substituting eqs. 2.21 and 3.1 into eq. 3.4 and using the result

$$\int_0^\infty \text{erfc}\left(\frac{x}{2\sqrt{D\tau}}\right) dx = 2\sqrt{\frac{D\tau}{\pi}} \tag{3.8}$$

it may be shown that

$$R_{\text{trans}}(t) = k_{\text{L}}^{\text{trans}}(t)\left[c_{\text{s}} - c_{\text{b}}\right] \tag{3.9}$$

where $k_{\text{L}}^{\text{trans}}(t)$ is the dynamic liquid-side dissolved-gas transfer coefficient given by

$$k_{\text{L}}^{\text{trans}}(t) = 2\sqrt{\frac{DS}{\pi(2a+1)}}\left[\frac{\Gamma(a+3/2) - \Gamma(a+3/2,(2a+1)St)}{\Gamma(a+1) - \Gamma(a+1,(2a+1)St)}\right] \tag{3.10}$$

It is evident that eqs. 3.6 and 3.10 have different mathematical structures. Substituting $a = 0$ and using the relation $\Gamma(3/2) = \sqrt{\pi}/2$ reduce eq. 3.10 to

$$k_{\text{L}}^{\text{trans}}(t) = \sqrt{DS}\frac{1 - \frac{2}{\sqrt{\pi}}\Gamma(3/2, St)}{1 - e^{-St}} \tag{3.10a}$$

which result was derived earlier by Mondal and Chatterjee (2017); this can be observed from eqs. 3.6, 3.8, and 3.13 of Case 1 in their work. As $t \rightarrow 0$, eq. 3.10 reduces to

$$k_{\text{L}}^{\text{trans}}(t \rightarrow 0) = 4S\left(\frac{a+1}{2a+3}\right)\sqrt{\frac{Dt}{\pi}} \tag{3.11}$$

according to which, during the initial moments of absorption, the dissolved-gas transfer coefficient, unlike the gas-absorption coefficient, is directly proportional to the surface renewal rate S.

As eqs. 3.7 and 3.11 show, for $t \rightarrow 0$, the gas-absorption coefficient $k_{\text{L}}^{\text{abs}}$ is inversely proportional to $\sqrt{t}$, whereas the dissolved-gas transfer coefficient $k_{\text{L}}^{\text{trans}}$ is directly proportional to it, that is, they have an inverse relationship.

As $t \rightarrow \infty$, that is, as steady state is approached, eqs. 3.6 and 3.10 simplify to

$$k_{\text{L}}^{\text{abs}}(t \rightarrow \infty) = k_{\text{L}}^{\text{trans}}(t \rightarrow \infty) = k_{\text{L}} = \frac{\Gamma(a+1/2)}{\Gamma(a+1)}\sqrt{\frac{(2a+1)DS}{\pi}} \tag{3.12}$$

which expression has also been presented before by Mondal and Chatterjee (2017). According to eq. 3.12, at steady state, the gas-absorption and dissolved-gas transfer coefficients coincide and, for $a = 0$ (conventional Danckwerts model), become equal to $\sqrt{DS}$, which is a well-known result.

Equation 3.12 can be expressed as

$$k_{\text{L}} = \alpha\sqrt{DS} \tag{3.13}$$

where

$$\alpha = \frac{\Gamma(a+1/2)}{\Gamma(a+1)}\sqrt{\frac{(2a+1)}{\pi}} \tag{3.14}$$

For $a = 0$, that is, the conventional Danckwerts age distribution, eq. 3.14 gives $\alpha = 1$, whereas for $a \rightarrow \infty$ (i.e., high wind speed), it can be shown that $\alpha \rightarrow \sqrt{2/\pi} \cong 0.797885$ [see also Table 1 in the Appendix (Electronic

Supplementary Material, S1) of the manuscript of Mondal and Chatterjee, 2017], and the surface age-distribution function $f(\tau, t \to \infty) \to S\delta(S\tau - 0.5)$, where $\delta(t)$ is the delta function. Hence, the value of α is bounded within these limits, and an average value can be used for all wind speeds:

$$\alpha \cong 0.5 + \sqrt{\frac{1}{2\pi}} \approx 0.9 \tag{3.15}$$

Thus, $k_L \approx 0.9\sqrt{DS}$ for the GD model, which can be compared to $k_L = \sqrt{DS}$ – the expression for the mass-transfer coefficient as predicted by the conventional Danckwerts model ($a = 0$). The uncertainty in the measured values of k_L, which can often be substantial, provides a justification for using the approximation $\alpha = 0.9$ in eq. 3.13. In summary, the GD model predicts a 10% lower value for the steady-state liquid-side mass-transfer coefficient k_L compared to the conventional Danckwerts model.

Figure 3.1 shows the behavior of the two mass-transfer coefficients, k_L^{abs} and k_L^{trans}, given by eqs. 3.6 and 3.10, respectively, as a function of process time t for the case of oxygen absorption into water at 25 °C, at the three different (experimental) wind speeds used by Garbe *et al.* (2002). The values of the parameters a and S were obtained from Table 2.1 (Chapter 2), whereas a value of 2.12×10^{-9} m^2 s^{-1} was used for the diffusion coefficient D of oxygen in water at 25 °C (Mondal and Chatterjee, 2017). The gas-absorption mass-transfer coefficient k_L^{abs} decays with t in an exponential-like manner, decreasing from an extremely high initial value and approaching a steady-state value asymptotically as t increases. In sharp contrast, the dissolved-gas mass-transfer coefficient k_L^{trans} increases in a smooth fashion from a value of zero and merges with the gas-absorption coefficient curve with approach toward steady state, as predicted by eq. 3.12. To our knowledge, Mondal and Chatterjee (2017) were the first to distinguish between these two mass-transfer coefficients in the gas-absorption literature.

The contrasting behavior of the gas-absorption and dissolved-gas transfer coefficients with process time t can be explained as follows. During the initial moments of absorption, the gas–liquid interface consists primarily of younger elements (i.e., small τ), which will have a high absorption rate (eq. 3.2). The gas-absorption coefficient k_L^{abs}, which is the average rate of absorption per unit concentration driving force (eq. 3.5), will therefore be high initially, as can be seen from eqs. 2.21, 3.2, and 3.3. However, younger elements will also have a lower dissolved-gas concentration (eq. 3.1). The dissolved-gas transfer coefficient k_L^{trans}, which represents the average rate of dissolved-gas transfer to the bulk liquid per unit concentration driving force (eq. 3.9), will therefore be low during the first moments of absorption (see eqs. 2.21, 3.1, and 3.4). However, as time progresses, the populations of young and old elements at the interface will more or less balance one another, as the age distribution evolves with the passage of time, and

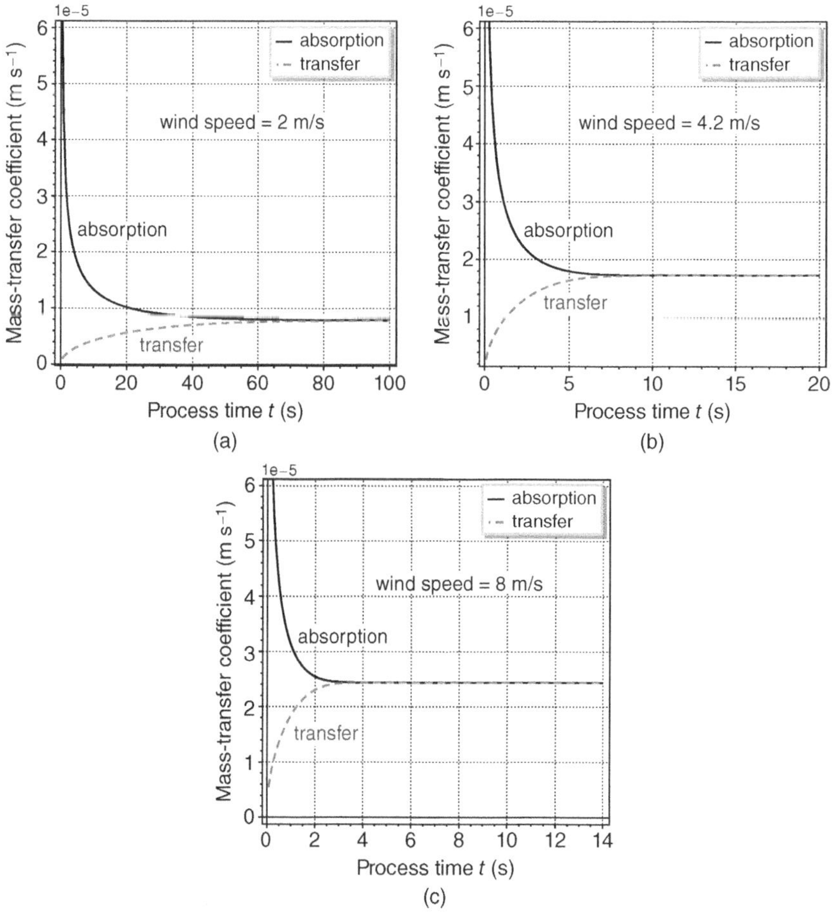

Figure 3.1 Behavior of the gas-absorption and dissolved-gas transfer coefficients (eqs. 3.6 and 3.10, respectively) with process time for oxygen absorption in water at 25 °C and wind speeds of (a) 2, (b) 4.2, and (c) 8 m s^{-1}. Values of the parameters a and S are given in Table 2.1 (Chapter 2) and $D = 2.12 \times 10^{-9}$ m^2 s^{-1}. Source: Horvath and Chatterjee (2018) / The Royal Society / CC BY 4.0.

the initial large divergence between these two coefficients will diminish as steady state is approached.

The difference between k_L^{abs} and k_L^{trans} signifies a heightened accumulation of dissolved gas in the interfacial liquid elements during the early period of absorption. This accumulation, which can last from a few seconds to a few minutes depending upon the intensity of turbulence, will be especially pronounced just after the beginning of absorption and will reduce to zero when steady state is reached.

Table 3.1 Gas-absorption and dissolved-gas mass-transfer coefficients for oxygen absorption in water at 25 °C calculated from eqs. 3.6, 3.10, and 3.12.

Wind speed m s^{-1}	Process time *t* s	$k_L^{abs}(t)$ m s^{-1} (eq. 3.6)	$k_L^{trans}(t)$ m s^{-1} (eq. 3.10)	Steady-state k_L^{abs} or k_L^{trans} (i.e., k_L) m s^{-1} (eq. 3.12)
2.0	150	7.87×10^{-6}	7.86×10^{-6}	7.87×10^{-6}
4.2	9	1.73×10^{-5}	1.73×10^{-5}	1.73×10^{-5}
8.0	4	2.44×10^{-5}	2.43×10^{-5}	2.44×10^{-5}

Values of the parameters a and S are reported in Table 2.1 (Chapter 2) and $D = 2.12\times10^{-9}$ m^2 s^{-1}.
Source: Horvath and Chatterjee (2018) / The Royal Society / CC BY 4.0.

Table 3.1 reports values of k_L^{abs} and k_L^{trans} for oxygen absorption in water at 25 °C for three different wind speeds (Table 2.1, Chapter 2) as determined from eqs. 3.6, 3.10, and 3.12. The second column of this table shows the process time at which both coefficients, calculated using eqs. 3.6 and 3.10, are virtually identical with one another and almost indistinguishable from the steady-state value predicted by eq. 3.12, which is shown in the last column of the table, that is, it is the time required to attain steady state practically. As can be concluded from Table 3.1, a fourfold increase in wind speed increases the steady-state value of k_L^{abs} or k_L^{trans} by a factor of 3.1 while decreasing the time required to attain steady state by a factor of 37. The values of the mass-transfer coefficients in this table are comparable to the values of 8.33×10^{-6} and 2.12×10^{-5} m s^{-1} for the liquid-side mass-transfer coefficient obtained experimentally by Hutchinson and Sherwood (1937) when pure oxygen was absorbed in water at 25 °C in a stirred flask (stirrer speed: 171 and 1025 revolutions per minute, respectively), with the stirred water surface being exposed to the gas.

3.3 Comparison of the Surface Renewal Model with Experimental Data

Krall (2013) conducted an extensive theoretical and experimental investigation on the transfer velocity (i.e., steady-state liquid-phase mass-transfer coefficient k_L) of a number of sparingly soluble gases in the Heidelberg Aeolotron and the Kyoto high-speed wind-wave tank. The fourth column of Table 3.2 shows some of Krall's experimental data on the transfer velocity for the nitrous oxide–water system at 20 °C at different wind speeds. These data are for a clean water surface (i.e., no surfactant) and are extracted from Figure 7.11 of Krall's dissertation, which reports experimental transfer velocities (measured in the Aeolotron) as a function of the reference wind speed, which can be converted to the usually

Table 3.2 Experimental and theoretical values of the steady-state liquid-side mass-transfer coefficient or transfer velocity k_L for the nitrous oxide – water system at 20 °C.

Wind speed (at 10 m) $m\,s^{-1}$	S s^{-1}	a	Experimental k_L $cm\,h^{-1}$	Theoretical k_L (eq. 3.12) $cm\,h^{-1}$	Absolute error (%)
1.7	0.014	0.21	2.06	1.71	17.1
2.5	0.078	1.03	3.26	3.77	15.5
3.4	0.145	1.84	5.10	5.00	1.96
4.9	0.250	2.97	7.40	6.45	12.9
7.1	0.374	4.00	13.73	7.83	43.0
					Average = 18.1%

Values of S and a as functions of the wind speed were obtained as explained in the text and $D = 1.88 \times 10^{-9}$ $m^2\,s^{-1}$. The experimental transfer velocities were extracted from Figure 7.11 in the dissertation of Krall (2013).
Source: Horvath and Chatterjee (2018) / The Royal Society / CC BY 4.0.

reported wind speed at a height of 10 m above the water surface by using Figure 7.9b in the same dissertation; these latter wind speeds are reported in the first column of Table 3.2. The second and third columns of this table show values of the two parameters of the GD age distribution, S and a, at these wind speeds; these were estimated by fitting quadratic expressions to the values of these parameters given in Table 2.1 (Chapter 2) as a function of wind speed (Horvath and Chatterjee, 2018). The diffusion coefficient of nitrous oxide in water at 25 °C is 1.91×10^{-9} $m^2\,s^{-1}$ (Krall, 2013), which when corrected to 20 °C (experimental temperature used by Krall, 2013) with the Wilke–Chang correlation (Perry *et al.*, 1984), gives a value of $D = 1.88 \times 10^{-9}$ $m^2\,s^{-1}$. Values of the theoretical transfer velocity (k_L), calculated by eq. 3.12, are provided in the fifth column of Table 3.2. It is seen that the agreement between the theoretical and experimental k_L is quite reasonable (average absolute error of prediction = 18.1%) when the following are taken into consideration: (1) the values of the two parameters of the GD age distribution, S and a, were estimated from the experimentally measured age distribution of surface renewal events, obtained in the context of heat exchange, on the Aeolotron as reported by Garbe *et al.* in (2002); (2) the experimental data on transfer velocities were reported by Krall in 2013, that is, about a decade separates these two investigations during which time period the Aeolotron was remodeled; and (3) there is some uncertainty in the experimental measurements of k_L, as can be observed in Figure 7.11 of Krall's thesis.

For large wind speeds, the transfer velocity determined from eq. 3.12 was significantly lower than that measured experimentally by Krall (2013). Two possible reasons for this are: (1) the correlations for S and a as functions of wind speed (see Figure 4a and b in Horvath and Chatterjee, 2018) are not valid since they only hold when the wind speed lies between 2 and 8 m s^{-1}; and (2) the air–water interface becomes broken with the appearance of bubbles and spray, with the surface renewal model no longer being able to provide an accurate picture of the ensuing complex mass-transfer process.

It was indicated in Chapter 1 that different mass-transfer theories predict that the liquid-side mass transfer coefficient $k_L \propto D^m$, where the value of the exponent m depends upon the mass-transfer model invoked. For instance, $m = 1$ for the film model, whereas $m = 1/2$ for the surface renewal model, as can be seen from eq. 3.12. Although there is some uncertainty, the data for the trace gases employed by Krall (2013) showed the exponent gradually changing from 2/3 (smooth surface) to 1/2 (wavy surface) as the wind speed increased.[1] For example, Krall's Fig. 7.14 shows that, for the clean air–water interface, m changes from 0.66 to 0.52 as the wind speed increases from 2.4 to 10.6 m s^{-1}, with the midpoint being at around a wind speed of 4.5 m s^{-1}. This behavior can be attributed to a preponderance of short-lived liquid elements (i.e., those that are not completely penetrated by the dissolved gas) at the air–water interface at high wind speeds when S will be high due to increased turbulence. The change in the value of m was observed even when there was a trace quantity of a surfactant (triton) present on the water surface. For a triton dosage of 0.26 μmol L^{-1}, the water surface was smooth until a wind speed of 7 m s^{-1} and became wavy at wind speeds greater than 11 m s^{-1}, with the midpoint of the transition being around a wind speed of 9 m s^{-1}. The surface renewal model is unable to capture this behavior of a changing m and therefore offers only a partial view into the nature of reality, which is a typical feature of scientific models (Bailer-Jones, 2009).

Notation

a	parameter of the generalized Danckwerts age distribution
$c(x, \tau)$	dissolved-gas concentration at location x in a surface element of age τ (kmol m^{-3})
c_b	dissolved-gas concentration in the bulk liquid (kmol m^{-3})
c_s	dissolved-gas concentration at the gas–liquid interface, that is, gas solubility (kmol m^{-3})
D	diffusion coefficient of the dissolved gas in the liquid (m^2 s^{-1})
$f(\tau, t)$	age distribution of surface elements at process time t (s^{-1})
k_L	steady-state liquid-side mass-transfer coefficient or transfer velocity (m s^{-1} or cm h^{-1})

1 According to Jähne (1980), experimentally determined values of m are never greater than 0.75 for air–water heat and gas transfer (see Figure 4.8 on p. 110 in his work).

$k_L^{abs}(t)$	liquid-side gas-absorption coefficient ($m\ s^{-1}$)
$k_L^{trans}(t)$	liquid-side dissolved-gas transfer coefficient ($m\ s^{-1}$)
m	exponent
$R_{abs}(t)$	average rate of gas absorption at process time t ($kmol\ m^{-2}\ s^{-1}$)
$R_{inst}(\tau)$	instantaneous rate of gas absorption in a surface element having an age of τ ($kmol\ m^{-2}\ s^{-1}$)
$R_{trans}(t)$	average rate of transfer of dissolved gas to the bulk liquid at process time t ($kmol\ m^{-2}\ s^{-1}$)
S	surface renewal frequency (Hz or s^{-1})
t	process time (s)
x	distance into a surface element measured from the gas–liquid interface (m)

Greek Letters

α	defined by eq. 3.14
$\Gamma(z, y)$	incomplete gamma function defined by eq. 2.20 (Chapter 2), with z and y being its arguments
$\delta(t)$	delta function
τ	age of a surface element (s)

Homework Exercises

i) Derive the expression for $k_L^{abs}(t)$ given by eq. 3.6. Show that eq. 3.6 reduces to eq. 3.7 as $t \to 0$.

ii) Derive the expression for $k_L^{trans}(t)$ given by eq. 3.10. Show that eq. 3.10 reduces to eq. 3.11 as $t \to 0$.

iii) Using eqs. 3.6a and 3.10a, develop plots showing the transient behavior of k_L^{abs} and k_L^{trans} similar to those shown in Figure 3.1 for the absorption of hydrogen in water for values of $S = 0.0187$ and $0.1760\ s^{-1}$. The diffusion coefficient D of hydrogen in water is $6 \times 10^{-9}\ m^2\ s^{-1}$ at 25 °C. Comment on your results.

References

Astarita, G. (1967). *Mass transfer with chemical reaction*. Amsterdam: Elsevier.

Bailer-Jones, D.M. (2009). *Scientific models in philosophy of science*. Pittsburgh, PA: University of Pittsburgh Press.

Danckwerts, P.V. (1970). *Gas-liquid reactions*. New York: McGraw-Hill.

Garbe, C.S., Jähne, B. and Haußecker, H. (2002). "Measuring the sea surface heat flux and probability distribution of surface renewal events," in *Gas transfer at water surfaces* (eds. Donelan, M.A., Drennan, W.M., Saltzman, E.S. and Anninkhof, R.W.). Washington, DC: American Geophysical Union, pp. 109–114.

Horvath, I.R. and Chatterjee, S.G. (2018). "A surface renewal model for unsteady-state mass transfer using the generalized Danckwerts age distribution function," *Royal Society Open Science*, 5(5), 172423. Available at: http://dx.doi.org/10.1098/rsos.172423.

Hutchinson, M.H. and Sherwood, T.K. (1937). "Liquid film in gas absorption," *Industrial & Engineering Chemistry*, 29(7), pp. 836–840. Available at: http://dx.doi.org/10.1021/ie50331a023.

Jähne, B. (1980). *On the parametrization of air-sea gas transfer with the aid of laboratory experiments*. Heidelberg: Dissertation, University of Heidelberg. Available at: http: //www.ub.uni-heidelberg.de/archiv/16796.

Krall, K. (2013). *Laboratory investigations of air-sea gas transfer under a wide range of water surface conditions*. Heidelberg: Doctoral dissertation, University of Heidelberg. Available at: http://www.ub.uni-heidelberg.de/archiv/14392.

Mondal, C. and Chatterjee, S.G. (2017). "A frequency quantum interpretation of the surface renewal model of mass transfer," *Royal Society Open Science*, 4(7), 170103. Available at: http://dx.doi.org/10.1098/rsos.170103.

Perry, R.H., Green, D.W. and Maloney, J.O. (eds.) (1984). *Perry's chemical engineers' handbook*. Sixth edition. New York: McGraw-Hill.

4

Physical Gas Absorption in a Stirred Batch Cell*

4.1 Introduction

The standard oxygen-transfer model used for the analysis of diffused and surface aerators under unsteady-state conditions is given by Tchobanoglous and Burton (1991)

$$\frac{\mathrm{d}c_\mathrm{b}}{\mathrm{d}t} = K_\mathrm{L}a\left[c_\mathrm{sat} - c_\mathrm{b}(t)\right] \tag{4.1}$$

where $c_\mathrm{b}(t)$ is the dissolved-oxygen (DO) concentration in the bulk liquid at time t, c_sat is the saturation DO concentration, K_L is an overall mass-transfer coefficient, and a is the gas–liquid interfacial contact area per unit volume of the liquid. The ASCE (American Society of Civil Engineers) standard model is based upon eq. 4.1, as are the improved two-zone oxygen-transfer models for diffused and surface aerators (McWhirter and Hutter, 1989; McWhirter *et al.*, 1995; Chern and Yu 1997).

It should be noted that in eq. 4.1, the rate of absorption is expressed as the product of a mass-transfer coefficient and a concentration driving force (Astarita, 1967; Danckwerts, 1970), which form can be theoretically justified with the film, penetration, surface renewal, and film-penetration models of mass transfer but only for steady-state conditions. The use of the same driving force to also describe absorption under unsteady-state conditions is open to theoretical inquiry (Chatterjee, 2013).

Recently, Madygulov and Vlasov (2023) have presented a new model of physical gas absorption in a stirred liquid that rests on a radically different conceptual foundation than the one used traditionally. Their model, which visualizes the absorption and desorption processes at the gas–liquid interface as a reversible

* The material in this chapter has been adapted from Huang, H. and Chatterjee, S.G. (2021) "Transient physical gas absorption in a stirred liquid using the surface renewal model of mass transfer," *Chemical Engineering Science*, 234, p. 116449. Available at: https://doi.org/10.1016/j.ces.2021.116449.

Applications of the Surface Renewal Model of Mass Transfer, First Edition.
Hengshuo Huang and Siddharth G. Chatterjee.

Companion website: www.wiley.com/go/huang/surfacerenewalmodel

first-order reaction with no concentration gradient of the dissolved gas existing in the bulk liquid because of intensive stirring, interprets the parameter K_L in eq. 4.1 as the rate constant of the desorption reaction with an Arrhenius temperature dependence. Experimental measurements of methane and carbon dioxide absorption in a stirred batch cell conducted at different temperatures and pressures were in excellent agreement with their model.

The first objective of this chapter is to develop a rigorous surface renewal model, whose chief elements have been discussed in earlier chapters, for modeling transient physical gas absorption in a stirred liquid of finite volume in the presence of zero and significant gas-phase mass-transfer resistance. Thus, the work presented here goes beyond that in Chapter 3, which investigated, from the surface renewal viewpoint, transient gas absorption in a large body of liquid (i.e., where the bulk-liquid concentration of dissolved gas is constant), assuming zero gas-phase resistance to mass transfer. The second objective is to provide a theoretical justification for eq. 4.1. The model will consider separately both the rate of absorption at the gas–liquid interface (R_{abs}) and the rate of transfer of dissolved gas to the bulk liquid (R_{trans}) that is caused by the surface renewal mechanism. The model will then be applied to study the absorption of oxygen and hydrogen in water.

4.2 Interfacial Mass-Transfer Model

4.2.1 Zero Gas-Phase Mass-Transfer Resistance

The analysis presented in this section is also valid for solid dissolution in a liquid. In the absorption of a pure gas (i.e., no gas-phase mass-transfer resistance), the interfacial concentration of dissolved gas in the liquid, c_s, will remain constant (equal to the solubility c_{sat} of the gas in the liquid) provided the gas-phase concentration of the species being absorbed is held constant. In contrast, the bulk-liquid concentration of dissolved gas can be a function of the process time t.

According to the surface renewal model, at any time t since the beginning of absorption, the surface of the liquid in which the gas is being absorbed is composed of a mosaic of chaotic eddies or liquid elements, which have ages ranging from zero to t. During its residence at the gas–liquid interface, a liquid element is assumed to be isolated from the bulk liquid. The unsteady-state diffusion of dissolved gas in a surface element, which is assumed to be of infinite depth, during its exposure time θ at the interface can be modeled by the diffusion equation, that is:

$$\frac{\partial c}{\partial \theta} = D\frac{\partial^2 c}{\partial x^2} \tag{4.2}$$

with initial and boundary conditions specified by

$$c(x, 0) = c_b(\theta = 0) \text{ at } \theta = 0 \text{ for } 0 \le x \le \infty \tag{4.3}$$

$$c(\infty, \theta) = c_b(\theta = 0) \text{ for } \theta > 0 \text{ as } x \to \infty \tag{4.4}$$

$$c(0, \theta) = c_s \text{ for } \theta > 0 \text{ at } x = 0 \tag{4.5}$$

In these equations, c and c_b are the concentrations of dissolved gas at location x (measured from the gas–liquid interface) in the surface element at time θ and that in the bulk liquid, respectively, and D is the diffusion coefficient of the dissolved gas in the liquid.

The solution of eq. 4.2, subject to eqs. 4.3–4.5, is (Carslaw and Jaeger, 1959; Crank, 1964)

$$c(x, \theta) = c_b(\theta = 0) + \left[c_s - c_b(\theta = 0)\right] \operatorname{erfc}\left(\frac{x}{2\sqrt{D\theta}}\right) \tag{4.6}$$

which, upon differentiating, gives the instantaneous gas-absorption rate $R_{\text{inst}}(\theta)$ at the surface of the liquid element as

$$R_{\text{inst}}(\theta) = -D\left(\frac{\partial c}{\partial x}\right)_{x=0} = \left[c_s - c_b(\theta = 0)\right]\sqrt{\frac{D}{\pi\theta}} \tag{4.7}$$

Consider a surface element with an age of τ at time t. Since θ is the time elapsed since the element first arrived at the gas–liquid interface, $\theta = 0$ corresponds to a time of $t - \tau$ whereas $\theta = \tau$ corresponds to a time of t. Thus, the concentration profile $c(x, \tau, t)$ and rate of absorption $R_{\text{inst}}(\tau, t)$ in the liquid element can be obtained from eqs. 4.6 and 4.7, respectively, with $\theta = 0$ being replaced with $t - \tau$ and θ with τ. We now assume that surface elements have ages that can be characterized by the unsteady-state form of the Danckwerts age-distribution function given by (see eq. 2.10 in Chapter 2)

$$f_D(\tau, t) = \frac{Se^{-S\tau}}{1 - e^{-St}} \tag{4.8}$$

We note that other surface age distributions are possible (Mondal and Chatterjee, 2017), which reflect different hypotheses about the initial state of the gas–liquid interface when the absorption process commences. Equation 4.8, which rests upon the hypothesis that the gas–liquid interface is instantaneously and completely formed with liquid elements flowing into it from the bulk liquid and departing from it to the bulk liquid at a constant rate when absorption begins, is used as a representative age-distribution function because of its simple form and also because it has been used in the literature to model crossflow ultrafiltration (Koltuniewicz and Noworyta, 1994). According to this equation, as the process time $t \to \infty$, the age distribution approaches the well-known steady-state form

$Se^{-S\tau}$, which was originally derived by Danckwerts (1951). The rate or frequency of surface renewal S of liquid elements at the gas–liquid interface is taken to be constant, that is, the level of turbulence or stirrer speed is assumed to be fixed.

The transient rate of absorption $R_{abs}(t)$ of the gas is then given by

$$R_{abs}(t) = \int_0^t R_{inst}(\tau, t) f_D(\tau, t)\, d\tau \tag{4.9}$$

From eqs. 4.7 to 4.9, it follows that

$$R_{abs}(t) = \frac{S\sqrt{D/\pi}}{1 - e^{-St}} \int_0^t \left[c_s - c_b(t-\tau)\right] \tau^{-1/2} e^{-S\tau}\, d\tau \tag{4.10}$$

which can be put into the form

$$R_{abs}(t) = \frac{\sqrt{DS/\pi}}{1 - e^{-St}} \int_0^{St} f\left(t - \frac{\alpha}{S}\right) \alpha^{-1/2} e^{-\alpha}\, d\alpha \tag{4.11}$$

where

$$\alpha = S\tau \tag{4.12}$$

and

$$f(t) = c_s - c_b(t) \tag{4.13}$$

As $St \to 0$, eq. 4.11 reduces to

$$R_{abs}(St \to 0) = 2f(0)\sqrt{\frac{D}{\pi t}} \tag{4.11a}$$

that is, the rate of absorption is unbounded during the initial moments of absorption ($t \to 0$) or for low surface renewal rates ($S \to 0$) and is independent of the surface renewal rate, S. Equation 4.11a is the well-known expression for the rate of absorption as predicted by the penetration or Higbie model (Danckwerts, 1970).

The convolution integral in eq. 4.11 can be evaluated by the following numerical procedure. Let

$$I^A = \int_0^{St} f\left(t - \frac{\alpha}{S}\right) \alpha^{-1/2} e^{-\alpha}\, d\alpha \tag{4.14}$$

At time $t = t_1 > 0$, eq. 4.14 can be approximated as

$$I_1^A \approx \frac{f(t_1) + f(0)}{2} \int_0^{St_1} \alpha^{-1/2} e^{-\alpha}\, d\alpha = \frac{f(t_1) + f(0)}{2} \left[\Gamma(1/2, 0) - \Gamma(1/2, St_1)\right] \tag{4.15}$$

where $\Gamma(y, z)$ is the incomplete gamma function defined by

$$\Gamma(y, z) = \int_z^{\infty} \alpha^{y-1} e^{-\alpha} d\alpha \tag{4.16}$$

It is to be noted that $\Gamma(1/2, 0) = \sqrt{\pi}$.

Similarly, at time $t = t_2 > t_1$, eq. 4.14 can be expressed as

$$I_2^{\mathrm{A}} \approx \frac{f(t_2) + f(t_2 - t_1)}{2} \int_0^{St_1} \alpha^{-1/2} e^{-\alpha} d\alpha + \frac{f(t_2 - t_1) + f(0)}{2} \int_{St_1}^{St_2} \alpha^{-1/2} e^{-\alpha} d\alpha - \frac{f(t_2) + f(t_2 - t_1)}{2} \left[\Gamma(1/2, 0) - \Gamma(1/2, St_1)\right] + \frac{f(t_2 - t_1) + f(0)}{2} \left[\Gamma(1/2, St_1) - \Gamma(1/2, St_2)\right] \tag{4.17}$$

Hence, by induction, it follows that at time $t = t_n > t_{n-1}$

$$I_n^{\mathrm{A}} \approx \frac{1}{2} \sum_{j=1}^{n} \left[f(t_n - t_{j-1}) + f(t_n - t_j)\right] \left[\Gamma(1/2, St_{j-1}) - \Gamma(1/2, St_j)\right] \tag{4.18}$$

where $t_0 = 0$. If the time levels t_1, t_2, and so on are separated by a uniform increment Δt, that is, $t_1 = \Delta t$, $t_2 = t_1 + \Delta t$, ..., $t_n = t_{n-1} + \Delta t$, it can be shown that eq. 4.18 reduces to

$$I_n^{\mathrm{A}} \approx \frac{1}{2} \sum_{j=1}^{n} \left[f(t_{n-j+1}) + f(t_{n-j})\right] \left[\Gamma(1/2, St_{j-1}) - \Gamma(1/2, St_j)\right] \tag{4.19}$$

In an ideal continuous-flow stirred-tank absorber (CFSTA), that is, one in which the bulk liquid is assumed to be completely mixed, c_b would eventually attain a constant value as $t \to \infty$, that is, when steady state is reached. Under this situation, eq. 4.10 approaches the conventional linear driving force (LDF) form

$$R_{\mathrm{abs}}(t \to \infty) = k_{\mathrm{L}}(c_{\mathrm{s}} - c_{\mathrm{b}}) \tag{4.20}$$

where k_{L} is the liquid-side mass-transfer coefficient given by

$$k_{\mathrm{L}} = \sqrt{DS} \tag{4.21}$$

which is a well-known expression of the surface renewal model (Danckwerts, 1970). This same form is also approached in unsteady-state physical gas absorption (e.g., in a stirred cell), provided the rate of surface renewal is sufficiently high, as can be seen by letting $S \to \infty$ in eq. 4.11.

The transient net rate of transfer $R_{\mathrm{trans}}(t)$ of dissolved gas to the bulk liquid for surface elements of infinite depth is given by (see Chapter 3)

$$R_{\mathrm{trans}}(t) = S \int_0^t \int_0^{\infty} \left[c(x, \tau, t) - c_{\mathrm{b}}(t - \tau)\right] f_{\mathrm{D}}(\tau, t)\, dx\, d\tau \tag{4.22}$$

where the first and second terms of the integrand on the right-hand side represent the two opposite directions of convective mass transfer due to the surface renewal mechanism, and $f_D(\tau, t)$ is given by eq. 4.8. Using eq. 4.6 (with $\theta = 0$ replaced with $t - \tau$ and θ replaced with τ) and eq. 4.8 in eq. 4.22, integrating with respect to x, and noting that

$$\int_0^{\infty} \operatorname{erfc}\left(\frac{x}{2\sqrt{D\theta}}\right) dx = 2\sqrt{\frac{D\tau}{\pi}} \tag{4.23}$$

yield

$$R_{\text{trans}}(t) = \frac{2\sqrt{DS/\pi}}{1 - e^{-St}} \int_0^{St} f\left(t - \frac{\alpha}{S}\right) \alpha^{1/2} e^{-\alpha} d\alpha \tag{4.24}$$

where α and $f(t - \alpha/S)$ are given by eqs. 4.12 and 4.13, respectively. The numerical approximation of the integral in eq. 4.24 is given by eq. 4.18 or eq. 4.19 with the argument ½ of the incomplete gamma function being replaced by 1½. We note that $\Gamma(3/2, 0) = 0.5\sqrt{\pi}$.

As $St \to 0$, eq. 4.24 reduces to

$$R_{\text{trans}}(St \to 0) = \frac{4}{3} f(0) \sqrt{\frac{DS}{\pi}} (St)^{1/2} \tag{4.24a}$$

that is, the rate of dissolved-gas transfer depends on the surface renewal rate S and approaches zero if either $S \to 0$ (low surface renewal) or $t \to 0$.

According to eqs. 4.11a and 4.24a, the rate of gas absorption as $t \to 0$ is inversely proportional to $\sqrt{t}$, whereas the rate of dissolved-gas transfer is directly proportional to it, that is, they have an inverse relationship, which has been pointed out previously by Mondal and Chatterjee (2017) for the case when the bulk-liquid concentration of dissolved gas is constant. As indicated in Chapter 1, this crucial difference between the absorption and transfer rates has been overlooked in much of the research work on gas absorption.

Upon comparing eq. 4.24 for R_{trans} with eq. 4.11 for R_{abs}, it becomes evident that their mathematical structures are different, and it is only under steady-state conditions ($t \to \infty$) or for a sufficiently high frequency of surface renewal ($S \to \infty$) that they both approach the LDF form given by eq. 4.20.

4.2.2 Finite Gas-Phase Mass-Transfer Resistance

The governing equations are eqs. 4.2–4.4 with the interfacial boundary condition given by

$$k_G(y_b - Hc) = -D \left.\frac{\partial c}{\partial x}\right|_{x=0} \quad \text{for } \theta > 0 \text{ at } x = 0 \tag{4.25}$$

where the bulk gas-phase concentration y_b of the solute gas is assumed to be constant, and H is Henry's constant. The solution of eq. 4.2, subject to eqs. 4.3, 4.4, and 4.25, is given by (Carslaw and Jaeger, 1959; Crank, 1964)

$$c(x,\theta) = c_b(\theta = 0) + \left(\frac{y_b}{H} - c_b(\theta = 0)\right) \times \left[\operatorname{erfc}\left(\frac{x}{2\sqrt{D\theta}}\right) - e^{hx+h^2D\theta}\operatorname{erfc}\left(h\sqrt{D\theta} + \frac{x}{2\sqrt{D\theta}}\right)\right] \tag{4.26}$$

where

$$h = \frac{k_G H}{D} \tag{4.27}$$

The instantaneous rate of absorption in the liquid element can be obtained from eq. 4.26:

$$R_{inst}(\theta) = -D\left(\frac{\partial c}{\partial x}\right)_{x=0} = Dh\left[y_b/H - c_b(\theta = 0)\right] e^{h^2D\theta}\operatorname{erfc}\left(h\sqrt{D\theta}\right) \tag{4.28}$$

The age-averaged transient rate of absorption at time t is obtained by substituting eq. 4.28 (with $\theta = 0$ replaced with $t - \tau$ and θ replaced with τ, as mentioned earlier) into eq. 4.9 and using eq. 4.8, which yields

$$R_{abs}(t) = \frac{DSh}{1 - e^{-St}} \int_0^t \left[\frac{y_b}{H} - c_b(t - \tau)\right] \operatorname{erfc}\left(h\sqrt{D\tau}\right) e^{-(S-h^2D)\tau} d\tau \tag{4.29}$$

Equation 4.29 can be recast as

$$R_{abs}(t) = \frac{Dh}{1 - e^{-St}} \int_0^{St} \left[g\left(t - \frac{\alpha}{S}\right)\right] \operatorname{erfc}\left(h\sqrt{\frac{D\alpha}{S}}\right) e^{-\left(1-\frac{h^2D}{S}\right)\alpha} d\alpha \tag{4.30}$$

where

$$g(t) = y_b/H - c_b(t) \tag{4.31}$$

As $St \to 0$, eq. 4.30 reduces to

$$R_{abs}(St \to 0) = Dg(0)h = k_G Hg(0) \tag{4.30a}$$

that is, the rate of absorption is finite if either $S \to 0$ or $t \to 0$, which is in sharp contrast to the case of no gas-phase mass-transfer resistance (see eq. 4.11a).

For a large rate of surface renewal, that is, as $S \to \infty$, that is, when the liquid-side mass-transfer resistance vanishes, eq. 4.29 reduces to the pseudo-steady-state LDF form given by

$$R_{abs}(t) \approx k_G H\left[\frac{y_b}{H} - c_b(t)\right] \tag{4.32}$$

which, as indicated earlier, is the standard form of the absorption rate used in wastewater engineering for evaluating aerators – see eq. 4.1. Here, $K_L = k_G H$ and the quantity y_b/H is the solubility c_{sat} of the gas in the liquid.

In an ideal CFSTA as $t \to \infty$, eq. 4.30 attains the conventional LDF steady-state form

$$R_{\text{abs}}(t \to \infty) = K_{\text{OL}}\left(\frac{y_{\text{b}}}{H} - c_{\text{b}}\right) \tag{4.33}$$

where

$$\frac{1}{K_{\text{OL}}} = \frac{1}{k_{\text{L}}} + \frac{1}{Hk_{\text{G}}} \tag{4.34}$$

with k_{L} given by eq. 4.21. In deriving eq. 4.33, we have made use of the relation (Abramowitz and Stegun, 1965)

$$\int_0^{\infty} \text{erfc}\left(h\sqrt{D\tau}\right) e^{-(S-h^2D)\tau} \text{d}\tau = \frac{1}{\sqrt{S}\left(h\sqrt{D} + \sqrt{S}\right)} \tag{4.35}$$

Therefore, we note that the law of addition of resistances (eq. 4.34), which is widely used in the field of gas absorption, strictly holds only under steady-state conditions.

Substituting eq. 4.26 into eq. 4.22, integrating with respect to x, and using eqs. 4.8 and 4.23 along with the result

$$\int_0^{\infty} e^{hx}\text{erfc}\left(\frac{x}{2\sqrt{D\tau}} + h\sqrt{D\tau}\right)\text{d}x = \frac{1}{h}\left[e^{-h^2D\tau} - \text{erfc}\left(h\sqrt{D\tau}\right)\right] \tag{4.36}$$

give

$$R_{\text{trans}}(t) = \frac{S}{1 - e^{-St}} \int_0^{St} g\left(t - \frac{\alpha}{S}\right) \times \left\{2\sqrt{\frac{D\alpha}{\pi S}} - \frac{1}{h}\left[1 - e^{h^2D\frac{\alpha}{S}}\text{erfc}\left(h\sqrt{\frac{D\alpha}{S}}\right)\right]\right\} e^{-\alpha}\text{d}\alpha \tag{4.37}$$

where $g(t)$ is defined by eq. (4.31). As $St \to 0$, eq. 4.37 approaches the form

$$R_{\text{trans}}(St \to 0) = \frac{4}{3}g(0)\sqrt{\frac{DS}{\pi}}(St)^{1/2} \tag{4.37a}$$

which can be compared to eq. 4.24a for the case of zero gas-phase mass-transfer resistance.

It is again observed by comparing eqs. 4.30 and 4.37 that in general, $R_{\text{abs}}(t) \neq R_{\text{trans}}$. Only for a sufficiently high rate of surface renewal ($S \to \infty$) or under steady-state conditions ($t \to \infty$), $R_{\text{trans}}(t)$ approaches $R_{\text{abs}}(t)$ and attains the LDF form given by eqs. 4.32 or 4.33.

The integral in eq. 4.30 can be evaluated by a similar procedure to that used for the zero gas-phase mass-transfer resistance case. Let

$$I^{\text{B}} = \int_0^{St} g(t - \alpha/S)\,\text{erfc}\left(h\sqrt{\frac{D\alpha}{S}}\right) e^{-(1-h^2D/S)\alpha}\text{d}\alpha \tag{4.38}$$

At time $t = t_n$ ($>t_{n-1}$)

$$I_n^B \approx \frac{1}{2} \sum_{j=1}^{n} \left[g\left(t_n - t_{j-1}\right) + g\left(t_n - t_j\right)\right] \left[J\left(St_j\right) - J\left(St_{j-1}\right)\right] \tag{4.39}$$

If the time levels t_1, t_2, and so on are separated by a uniform increment Δt, then

$$I_n^B \approx \frac{1}{2} \sum_{j=1}^{n} \left[g\left(t_{n-j+1}\right) + g\left(t_{n-j}\right)\right] \left[J\left(St_j\right) - J\left(St_{j-1}\right)\right] \tag{4.40}$$

where the function $J(\beta)$ is defined by

$$J(\beta) = \int_0^{\beta} e^{-(1-h^2D/S)\alpha} \operatorname{erfc}\left(h\sqrt{\frac{D\alpha}{S}}\right) d\alpha = \frac{1}{h\sqrt{D/S}+1} + \frac{1}{h^2D/S-1}$$
$$\times \left[e^{-(1-h^2D/S)\beta} \operatorname{erfc}\left(h\sqrt{D\beta/S}\right) - h\sqrt{\frac{D}{\pi S}}\Gamma\left(\frac{1}{2},\beta\right)\right] \tag{4.41}$$

The approximate representation of the integral in eq. 4.37 is given by eq. 4.39 or eq. 4.40 with the function $J(\beta)$ replaced with another function $K(\beta)$, which is defined by

$$K(\beta) = \int_0^{\beta} \left\{2\sqrt{\frac{D\alpha}{\pi S}} - \frac{1}{h}\left[e^{h^2D\alpha/S} \operatorname{erfc}\left(h\sqrt{D\alpha/S}\right)\right]\right\} e^{-\alpha} d\alpha$$
$$= 2\sqrt{\frac{D}{\pi S}}\left[\frac{\sqrt{\pi}}{2} - \Gamma\left(\frac{3}{2},\beta\right)\right] + \frac{1}{h}\left[J(\beta) + e^{-\beta} - 1\right] \tag{4.42}$$

where $J(\beta)$ is given by eq. 4.41.

4.3 Philosophical Interlude

The rates of absorption and dissolved-gas transfer, as seen from eqs. 4.11 and 4.24 for the zero gas-phase mass-transfer resistance case and eqs. 4.30 and 4.37 for the finite gas-phase mass-transfer resistance case, are expressed in terms of convolution integrals. According to these equations, the rates of absorption and dissolved-gas transfer at any time t contain the entire history of the concentration driving force (multiplied by weighting factors) starting from time $t = 0$ onward. If $f(t)$ in eqs. 4.11 and 4.24 or $g(t)$ in eqs. 4.30 and 4.37 is considered to be amplitudes of waves at time t, then all such waves, whose amplitudes vary from $f(0)$ to $f(t)$ or $g(0)$ to $g(t)$, are transformed or enfolded into the instantaneous or point value of the absorption rate $R_{\text{abs}}(t)$ or transfer rate $R_{\text{trans}}(t)$ at process time t. This is analogous to a hologram in which every point of the image enfolds or carries information about the entire object. Bohm (1980), who used the example of the

Green's function, referred to such transformations as metamorphoses.[1] Only for a sufficiently high rate of surface renewal ($S \to \infty$) or under steady-state conditions ($t \to \infty$) when the past history becomes insignificant, the holographic form of these equations reduces to the familiar LDF form (see eqs. 4.20, 4.32, and 4.33), which can be compared metaphorically to the ordinary point-to-point correspondence between a structure and its image on a photographic plate, that is, there is a one-to-one mapping between the instantaneous concentration driving force and the absorption rate or dissolved-gas transfer rate at that moment. Thus, the holographic form of these equations collapses into the appropriate photographic form as $S \to \infty$ or as $t \to \infty$.

The equations for R_{abs} and R_{trans} given by 4.11, 4.24, 4.30, and 4.37 reflect the philosophical principle of undivided wholeness, according to which the part contains the whole or is an aspect of the whole. This can be contrasted with the Cartesian mechanistic philosophical framework of modern science, which believes that an entity or a process is composed of a number of fundamental and unchanging parts and their pre-assigned interactions from which the whole can be derived. For more details on the profound differences between these two worldviews, the reader is referred to Bohm (2003) and Chatterjee (2013).

4.4 Bulk Mass-Transfer Model

Let us consider the unsteady-state batch operation of a stirred-tank gas–liquid absorber in which there is no inflow or outflow of liquid. The mass balance for dissolved gas in the bulk liquid is given by

$$\frac{\mathrm{d}c_{\text{b}}}{\mathrm{d}t} = aR_{\text{trans}}(t) \tag{4.43}$$

with

$$c_{\text{b}} = c_{\text{b,init}} \text{ at } t = 0 \tag{4.44}$$

Here, $c_{\text{b,init}}$ is the initial dissolved-gas concentration in the bulk liquid and, as mentioned earlier, a is the gas–liquid interfacial contact area per unit volume of liquid, whose value is either known or can be obtained from an empirical correlation (see Section 5.3 in Chapter 5).

1 According to Bohm (1980), a metamorphosis refers to a transformation "in which everything alters in a thoroughgoing manner while some subtle and highly implicit features remain invariant." This change, for example, between an object and its holographic image, is much more radical than a geometric transformation (i.e., translation, rotation, or dilation).

4.4.1 Zero Gas-Phase Mass-Transfer Resistance

Substituting eq. 4.24 into eq. 4.43 yields the following integro-differential equation that describes the change of the bulk liquid dissolved-gas concentration with process time:

$$\frac{dc_b}{dt} = -\frac{df}{dt} = \frac{K_1}{1 - e^{-St}} \int_0^{St} f(t - \alpha/S)\,\alpha^{1/2} e^{-\alpha}\,d\alpha \tag{4.45}$$

which can be compared to the LDF model expressed by eq. 4.1. Here

$$K_1 = 2a\sqrt{DS/\pi} \tag{4.46}$$

For $t \to 0$, substituting eq. 4.24a into eq. 4.43 and integrating with respect to t yield

$$c_b(t \to 0) = \frac{4}{9} K_1 \sqrt{S} f(0)\, t^{3/2} + c_{b,init} \tag{4.47}$$

where

$$f(0) = c_{sat} - c_{b,init} \tag{4.48}$$

which follows from eq. 4.13. Eq. 4.47 can be used to calculate the value of c_b at the first time level $t_1 = \Delta t$, that is, $c_b(t_1)$ after which $f(t_1)$ may be obtained from eq. (4.13):

$$f(t_1) = c_{sat} - c_b(t_1) \tag{4.49}$$

For all subsequent time levels, that is, $t_2 = t_1 + \Delta t, \ldots, t_n = t_{n-1} + \Delta t$, the following procedure can be used to calculate $f(t_{n+1})$, given the values of f at all earlier time levels. Representing the time derivative in eq. 4.45 by a centered-difference approximation and using the numerical approximation of the convolution integral on its right-hand side, as discussed earlier, yield

$$f(t_{n+1}) = f(t_{n-1}) - \frac{K_1 \Delta t}{1 - e^{-St_n}} \sum_{j=1}^{n} \left[f(t_{n-j+1}) + f(t_{n-j}) \right] \left[\Gamma(3/2, St_{j-1}) - \Gamma(3/2, St_j) \right] \tag{4.50}$$

where $n = 1, 2, 3, \ldots$ The values of c_b at these time levels can then be calculated using eq. 4.13.

Once $c_b(t)$ is known, the gas absorption and transfer rates can be obtained from eqs. 4.11 and 4.24, respectively, by employing the numerical procedure given in Section 4.2.1.

4.4.2 Finite Gas-Phase Mass-Transfer Resistance

The mass balance for dissolved gas in the bulk liquid is given by eq. 4.43 with R_{trans} now to be obtained from eq. 4.37. The procedure for calculating $c_b(t)$ is similar to that for the zero gas-phase mass-transfer resistance case. For the first time level t_1, the value of $c_b(t_1)$ can be calculated by eq. 4.47 with $f(0)$ replaced by $g(0)$, where (see eq. 4.31)

$$g(0) = y_b/H - c_{b,init} \tag{4.51}$$

For all subsequent time levels, the following equation is used to determine $g(t_{n+1})$:

$$g\left(t_{n+1}\right) = g\left(t_{n-1}\right) - \frac{K_2 \Delta t}{1 - e^{-St_n}} \sum_{j=1}^{n} \left[g\left(t_{n-j+1}\right) + g\left(t_{n-j}\right)\right] \left[K\left(St_j\right) - K\left(St_{j-1}\right)\right] \tag{4.52}$$

where the function K is defined by eq. 4.42, $n = 1, 2, 3, \ldots$, and $K_2 = aS$. The value of c_b at these time levels can then be obtained from eq. 4.31.

After $c_b(t)$ has been determined, $R_{abs}(t)$ and $R_{trans}(t)$ can be obtained from eqs. 4.30 and 4.37, respectively, using the numerical method described earlier in Section 4.2.2.

4.5 Analytical Expressions for the Rates of Absorption and Dissolved-Gas Transfer Using the Laplace Transform Method

The Laplace transform technique can be used to develop explicit mathematical formulas for the rates of gas absorption and dissolved-gas transfer, $R_{abs}(t)$ and $R_{trans}(t)$, respectively, which is not possible with the numerical method presented in Sections 4.2.1 and 4.2.2. The Gaver–Stehfest numerical inversion procedure (Abate and Whitt, 2006) is employed to determine $c_b(t)$ from its Laplace transform.

4.5.1 Gaver–Stehfest Algorithm

Consider a function $p(t)$ whose Laplace transform is $P(z)$, where z is the Laplace transform parameter. If $P(z)$ is known, $p(t)$ can be determined from the formula

$$p(t) = \frac{\ln 2}{t} \sum_{i=1}^{2M} w_i \cdot P\left(\frac{i \ln 2}{t}\right) \tag{4.53}$$

where the weight w_i is obtained from

$$w_i = \frac{(-1)^{M+i}}{M!} \sum_{j=\left\lfloor \frac{i+1}{2} \right\rfloor}^{\min\{i,M\}} j^{M+1} C_j^M \, C_j^{2j} \, C_{i-j}^j \tag{4.54}$$

with $\lfloor y \rfloor$ being the greatest integer less than or equal to y and C_l^m being the binomial coefficient, which is defined by

$$C_l^m = \frac{m!}{l!\,(m-l)!} \tag{4.55}$$

The quantity $2M$ in eq. 4.53 equals the number of transform evaluations required, $2.2M$ is the required system precision in decimal digits, and $0.90M$ is the number of significant digits produced for good transforms (Abate and Whitt, 2006). Therefore, the absolute fractional relative error in the evaluation of the inverse transform is estimated to be $10^{-0.90M}$.

4.5.2 Zero Gas-Phase Mass-Transfer Resistance

Equation 4.24 may be rewritten as

$$R_{\text{trans}}(t) = \frac{2S^2}{1-e^{-St}}\sqrt{\frac{D}{\pi}}\int_0^t f(t-\tau)\,\tau^{\frac{1}{2}}e^{-S\tau}\,d\tau \tag{4.56}$$

For $St > 4$, $1-e^{-St} > 0.9817 \approx 1$, and so to a good approximation, eq. 4.56 can be expressed as

$$R_{\text{trans}}(t) \approx 2S^2\sqrt{\frac{D}{\pi}}\int_0^t f(t-\tau)\cdot\tau^{\frac{1}{2}}\,e^{-S\tau}\,d\tau \tag{4.57}$$

Upon substitution of eq. 4.57 into eq. 4.43, using eq. 4.13, and taking the Laplace transform of both sides of the resulting equation, we obtain

$$\mathcal{L}\left[-\frac{df(t)}{dt}\right] = 2aS^2\sqrt{\frac{D}{\pi}}\cdot\mathcal{L}\left[\int_0^t f(t-\tau)\cdot\tau^{\frac{1}{2}}\,e^{-S\tau}\,d\tau\right] \tag{4.58}$$

where $\mathcal{L}$ denotes the Laplace transform operator. Applying the derivative and convolution theorems of the Laplace transformation to eq. 4.58 gives

$$-\left[zF(z)-f(0)\right] = 2aS^2\sqrt{\frac{D}{\pi}}\cdot F(z)\cdot\frac{\sqrt{\pi}}{2}\frac{1}{(z+S)^{\frac{3}{2}}} \tag{4.59}$$

with $F(z)$ being the Laplace transform of $f(t)$. Solving eq. 4.59 for $F(z)$ yields

$$F(z) = f(0)\,F_1(z) \tag{4.60}$$

where $f(0)$ is given by eq. 4.48 and

$$F_1(z) = \frac{(z+S)^{\frac{3}{2}}}{z(z+S)^{\frac{3}{2}} + aS^2\sqrt{D}} \tag{4.61}$$

It follows from eqs. 4.48, 4.53, and 4.60 that

$$c_b(t) = c_{\text{sat}}\left[1 - \frac{\left(c_{\text{sat}}-c_{\text{b,init}}\right)}{c_{\text{sat}}}\frac{\ln 2}{t}\sum_{i=1}^{2M} w_i\,F_1\left(\frac{i\ln 2}{t}\right)\right] \tag{4.62}$$

where the weight w_i and function $F_1(z)$ are given by eqs. 4.54 and 4.61, respectively.

The rates of absorption and dissolved-gas transfer can be obtained by first using the method of nonlinear least squares to fit numerical values of $c_b(t)$ calculated using eq. 4.62 with the following expression obtained from the conventional LDF pseudo-steady-state model, which is presented in Section 4.6 (see eq. 4.73):

$$c_b(t) = c_{sat}\left(1 - e^{-k_L a t}\right) + c_{b,init} e^{-k_L a t} \tag{4.63}$$

where $k_L a$ is to be considered a fitting parameter. Substitution of eq. 4.63 into eqs. 4.10 and 4.24 then yields the following explicit formulas for the gas absorption and dissolved-gas transfer rates:

$$R_{abs}(t) = c_{sat}\frac{\sqrt{D}\,S}{\left(1 - e^{-St}\right)}\cdot\frac{e^{-k_L a t}}{\sqrt{S - k_L a}}\operatorname{erf}\left(\sqrt{\left(S - k_L a\right)t}\right) \tag{4.64}$$

and

$$R_{trans}(t) = 2c_{sat}\sqrt{\frac{D}{\pi}}\cdot\frac{S^2}{\left(1 - e^{-St}\right)}\left[\frac{\sqrt{\pi}\operatorname{erf}\left(\sqrt{\left(S - k_L a\right)t}\right)e^{-k_L a t}}{2\left(S - k_L a\right)^{\frac{3}{2}}} - \frac{\sqrt{t}\,e^{-St}}{S - k_L a}\right] \tag{4.65}$$

where it has been assumed that $c_{b,init} = 0$. Equations 4.64 and 4.65 can be expressed in an alternative fashion by using the extended Euler gamma function (see eq. 4.16) as

$$R_{abs}(t) = c_{sat}\sqrt{\frac{D}{\pi}}\cdot\frac{S}{\left(1 - e^{-St}\right)}\cdot\frac{e^{-k_L a t}}{\sqrt{S - k_L a}}\left[\Gamma\left(\frac{1}{2}, 0\right) - \Gamma\left(\frac{1}{2}, \left(S - k_L a\right)t\right)\right] \tag{4.66}$$

and

$$R_{trans}(t) = 2c_{sat}\sqrt{\frac{D}{\pi}}\cdot\frac{e^{-k_L a t}}{\left(1 - e^{-St}\right)}\cdot\frac{S^2}{\left(S - k_L a\right)^{\frac{3}{2}}}\left[\Gamma\left(\frac{3}{2}, 0\right) - \Gamma\left(\frac{3}{2}, \left(S - k_L a\right)t\right)\right] \tag{4.67}$$

Equations 4.64 and 4.65 or their equivalents (eqs. 4.66 and 4.67) show that the rates of absorption and dissolved-gas transfer depend upon the following four parameters: c_{sat}, D, S, and $k_L a$ (volumetric mass-transfer coefficient).

4.5.3 Finite Gas-Phase Mass-Transfer Resistance

By following a similar procedure to that used for the case of zero gas-phase mass-transfer resistance, it may be shown that

$$\mathcal{L}\left[c_{\mathrm{b}}(t)\right]=\left(\frac{y_{\mathrm{b}}}{H}-c_{\mathrm{b,init}}\right)\left[1-\frac{1}{a\,S^{2}R(z)+z}\right] \tag{4.68}$$

where the function $R(z)$ is given by

$$R(z)=\frac{\sqrt{D}}{(z+S)^{\frac{3}{2}}}-\frac{1}{h(z+S)}+\frac{1}{h}\cdot\frac{1}{\sqrt{z+S}\cdot\left(\sqrt{z+S}+h\sqrt{D}\right)} \tag{4.69}$$

To obtain the rates of absorption and dissolved-gas transfer, the values of $c_{\mathrm{b}}(t)$, obtained from the numerical inversion of eq. 4.68 by the Gaver–Stehfest algorithm, are fitted with the pseudo-steady-state expression (see eq. 4.73)

$$c_{\mathrm{b}}(t)=\frac{y_{\mathrm{b}}}{H}\left(1-e^{-K_{\mathrm{OL}}at}\right)+c_{\mathrm{b,init}}e^{-K_{\mathrm{OL}}at} \tag{4.70}$$

where $K_{\mathrm{OL}}a$ is a fitting parameter. Substitution of eq. 4.70 into eqs. 4.30 and 4.37 then yields the following equations (assuming $c_{\mathrm{b,init}}=0$):

$$R_{\mathrm{abs}}(t)=\frac{Dhy_{\mathrm{b}}}{H}\cdot\frac{S}{\left(1-e^{-St}\right)}\cdot\left(\frac{e^{-K_{\mathrm{OL}}at}}{h^{2}D+K_{\mathrm{OL}}a-S}\right)\times\left[\operatorname{erfc}\left(h\sqrt{Dt}\right)e^{\left(h^{2}D+K_{\mathrm{OL}}a-S\right)t}+\frac{h\sqrt{D}\operatorname{erf}\left(\sqrt{\left(S-K_{\mathrm{OL}}a\right)t}\right)}{\sqrt{S-K_{\mathrm{OL}}a}}-1\right] \tag{4.71}$$

and

$$R_{\mathrm{trans}}(t)=\frac{S^{2}y_{\mathrm{b}}}{H}\cdot\frac{e^{-K_{\mathrm{OL}}at}}{\left(1-e^{-St}\right)}\times\left\{\begin{array}{l}-\dfrac{2\sqrt{Dt}\,e^{\left(K_{\mathrm{OL}}a-S\right)t}}{\sqrt{\pi}\left(S-K_{\mathrm{OL}}a\right)}+\dfrac{\sqrt{D}\operatorname{erf}\left(\sqrt{\left(S-K_{\mathrm{OL}}a\right)t}\right)}{\left(S-K_{\mathrm{OL}}a\right)^{\frac{3}{2}}}+\dfrac{e^{\left(K_{\mathrm{OL}}a-S\right)t}-1}{h\left(S-K_{\mathrm{OL}}a\right)}+\dfrac{1}{h\left(h^{2}D+K_{\mathrm{OL}}a-S\right)}\\ \times\left[\operatorname{erfc}\left(h\sqrt{Dt}\right)e^{\left(h^{2}D+K_{\mathrm{OL}}a-S\right)t}+\dfrac{h\sqrt{D}\operatorname{erf}\left(\sqrt{\left(S-K_{\mathrm{OL}}a\right)t}\right)}{\sqrt{S-K_{\mathrm{OL}}a}}-1\right]\end{array}\right\} \tag{4.72}$$

Since $c_{\mathrm{sat}}=y_{\mathrm{b}}/H$, it follows from eqs. 4.27, 4.71, and 4.72 that R_{abs} and R_{trans} depend on the following six parameters: c_{sat}, D, k_{G}, H, S, and $K_{\mathrm{OL}}a$ (volumetric mass-transfer coefficient).

The Laplace transform technique discussed in Sections 4.5.2 and 4.5.3 used the approximation $1-e^{-St}>0.9817\approx1$ for $St>4$ (e.g., see eqs. 4.56 and 4.57). A more rigorous method, which does not use this approximation, is presented in Appendix B for the cases of both zero and finite gas-phase mass-transfer resistance.

4.6 Conventional Pseudo-Steady-State LDF Model

This simple model, which does not distinguish between the gas absorption and dissolved-gas transfer rates (i.e., it tacitly assumes that they are the same) in contrast to the rigorous surface renewal model presented earlier, is obtained by integrating eq. 4.1 using the initial condition specified by eq. 4.44:

$$c_{\rm b}(t) = c_{\rm sat}\left(1 - e^{-K_{\rm L}at}\right) + c_{\rm b,init}e^{-K_{\rm L}at} \tag{4.73}$$

where $K_{\rm L} = k_{\rm L}$ (see eq. 4.21) for no gas-phase mass-transfer resistance, and $K_{\rm L} = K_{\rm OL}$ (see eq. 4.34) and $c_{\rm sat} = y_{\rm b}/H$ for the case of finite gas-phase mass-transfer resistance. From eq. 4.73, it follows that

$$R_{\rm abs}(t) = R_{\rm trans}(t) = K_{\rm L}\left[c_{\rm sat} - c_{\rm b}(t)\right] \tag{4.74}$$

It is clear from eq. 4.74 that the rate of absorption or dissolved-gas transfer is maximum but finite when $t = 0$ and thereafter declines monotonically with process time as $c_{\rm b}(t)$ increases. It also follows from eq. 4.73 that $c_{\rm b}(t)$ is governed by $c_{\rm sat}$, $c_{\rm b,init}$, and the volumetric mass-transfer coefficient $K_{\rm L}a$, whose magnitude can be determined by fitting eq. 4.73 to transient dissolved-gas concentration measurements, as is common practice. However, the rate of absorption or transfer, as can be seen from eq. (4.74), is governed by only two parameters, that is, $c_{\rm sat}$ and the overall mass-transfer coefficient, $K_{\rm L}$.

4.7 Absorption of Oxygen and Hydrogen in Water

This section illustrates the application of the surface renewal model to describe the absorption of oxygen and hydrogen in water contained in a stirred batch cell. A value of $M = 6$ in the Gaver–Stehfest algorithm (Section 4.5.1) is used, and the initial liquid-phase concentration of dissolved gas, $c_{\rm b,init}$, is taken to be zero in all calculations.

The absorption of eight different pure gases at 25 °C and atmospheric pressure in a stirred flask containing water, with the water surface being exposed to the gas, was studied by Hutchinson and Sherwood (1937) who calculated values of the liquid-side mass-transfer coefficient $k_{\rm L}$ from transient dissolved-gas concentration measurements and the integrated form of eq. 1 (see eq. 4.73) at different stirrer speeds. For demonstration purposes, we will first consider their data for oxygen absorption at a stirrer speed of 171 RPM (revolutions per minute). The value of D (diffusion coefficient of DO in water) at 25 °C was estimated from the diffusivity value at 20 °C, reported in Table III of their paper (Hutchinson and Sherwood, 1937), and by using the Wilke–Chang correlation (Perry *et al.*, 1984) to correct it to

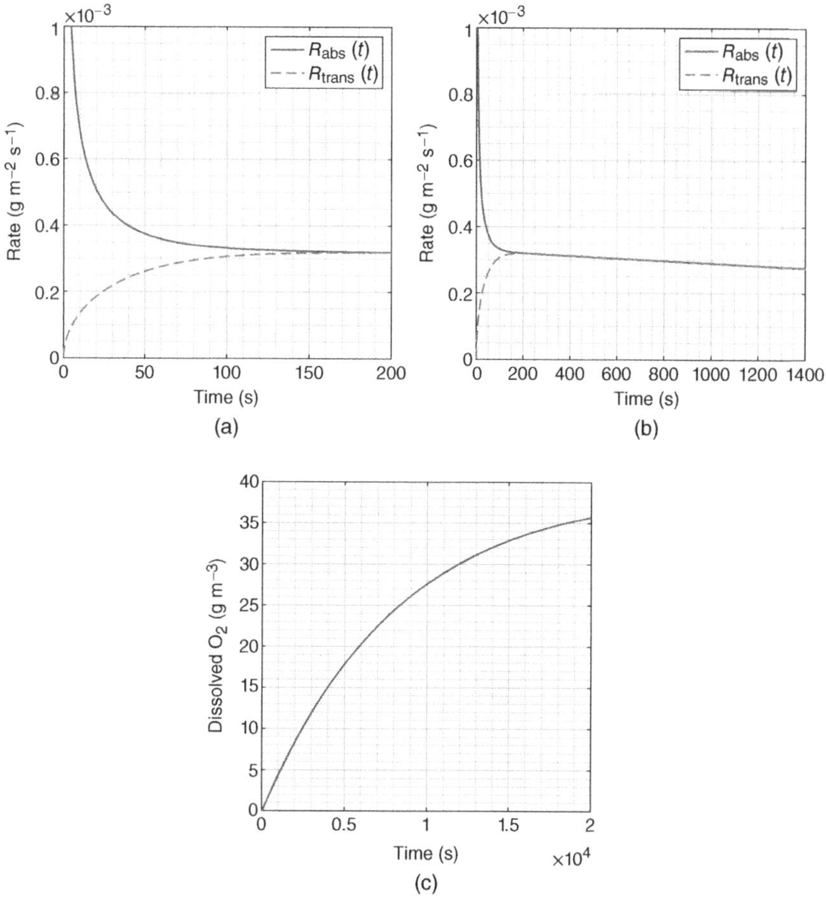

Figure 4.1 Variation of the rates of gas absorption and dissolved-gas transfer (a, b) and bulk-liquid concentration of dissolved gas (c) with process time in the absorption of pure oxygen into water contained in a stirred flask at 25 °C and atmospheric pressure for the case of zero gas-phase mass-transfer resistance. Parameter values used in the calculations, which were performed with the equations developed in Sections 4.2.1 and 4.4.1, are reported in Table 4.1. Source: Huang and Chatterjee (2021) / with permission of Elsevier.

25 °C. Values of 1, 2, 10, and 100 s were used for the time step Δt in the calculations, which were performed with the equations developed in Sections 4.2 and 4.4.

Figure 4.1 shows results for the absorption of pure oxygen (i.e., zero gas-phase mass-transfer resistance) in a stirred flask of water using the surface renewal model that was presented in Sections 4.2.1 and 4.4.1. Table 4.1 reports values of the parameters that were used in the model. The solubility of pure oxygen in water

Table 4.1 Parameter values for the transient absorption of pure oxygen (i.e., zero gas-phase mass-transfer resistance) into water contained in a stirred flask with oxygen above the water surface.

Specified parameters	Value
temperature[a]	25 °C
pressure[a]	1 atm
solubility of pure oxygen in water (c_{sat})[b]	$39.04\,g\,m^{-3}$
diffusion coefficient of oxygen in water (D; estimated from the value at 20 °C[a] and the Wilke–Chang correlation[c])	$2.14 \times 10^{-9}\,m^2\,s^{-1}$
liquid-phase mass-transfer coefficient (k_L)[a]	$8.33 \times 10^{-6}\,m\,s^{-1}$ ($3\,cm\,h^{-1}$; stirrer speed = 171 RPM)
gas–liquid interfacial area of contact (A)[a]	$7.35 \times 10^{-4}\,m^2$ ($7.35\,cm^2$)
volume of liquid in flask (V)[a]	$5 \times 10^{-5}\,m^3$ ($50\,cm^3$)
Calculated parameters	**Value**
gas–liquid interfacial area per unit volume of liquid ($a = A/V$)	$14.7\,m^{-1}$
surface renewal frequency ($S = k_L{}^2/D$; estimated from eq. 4.21)	0.0325 Hz (s^{-1})

a) Hutchinson and Sherwood (1937).
b) Xing *et al.* (2014).
c) Perry *et al.* (1984).
Source: Huang and Chatterjee (2021) / with permission of Elsevier.

at one atmosphere and 25 °C was obtained from Xing *et al.* (2014), whereas the values of the remaining specified parameters are those reported by Hutchinson and Sherwood (1937). The magnitudes of the gas–liquid interfacial contact area a and surface renewal rate S were estimated as shown in the table. The transient behavior of the rates of gas absorption (R_{abs}) and dissolved-gas transfer (R_{trans}) to the bulk liquid is depicted in Figure 4.1a for short times ($t = 0$ to 200 s) after the start of absorption. It is observed that R_{abs} decays in an exponential-like manner with process time t, whereas, in sharp contrast, R_{trans} increases in a smooth fashion from a value of zero, joining the absorption rate curve at about $t = 150$ s (2.5 min) when these two rates are approximately $3.2 \times 10^{-4}\,g\,m^{-2}\,s^{-1}$. After this initial time period, the absorption and transfer rates virtually merge into a single curve and decline with process time as the water becomes increasingly saturated with oxygen (Figure 4.1b). This behavior can be contrasted with the case of transient absorption of oxygen in a large body of water that was analyzed with the surface renewal model in Chapter 3. In that case, the absorption and transfer coefficients, which are equal to the respective rates divided by a constant concentration driving force, although distinct initially, merged together after a time period ranging from a few seconds to a few minutes, and subsequently remained constant with process time (see Figure 3.1 in Chapter 3). That is, there was no decline in these rates with time unlike here, which is concerned with physical gas absorption in a finite volume of liquid. The variation of DO in the bulk liquid with time shown in Figure 4.1c follows the typical pattern that is observed experimentally. Initially, the DO increases rapidly and thereafter increasingly slowly as saturation is approached. At $t = 20{,}000$ s (about 5.6 h), the DO level reaches 35.72 g m^{-3}, which is 91.5% of the saturation value (c_{sat}), as shown in Table 4.1.

Figure 4.2 shows results for the absorption of oxygen from air into water when the gas-phase mass-transfer resistance is finite (Sections 4.2.2 and 4.4.2); the parameter values used in the calculations are provided in Table 4.2. The value of Henry's law coefficient H for oxygen in water was derived from the compilation of Sander (2015). The results are qualitatively similar to those for the case of zero-gas phase mass-transfer resistance (i.e., absorption of pure oxygen) except that the magnitudes of the absorption and transfer rates and the DO concentration in the bulk liquid are much lower because of the presence of gas-phase mass-transfer resistance and the molar concentration of oxygen in air being only 21%.

We now look at results obtained with the Laplace transform method for the case of finite gas-phase mass-transfer resistance (Section 4.5.3). Figure 4.3 shows the behavior of the bulk-liquid DO concentration with process time over a seven order-of-magnitude change in k_G. As k_G decreases from 3×10^{-2} to 3×10^{-5} m s^{-1}, the DO curve changes insignificantly because the mass-transfer resistance lies almost entirely in the liquid. When $k_G = 3 \times 10^{-6}$ m s^{-1}, about 8% of the total mass-transfer resistance, estimated from the law of addition of resistances (eq. 4.34), which, as mentioned earlier, is strictly valid only under steady-state conditions, lies in the gas phase with the remaining 92% residing in the liquid phase, causing a slight depression in the DO level. As k_G decreases to 3×10^{-7} m s^{-1} and then further to 3×10^{-8} m s^{-1}, the gas-phase resistance starts becoming important, constituting 31% and 47% of the total resistance, respectively, which produces a significant depression in the DO level. A further drop by one order of magnitude in k_G shifts a major part of the overall mass-transfer resistance to the gas phase and leads to a great reduction in the DO level and in its rate of increase with time (Figure 4.3).

When there is no gas-phase mass-transfer resistance, the value of k_L obtained with the Laplace transform method (Section 4.5.2) is equal to 8.28×10^{-6} m s^{-1}, which is in very good agreement with the value of 8.33×10^{-6} m s^{-1} for k_L shown in Table 4.1. Likewise, in the presence of finite gas-phase mass-transfer resistance, the value of K_{OL} obtained with the Laplace transform method (Section 4.5.3) and that calculated from the steady-state law of addition of resistances (eq. (4.34))

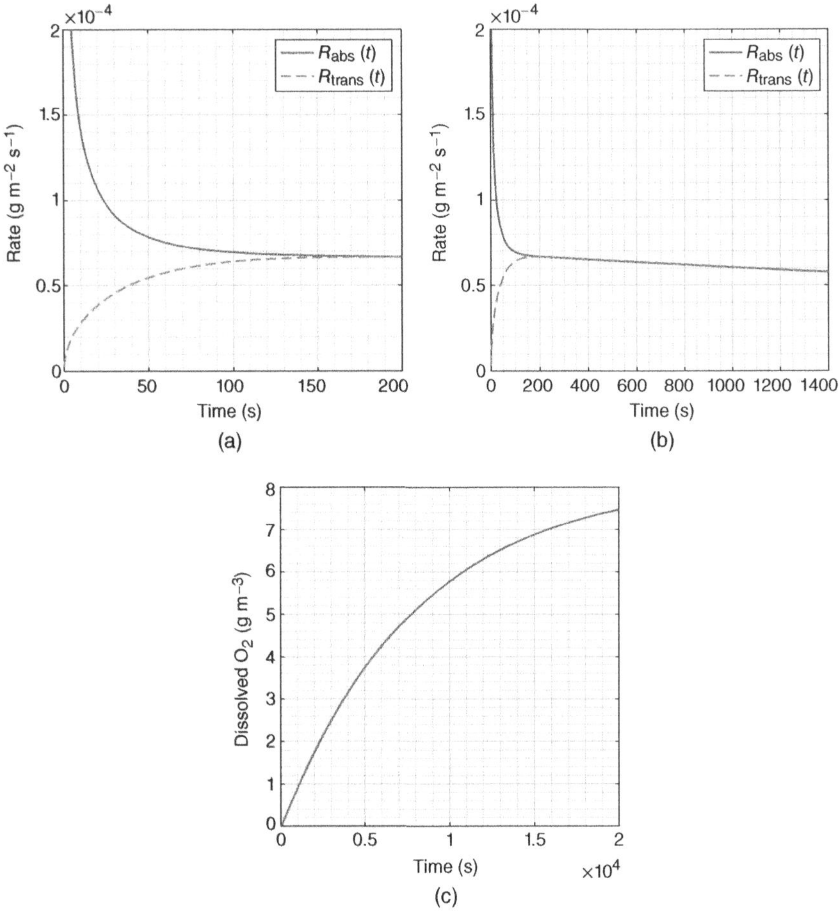

Figure 4.2 Variation of the rates of gas absorption and dissolved-gas transfer (a, b) and bulk-liquid concentration of dissolved gas (c) with process time for oxygen absorption from air at 25 °C and atmospheric pressure into water contained in a stirred flask for the case of finite gas-phase mass-transfer resistance. Parameter values used in the calculations, which were performed with the equations developed in Sections 4.2.2 and 4.4.2, are given in Table 4.2. Source: Huang and Chatterjee (2021) / with permission of Elsevier.

using the parameter values given in Table 4.2, are in excellent agreement with one another, as Table 4.3 shows. This suggests that, even under transient conditions, the conventional pseudo-steady-state model for the dissolved-gas bulk concentration $c_b(t)$ as given by eq. 4.73 with k_L obtained from eq. 4.21 or K_{OL} determined from the (steady-state) law of addition of resistances (eq. 4.34) will be a very good approximation of the theoretical $c_b(t)$ predicted by the rigorous surface-renewal

Table 4.2 Parameter values for the transient oxygen absorption from air into water contained in a stirred flask with air above the water surface for the finite gas-phase mass-transfer resistance case.

Specified parameters	**Value**
temperature[a)]	25 °C
pressure[a)]	1 atm
bulk gas-phase oxygen concentration (y_b, corresponds to air containing 21 mol percent oxygen)	274.62 g m^{-3}
gas-phase mass-transfer coefficient (k_G)	Assumed to be equal to 3×10^{-2} m s^{-1} (Figures 4.2 and 4.5a), 3×10^{-2} to 3×10^{-9} m s^{-1} (Figures 4.3 and 4.4), and 3×10^{-7} m s^{-1} (Figure 4.5b)
Henry's law solubility coefficient for oxygen in pure water (H)[b)]	33.618
diffusion coefficient of oxygen in water (D; estimated from the value at 20 °C[a)] and the Wilke–Chang correlation[c)])	2.14×10^{-9} m^2 s^{-1}
liquid-phase mass-transfer coefficient (k_L)[a)]	8.33×10^{-6} m s^{-1} (3 cm h^{-1}; stirrer speed = 171 RPM)
gas–liquid interfacial area of contact (A)[a)]	7.35×10^{-4} m^2 (7.35 cm^2)
volume of liquid in flask (V)[a)]	5×10^{-5} m^3 (50 cm^3)
Calculated parameters	**Value**
gas–liquid interfacial area per unit volume of liquid ($a = A/V$)	14.7 m^{-1}
surface renewal frequency ($S = k_L{}^2/D$; estimated from eq. 4.21)	0.0325 Hz (s^{-1})

a) Hutchinson and Sherwood (1937).
b) Sander (2015).
c) Perry *et al.* (1984).
Source: Huang and Chatterjee (2021) / with permission of Elsevier.

model (except for very short values of the process time t, as will be discussed later).

Figure 4.4 compares the gas absorption and dissolved-gas transfer rates (i.e., R_{abs} and R_{trans}) for the case of finite-gas-phase mass-transfer resistance as predicted by the Laplace transform (LT) method (eqs. 4.71 and 4.72 in Section 4.5.3) with those calculated using the original method (OM), which was presented in Sections 4.2.2 and 4.4.2, over a seven order-of-magnitude change in k_G for short values

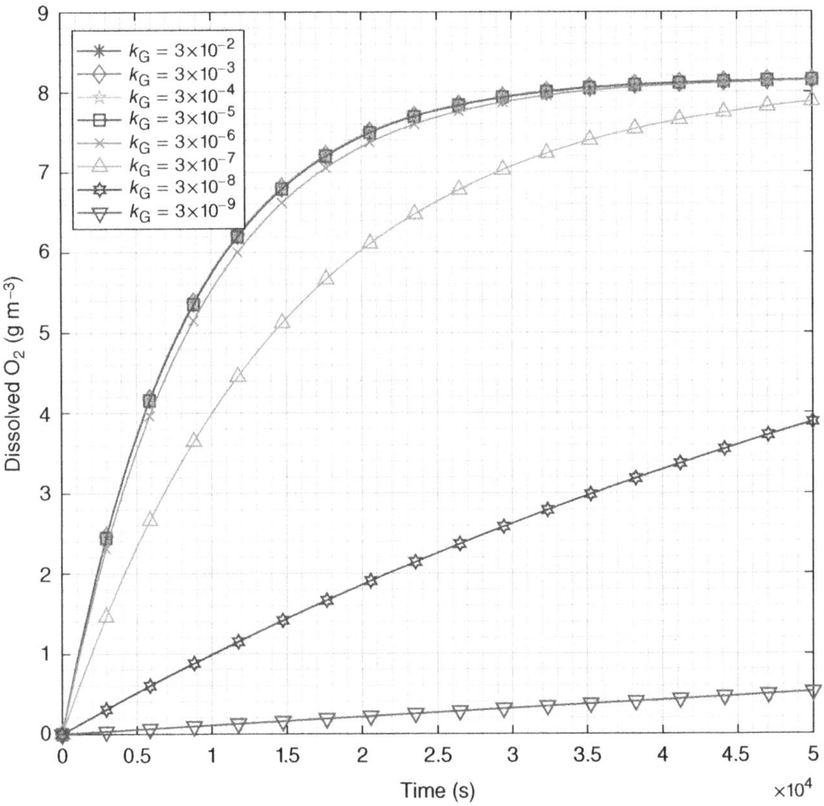

Figure 4.3 Variation of the bulk-liquid concentration of dissolved gas with process time for oxygen absorption from air at 25 °C and atmospheric pressure into water contained in a stirred flask for the case of finite gas-phase mass-transfer resistance as a function of the gas-phase mass-transfer coefficient, k_G. The Laplace transform method, presented in Sections 4.5.1 and 4.5.3, was used for the calculations with the parameter values reported in Table 4.2. The units of k_G are m s^{-1}. Source: Huang and Chatterjee (2021) / with permission of Elsevier.

of the process time t. Three conclusions follow from this figure: (1) The OM and LT methods are in excellent agreement; (2) below a k_G value of 3×10^{-7} m s^{-1}, the rates of absorption and transfer are severely depressed; (3) in all eight cases, the absorption and transfer curves virtually merge into a single curve within the first 150–200 s (2.5–3.3 min) of absorption, indicating that the period of accumulation of dissolved gas in the surface elements lasts only for a short time duration.

We now compare predictions of the rigorous surface renewal OM model (Sections 4.4.1 and 4.4.2) using the parameter values given in Tables 4.1 and 4.2

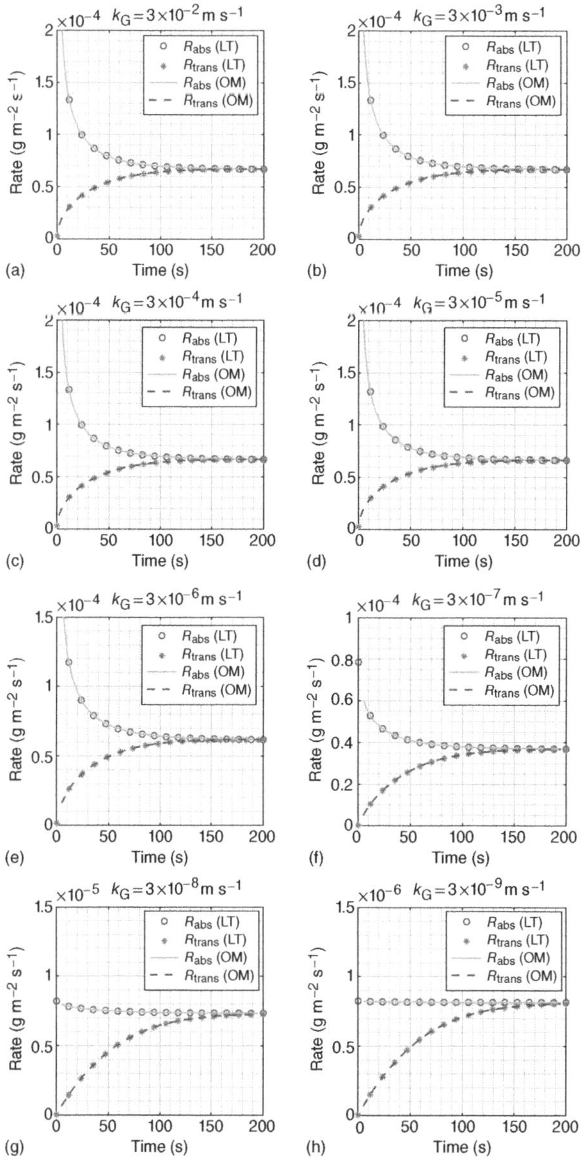

Figure 4.4 Comparison of the rates of gas absorption and dissolved-gas transfer with process time for oxygen absorption from air at 25 °C and atmospheric pressure into water contained in a stirred flask for the case of finite gas-phase mass-transfer resistance as predicted by the rigorous surface renewal model using two methods of calculation. OM = original method (Sections 4.2.2 and 4.4.2) and LT = Laplace transform method (eqs. 4.71 and 4.72 of Section 4.5.3). The parameter values used in the calculations are reported in Table 4.2. Source: Huang and Chatterjee (2021) / with permission of Elsevier.

Table 4.3 Comparison of the values of the overall mass-transfer coefficient K_{OL} obtained with two different methods for transient oxygen absorption from air into water contained in a stirred flask at 25 °C and atmospheric pressure, with air above the water surface for the finite gas-phase mass-transfer resistance case. Values of k_L and H are given in Table 4.2.

k_G (m s^{-1})	K_{OL} (Laplace transform method, Section 4.5.3) (m s^{-1})	K_{OL} from eq. (4.34) (m s^{-1})
3×10^{-2}	8.28×10^{-6}	8.33×10^{-6}
3×10^{-3}	8.28×10^{-6}	8.33×10^{-6}
3×10^{-4}	8.28×10^{-6}	8.32×10^{-6}
3×10^{-5}	8.22×10^{-6}	8.26×10^{-6}
3×10^{-6}	7.67×10^{-6}	7.69×10^{-6}
3×10^{-7}	4.56×10^{-6}	4.56×10^{-6}
3×10^{-8}	8.78×10^{-7}	9.00×10^{-7}
3×10^{-9}	9.92×10^{-8}	9.96×10^{-8}

Source: Huang and Chatterjee (2021) / with permission of Elsevier.

with those of the conventional pseudo-steady-state model (eq. 4.73). For both cases of zero and finite gas-phase mass-transfer resistance, the DO level of the bulk liquid as predicted by the pseudo-steady-state LDF and rigorous surface renewal OM models are in excellent agreement, that is, they cannot be visually distinguished in plots like Figures 4.1c and 4.2c except at short times when there is a large discrepancy. For example, Figure 4.5a and b compares the predictions of the rigorous surface renewal OM model (Section 4.4.2) with those of the conventional pseudo-steady-state model during the initial moments of absorption and also shows the effect of gas-phase mass-transfer resistance. The approximate model predicts a higher DO level in both cases. When $k_G = 3 \times 10^{-2}$ m s^{-1} (Figure 4.5a), the difference between the two models is 5.4% at $t = 500$ s (8.3 min) and decreases to 2.2% as the time increases to 1000 s (16.7 min). Reducing the value of k_G by five orders of magnitude (Figure 4.5b) depresses the DO level and also widens the discrepancy between the two models during the initial moments of absorption. Thus, at $t = 500$ s, the difference is 8.1%, which reduces to 3.6% at $t = 1000$ s. One other difference between the LDF and rigorous surface renewal models is noticeable in Figure 4.5a and b. As $t \to 0$ (i.e., for short times), the approximate model predicts a linear variation of the DO concentration with process time (see eq. 4.73), whereas the rigorous model exhibits a distinct curvature, which is likely due to the influence of mass transfer.

The results presented so far suggest a practical procedure for performing unsteady-state physical absorption calculations: (1) Fit experimental

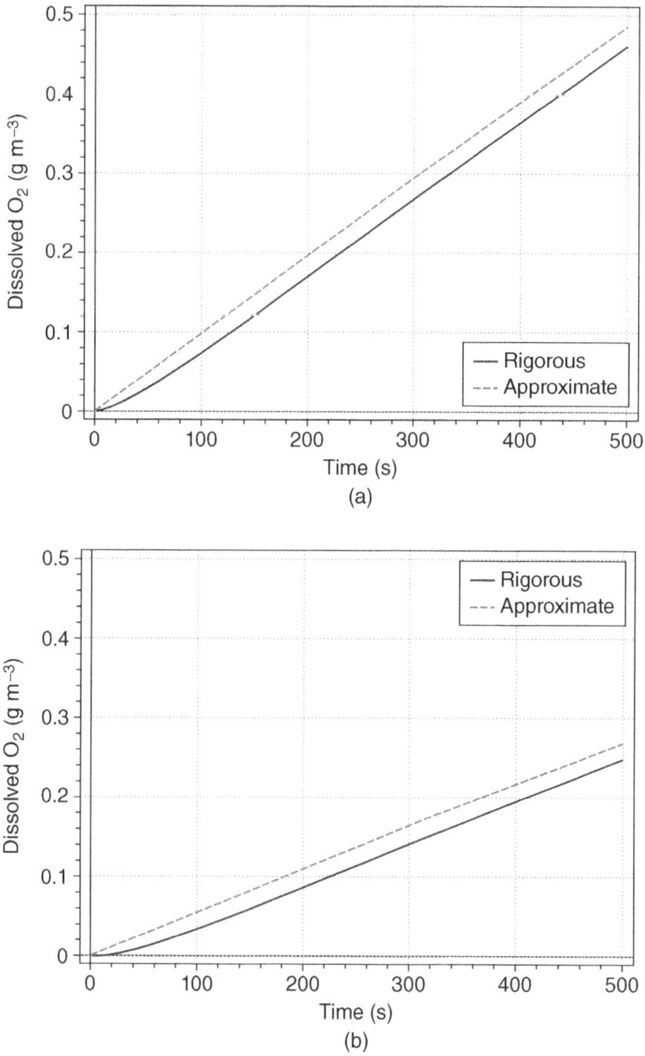

Figure 4.5 Effect of gas-phase mass-transfer resistance on the variation of dissolved-gas bulk-liquid concentration with process time for oxygen absorption from air at 25 °C and atmospheric pressure into water in a stirred flask. (a) $k_G = 3 \times 10^{-2}$ m s^{-1} and (b) $k_G = 3 \times 10^{-7}$ m s^{-1}. Parameter values used in the calculations, which were performed with the rigorous surface renewal OM model (Sections 4.2.2 and 4.4.2) and the approximate pseudo-steady-state LDF model (eqs. 4.34 and 4.73), are reported in Table 4.2. Source: Huang and Chatterjee (2021) / with permission of Elsevier.

measurements of the transient bulk-liquid dissolved-gas concentration with the pseudo-steady-state LDF model (eq. 4.73), which will yield values of $k_L a$ or $K_{OL} a$. (2) Assuming that the gas–liquid contact area a is known or can be estimated from an appropriate correlation, divide the volumetric mass-transfer coefficient ($k_L a$ or $K_{OL} a$) by it in order to obtain the value of the mass-transfer coefficient (k_L or K_{OL}). (3) Calculate the surface renewal rate S from either eq. 4.21 (zero gas-phase mass-transfer resistance) or eqs. 4.21 and 4.34 (finite gas-phase mass-transfer resistance). (4) Determine the gas absorption and dissolved-gas transfer rates from either eqs. 4.64 and 4.65 (or their equivalents, eqs. 4.66 and 4.67) or eqs. 4.71 and 4.72.

We saw in Section 4.2.2 that the LDF form of the absorption rate (eqs. 4.32 and 4.33) emerged out of the more general convoluted form (eq. 4.30) under certain conditions. The general form enfolds the entire history of the concentration driving force since the beginning of absorption, and it is only under special conditions, that is, as $S \to \infty$ (insignificant liquid-phase mass-transfer resistance) or $t \to \infty$ (steady-state conditions) when the history of the driving force becomes unimportant that the LDF form crystallizes out of the general form. The same holds for the rate of dissolved-gas transfer (eq. 4.37). The LDF form is also valid when the mass-transfer driving force is constant, as in gas absorption in a large body of liquid, in which case the mass-transfer coefficients become dynamic, that is, functions of process time, as can be observed in Figure 3.1 of Chapter 3. Since the method of calculation proposed in the previous paragraph for performing unsteady-state calculations is a hybrid of the simple pseudo-steady-state LDF model and the rigorous unsteady-state surface renewal LT model, it is inconsistent from a theoretical viewpoint. Nevertheless, most scientific models contain contradictions within themselves that arise from conflicting assumptions made in order to simplify theoretical complexity (Chatterjee, 2012).

It was also seen in Chapter 3 that an increase in the rate of surface renewal S reduced the initial period of separation between the gas-absorption and dissolved-gas transfer rate curves in a dramatic manner. To confirm this phenomenon, the surface renewal model is applied to the absorption of pure hydrogen in water at 25 °C and atmospheric pressure, a gas–liquid system which was also investigated by Hutchinson and Sherwood (1937). Figure 4.6 shows the rates of absorption and dissolved-gas transfer for this gas–liquid system for the case of zero gas-phase mass-transfer resistance using the LT method (Sections 4.5.1 and 4.5.2) at two different stirrer speeds (171 and 1025 RPM) for the first 250 s after the beginning of absorption. Table 4.4 reports values of the parameter values used in these calculations. As the stirrer speed rises from 171 to 1025 RPM, the value of the surface renewal rate S increases from 0.0187 to 0.176 s^{-1} (Table 4.4). Thus, a 500% rise in stirrer speed increases S by 840% and leads to a decrease in the initial period of separation between the gas-absorption and dissolved-gas transfer

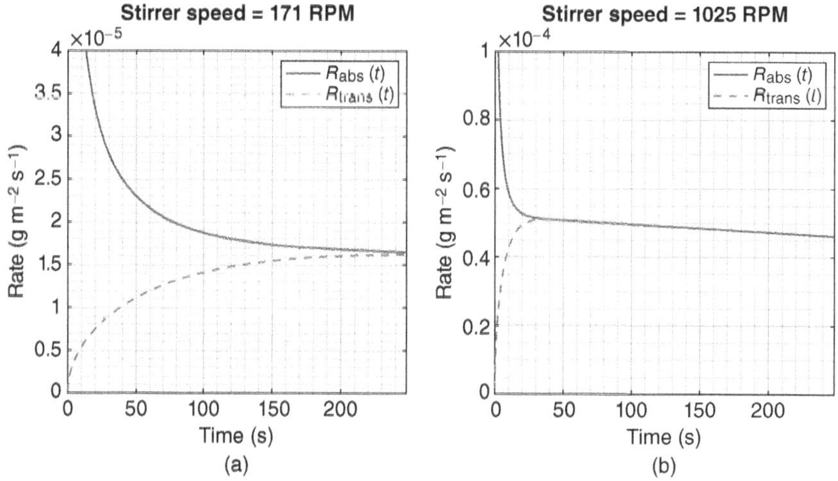

Figure 4.6 Variation of the rates of gas absorption and dissolved-gas transfer with process time in the absorption of pure hydrogen into water contained in a stirred flask at 25 °C and atmospheric pressure for the case of zero gas-phase mass-transfer resistance at stirrer speeds of (a) 171 RPM and (b) 1025 RPM. Parameter values used in the calculations, which were performed with the equations developed in Sections 4.5.1 and 4.5.2 (LT method), are given in Table 4.4. Source: Huang and Chatterjee (2021) / with permission of Elsevier.

rate curves from approximately 250 s (4.2 min) to 35 s (0.6 min) – a reduction of 86% (Figure 4.6). It can be concluded from these results that stirring has a great influence on mass transfer.

The surface renewal rate S, which is believed to be a purely hydrodynamic parameter, can be estimated from direct measurements of the surface-age distribution (see Mondal and Chatterjee, 2017 and Horvath and Chatterjee, 2018, and the references cited therein). However, in mass-transfer studies, it is often calculated from values of the mass-transfer coefficient as done in this work (see eq. 4.21). At a stirrer speed of 171 RPM, the value of S reported in Table 4.4 for hydrogen absorption in water is approximately 58% of that given in Table 4.1 for oxygen absorption. Errors made in estimating the diffusion coefficient of the dissolved gas may be one possible reason for this discrepancy. According to Tables 4.4 and 4.1, the solubility of hydrogen in water is only about 4% as that of oxygen. Since hydrogen is a smaller molecule than oxygen, it has a diffusion coefficient in water, which is about three times larger. These lead to the rate of oxygen absorption in water to be about an order of magnitude higher than that of hydrogen absorption in the time range of 200–250 s at a stirrer speed of 171 RPM, which can be seen by comparing Figure 4.1a and b with Figure 4.6a and b.

Table 4.4 Parameter values for the transient absorption of pure hydrogen into water contained in a stirred flask with hydrogen above the water surface for the case of zero gas-phase mass-transfer resistance.

Specified parameters	Value
temperature[a]	25 °C
pressure[a]	1 atm
solubility of pure hydrogen in water (c_{sat})[a]	$1.6\ g\ m^{-3}$
diffusion coefficient of hydrogen in water (D; estimated from the value at 20 °C[a] and the Wilke–Chang correlation[b])	$6.00 \times 10^{-9}\ m^2\ s^{-1}$
Liquid-phase mass-transfer coefficient (k_L)[a]	$1.06 \times 10^{-5}\ m\ s^{-1}$ (stirrer speed = 171 RPM) $3.25 \times 10^{-5}\ m\ s^{-1}$ (stirrer speed = 1025 RPM)
gas–liquid interfacial area of contact (A)[a]	$7.35 \times 10^{-4}\ m^2$ (or $7.35\ cm^2$)
volume of liquid in flask (V)[a]	$5 \times 10^{-5}\ m^3$ (or $50\ cm^3$)
Calculated parameters	**Value**
gas–liquid interfacial area per unit volume of liquid ($a = A/V$)	$14.7\ m^{-1}$
surface renewal frequency ($S = k_L^2/D$; estimated from eq. 4.21)	0.0187 Hz (s^{-1}) (stirrer speed = 171 RPM) 0.1760 Hz (s^{-1}) (stirrer speed = 1025 RPM)

a) Hutchinson and Sherwood (1937).
b) Perry *et al.* (1984).
Source: Huang and Chatterjee (2021) / with permission of Elsevier.

We conclude by noting that the values of the liquid-phase mass-transfer coefficient k_L shown in Tables 4.1, 4.2, and 4.4 for oxygen and hydrogen absorption in water are of the same order of magnitude as those of the desorption rate constant reported by Madygulov and Vlasov (2023) in the absorption of pure methane and carbon dioxide into water – see Tables 1 and 2 in their paper.

Notation

a	gas–liquid interfacial contact area per unit volume of liquid (m^{-1})
A	gas–liquid contact area (m^2)
c	concentration of solute gas at location x in a liquid element at time θ ($kmol\ m^{-3}$ or $g\ m^{-3}$)
c_b	concentration of solute gas in the bulk liquid ($kmol\ m^{-3}$ or $g\ m^{-3}$)

$c_{b,init}$	initial concentration of solute gas in the bulk liquid (kmol m^{-3} or g m^{-3})
c_s	interfacial (i.e., at x 0) concentration of solute gas in the liquid (kmol m^{-3} or g m^{-3})
c_{sat}	solubility of solute gas in the liquid or equal to y_b/H (kmol m^{-3} or g m^{-3})
C_l^m	binomial coefficient; defined by eq. 4.55
D	diffusion coefficient of solute gas in the liquid (m^2 s^{-1})
$f(t)$	$c_s - c_b(t)$ (kmol m^{-3} or g m^{-3})
$f_D(\iota, t)$	Danckwerts surface age-distribution function; given by eq. 4.8 (s^{-1})
$F(z)$	Laplace transform of $f(t)$
$F_1(z)$	given by eq. 4.61
$g(t)$	$y_b/H - c_b(t)$ (kmol m^{-3} or g m^{-3})
h	defined by eq. 4.27 (m^{-1})
H	Henry's constant
I^A	integral defined by eq. 4.14 (kmol m^{-3} or g m^{-3})
I^B	integral defined by eq. 4.38 (kmol m^{-3} or g m^{-3})
$I_1{}^A$, $I_2{}^A$, and $I_n{}^A$	values of I^A at values of t equal to t_1, t_2, and t_n, respectively (kmol m^{-3} or g m^{-3})
$I_n{}^B$	value of I^B at t equal to t_n (kmol m^{-3} or g m^{-3})
i	index of summation
j	index of summation
$J(\beta)$	function defined by eq. 4.41
$K(\beta)$	function defined by eq. 4.42
k_G	gas-phase mass-transfer coefficient (m s^{-1})
k_L	liquid-phase mass-transfer coefficient (m s^{-1})
K_L	overall mass-transfer coefficient in eq. 4.1 (m s^{-1})
K_{OL}	overall mass-transfer coefficient given by eq. 4.34 (m s^{-1})
K_1	defined by eq. 4.46
K_2	aS (m^{-1} s^{-1})
l	integer
$\mathcal{L}$	Laplace transform operator
m, M, n	integers
$p(t)$	function of t
$P(z)$	Laplace transform of $p(t)$
R_{abs}	age-averaged rate of absorption of solute gas (kmol m^{-2} s^{-1} or g m^{-2} s^{-1})
R_{inst}	instantaneous rate of absorption of solute gas in a liquid element (kmol m^{-2} s^{-1} or g m^{-2} s^{-1})

R_{trans}	age-averaged net rate of transfer of dissolved gas to the bulk liquid due to the surface renewal mechanism (kmol m^{-2} s^{-1} or g m^{-2} s^{-1})
S	frequency or rate of renewal of liquid elements at the gas–liquid interface (Hz or s^{-1})
t	process time (s)
t_0	value of t at zero time level; equal to zero (s)
t_1 and t_2	values of t at time levels 1 and 2, respectively (s)
t_{j-1} and t_j	values of t at time levels $j-1$ and j, respectively (s)
t_{n-1}	value of t at time level $n-1$ (s)
V	volume of liquid (m^3)
w_i	weight; defined by eq. 4.54
x	distance into a liquid element measured from the gas–liquid interface (m)
y	argument of the incomplete gamma function or a variable
y_b	concentration of solute gas in the bulk gas phase (kmol m^{-3} or g m^{-3})
z	argument of the incomplete gamma function or Laplace transform parameter

Greek Letters

α	defined by eq. 4.12
β	argument in eqs. 4.41 and 4.42
Δt	uniform time increment (s)
$\Gamma(y, z)$	incomplete gamma function; defined by eq. 4.16
θ	time of exposure of a liquid element at the gas–liquid interface (s)
τ	age of a liquid element at the gas–liquid interface at time t (s)

Homework Exercises

i) Show that eq. 4.11a is obtained as $St \rightarrow 0$ in eq. 4.11.

ii) Show that eq. 4.24a results as $St \rightarrow 0$ in eq. 4.24.

iii) Show that for a large rate of surface renewal, that is, as $S \rightarrow \infty$, eq. 4.29 reduces to the pseudo-steady-state LDF form given by eq. 4.32.

iv) Using eqs. 4.64 and 4.65 (or eqs. 4.66 and 4.67) develops plots showing the transient behavior of R_{abs} and R_{trans} when pure hydrogen under atmospheric pressure is absorbed in a stirred flask containing 85 g of water at 25 °C, which initially contains no dissolved hydrogen. Use values of $k_L = 5.93 \times 10^{-5}$ m s^{-1}

and $A = 34\ \text{cm}^2$. The diffusion coefficient D and solubility of hydrogen c_{sat} in water are $6 \times 10^{-9}\ \text{m}^2\ \text{s}^{-1}$ and $1.6\ \text{g m}^{-3}$, respectively.[2] Also, plot the dissolved hydrogen concentration in the bulk liquid as a function of process time using the equations developed in Section 4.4.1. Comment on your results.

References

Abate, J. and Whitt, W. (2006). "A unified framework for numerically inverting Laplace transforms," *INFORMS Journal on Computing*, 18(4), pp. 408–421. Available at: https://doi.org/10.1287/ijoc.1050.0137. See also http://www.columbia.edu/~ww2040/JoC.pdf.

Abramowitz, M. and Stegun, I.A. (1965). *Handbook of mathematical functions—with formulas, graphs, and mathematical tables*. New York: Dover Publications.

Astarita, G. (1967). *Mass transfer with chemical reaction*. Amsterdam: Elsevier.

Bohm, D. (1980). *Wholeness and the implicate order*. London: Routledge & Kegan Paul.

Bohm, D. (2003). "The causal–ontological interpretation and implicate orders," in *The essential David Bohm* (ed. Nichol, L.). London: Routledge, pp.183–198.

Carslaw, H.S. and Jaeger, J.C. (1959). *Conduction of heat in solids*. London: Oxford University Press.

Chatterjee, S.G. (2012). "The nature of scientific theory," *Current Science*, 102(3), pp. 386–388. Available at: https://www.jstor.org/stable/24083882.

Chatterjee, S.G. (2013). "Two contrasting paradigms of science: Cartesian mechanism and undivided wholeness," *The International Journal of Science in Society*, 4(3), pp. 75–93. Available at: https://doi.org/10.18848/1836-6236/CGP/v04i03/51394.

Chern, J.-M. and Yu, C.-F. (1997). "Oxygen transfer modeling of diffused aeration systems," *Industrial & Engineering Chemistry Research*, 36(12), pp. 5447–5453. Available at: https://doi.org/10.1021/ie9700627.

Crank, J. (1964). *The mathematics of diffusion*. London: Oxford University Press.

Danckwerts, P.V. (1951). "Significance of liquid-film coefficients in gas absorption," *Industrial & Engineering Chemistry*, 43(6), pp. 1460–1467. Available at: https://doi.org/10.1021/ie50498a055.

Danckwerts, P.V. (1970). *Gas-liquid reactions*. New York: McGraw-Hill.

Horvath, I.R. and Chatterjee, S.G. (2018). "A surface renewal model for unsteady-state mass transfer using the generalized Danckwerts age distribution function," *Royal Society Open Science*, 5(5), 172423. Available at: https://doi.org/10.1098/rsos.172423.

2 The parameter values specified in this problem are derived from Hutchinson and Sherwood (1937) and Madygulov and Vlasov (2023).

Huang, H. and Chatterjee, S.G. (2021). “Transient physical gas absorption in a stirred liquid using the surface renewal model of mass transfer,” *Chemical Engineering Science*, 234, 116449. Available at: https://doi.org/10.1016/j.ces.2021.116449.

Hutchinson, M.H. and Sherwood, T.K. (1937). “Liquid film in gas absorption,” *Industrial & Engineering Chemistry*, 29(7), pp. 836–840. Available at: https://doi.org/10.1021/ie50331a023.

Koltuniewicz, A. and Noworyta, A. (1994). “Dynamic properties of ultrafiltration systems in light of the surface renewal theory,” *Industrial & Engineering Chemistry Research*, 33(7), pp. 1771–1779. Available at: https://doi.org/10.1021/ie00031a016.

Madygulov, M.S. and Vlasov, V.A. (2023). “Physical gas absorption into a stirred liquid: a new kinetic model and experimental studies,” *Heat and Mass Transfer*, 59(8), pp. 1397–1408. Available at: https://doi.org/10.1007/s00231-023-03346-y.

McWhirter, J.R. and Hutter, J.C. (1989). “Improved oxygen mass transfer modeling for diffused/subsurface aeration systems,” *AIChE Journal*, 35(9), pp. 1527–1534. Available at: https://doi.org/10.1002/aic.690350913.

McWhirter, J.R., Chern, J.-M. and Hutter, J.C. (1995). “Oxygen mass transfer fundamentals of surface aerators,” *Industrial & Engineering Chemistry Research*, 34(8), pp. 2644–2654. Available at: https://doi.org/10.1021/ie00047a013.

Mondal, C. and Chatterjee, S.G. (2017). “‘A frequency quantum interpretation of the surface renewal model of mass transfer’, *Royal Society Open*,” *Science*, 4(7), 170103. Available at: https://doi.org/10.1098/rsos.170103.

Perry, R.H., Green, D.W. and Maloney, J.O. (eds.) (1984). *Perry's chemical engineers' handbook*. Sixth edition. New York: McGraw-Hill.

Sander, R. (2015). “Compilation of Henry's law constants (version 4.0) for water as solvent,” *Atmospheric Chemistry and Physics*, 15(8), pp. 4399–4981. Available at: https://doi.org/10.5194/acp-15-4399-2015.

Tchobanoglous, G. and Burton, F.L. (1991). *Wastewater engineering: treatment, disposal, and reuse. Third*. New York: McGraw-Hill.

Xing, W., Yin, M., Lv, Q., Hu, Y., Liu, C. and Zhang, J. (2014). “Oxygen solubility, diffusion coefficient, and solution viscosity,” in *Rotating Electrode Methods and Oxygen Reduction Electrocatalysts*. Elsevier, pp.1–31. Available at: https://doi.org/10.1016/B978-0-444-63278-4.00001-X.

5

Gas Absorption with First-Order Reaction in a Stirred Batch Cell*

5.1 Introduction

The absorption of a gas into a liquid where it undergoes a first-order chemical reaction with a liquid-phase reagent has been the subject of numerous investigations. Examples of gas–liquid systems that exhibit first or pseudo-first-order behavior, under certain conditions, are the absorption of CO_2 into alkaline and carbonate–bicarbonate buffer solutions (Danckwerts *et al.*, 1963; Richards *et al.*, 1964; Danckwerts, 1965; Cents *et al.*, 2001). Another example is the reaction between dissolved oxygen (DO) and sodium sulfite in aqueous medium in the presence or absence of a catalyst. Hartridge and Roughton (1923) utilized the rapid-mixing continuous flow method to measure the homogeneous rate of the oxidation of aqueous solutions of sodium sulfite with cobalt ion catalyst, which greatly abated the effect of mass transfer of oxygen between phases because of the participation of the DO in the reaction. According to de Waal and Beek (1967), the oxidation of sodium sulfite solutions with cobaltous ions was first order in oxygen over a wide range of sulfite concentrations, and the reaction rate constant could be varied from about $40\,s^{-1}$ up to about $5000\,s^{-1}$ by changing the catalyst concentration. However, in the same paper, they mentioned (citing another investigation) that the reaction could be second order in oxygen. Reith and Beek (1973), citing literature data on the oxidation of sodium sulfite solutions, indicated that the reaction was first order only in the case of pure oxygen at atmospheric pressure but second order when oxygen was absorbed from a gas mixture. Srivastava *et al.* (1968) found their data to be most consistent with first-order kinetics with respect to oxygen when the concentration of sulfite was significantly higher than

*The material in this chapter has been adapted from Huang, H. and Chatterjee, S.G. (2023) 'A rigorous surface renewal model for transient gas absorption with first-order chemical reaction in a stirred liquid', *Chemical Engineering Science*, 268, p. 118393. Available at: https://doi.org/10.1016/j.ces.2022.118393.

Applications of the Surface Renewal Model of Mass Transfer, First Edition.
Hengshuo Huang and Siddharth G. Chatterjee.

Companion website: www.wiley.com/go/huang/surfacerenewalmodel

0.03 mol L^{-1}. However, Chen and Barron (1972) found that this reaction system was zero order with respect to oxygen at about 0.015 mol L^{-1} of sulfite. Sivaji and Murty (1982) reported that the sulfite oxidation reaction was fast and second order with respect to oxygen at a sulfite concentration of 0.39 mol $L^{-1}$and was first order with respect to both oxygen and the reacting species at a sulfite level of 0.15 mol L^{-1}. However, it likely changed to a fast pseudo-first-order reaction below about 0.16 mol L^{-1} of sulfite and in the presence of a catalyst at concentrations of about 10^{-5} mol L^{-1}. It was found by Wilkinson *et al.* (1993) that in uncatalyzed sodium sulfite oxidation, the first-order reaction rate constant varied from 0.4 s^{-1} to 20 s^{-1}, which positively depended on the concentration of added sulfuric acid varying from 0.001 to 0.2 mol L^{-1}. Thus, it can be concluded that the oxidation of sodium sulfite solutions can be first or second order in oxygen depending on the experimental conditions.

Danckwerts (1950, 1951, 1970) derived theoretical expressions for the rate of absorption of a gas undergoing a first-order reaction in a liquid using the penetration and surface renewal models. Kayihan and Sandall (1974) proposed an eddy-diffusivity model for gas absorption accompanied by a first-order reaction in a turbulent liquid film flowing along a wall and developed a correlation for the Sherwood number as a function of a dimensionless diffusion parameter and the dimensionless distance along the surface of the film. Using the film and penetration models, Lightfoot (1958, 1962) analyzed the absorption of a gas accompanied by a simultaneous first-order irreversible chemical reaction in an agitated liquid contained in a tank and developed expressions for the ratio of rate of absorption with reaction to that of physical absorption, assuming zero bulk-liquid concentration of the dissolved gas. He found this ratio to be virtually independent of the interfacial mass-transfer model used from which he drew the conclusion that the effect of chemical reaction could be predicted without detailed knowledge of the gas–liquid hydrodynamics in the tank.

As pointed out by Chatterjee (2010) and Huang and Chatterjee (2021), the interfacial and bulk-liquid mass balances for the dissolved gas are linked because the bulk-liquid concentration of dissolved gas c_b appears in the interfacial mass-transfer model either as a boundary condition (film model) or as an initial condition and a boundary condition (surface renewal model). Both mass balances have to be solved simultaneously in order to obtain an integrated view of the absorption process, with the additional consideration of gas-phase resistance to mass transfer. However, as mentioned in Chapter 1, theoretical modeling in the literature on gas absorption has often focused on interfacial mass transport, for example, on obtaining analytical and numerical approximations for the enhancement factor for different types of gas–liquid reactions in which the reaction regime in the bulk liquid is assumed a priori (e.g., equilibrium, slow or fast reaction, and zero concentration of dissolved gas in the bulk liquid). For example,

Onda *et al.* (1968) in their work on CO_2 absorption into aqueous NaOH solutions in a packed column calculated the overall gas-phase mass-transfer coefficient from an equation that assumed that the bulk-liquid concentration of dissolved CO_2 was zero (see eq. 4 in their paper). DeCoursey (1974) also assumed zero bulk concentration of the solute gas while deriving an approximate expression for the enhancement factor for a second-order reaction using the Danckwerts model. However, this assumption will only hold for a fast reaction (e.g., Hatta number ≥ 0.5), as shown by Chatterjee (2010) in his analysis of a gas–liquid reactor with the film and surface renewal models, but not for a slow one, as will be seen later in this chapter.

As indicated in Chapter 1, the rate of absorption at the gas–liquid interface R_{abs} and the rate of dissolved-gas transfer to the bulk liquid R_{trans} are not the same in reactive gas absorption. Both of these rates need to be considered simultaneously for rational analysis and design of gas–liquid contacting equipment like stirred tanks and packed columns. For a first-order reaction between the dissolved gas and a liquid-phase reagent, analytical expressions for R_{trans} have been provided by Danckwerts (1970) for the film model and Chatterjee (2010) in the case of the surface renewal model for steady-state conditions. However, these two rates have often been conflated in the literature (Lightfoot, 1958; Tontiwachwuthikul *et al.*, 1992; Kucka *et al.*, 2003). In the case of the film model, R_{trans} can be obtained in a relatively straightforward fashion by evaluating the concentration gradient of the dissolved gas at the inner edge of the film either analytically or numerically. However, for the more sophisticated surface renewal model, which is believed to be more realistic than the stagnant film model, it is not elementary to obtain R_{trans} either analytically or numerically since that would entail the evaluation of a double integral over time and space, as shown by eqs. 3.4 and 4.22 in Chapters 3 and 4, respectively.

It should be noted that van Elk *et al.* (2000) and van Elk (2001) presented a comprehensive numerical model for a gas–liquid continuous-flow stirred-tank reactor (CFSTR) using the penetration model of Higbie. The two primary features of their numerical model of the gas–liquid contactor are the micro- and macromodels, respectively, with the former depicting interfacial mass transfer and the latter the mixing process within the individual gas and liquid phases (such as material conservation of a component in one given phase). van Elk and coworkers developed a novel method for achieving the simultaneous solution of the penetration model and the macromodel for predicting the dynamic behavior of an ideal CFSTR for a variety of reversible and irreversible gas–liquid reactions. They achieved this by assuming that there was instantaneous mixing between a surface element after its residence at the gas–liquid interface and the bulk liquid, which effectively reduced the mass balances of the macromodel to those of a batch reactor. However, it is commonly held that the penetration model, which assumes the same residence time for all surface elements at the gas–liquid interface, is less close to reality compared to the surface renewal model, which postulates a distribution of residence times, and which has been confirmed experimentally (Lamb *et al.*, 1969; Lesage *et al.*, 2002).

The chief intent of this chapter, which is an extension of the work presented in Chapter 4, is to develop a rigorous surface renewal model for unsteady-state gas absorption in a stirred liquid, in which the dissolved gas undergoes a first-order chemical reaction with a liquid-phase reactant in the absence and presence of significant gas-phase mass-transfer resistance. The model will take into consideration both R_{abs} and R_{trans} and is a preliminary step toward using the surface renewal model for analyzing chemical gas absorption in a comprehensive manner. It will be illustrated by applying it to the absorption of oxygen in water containing a reagent like sodium sulfite.

5.2 Interfacial Mass-Transfer Model

5.2.1 Zero Gas-Phase Mass-Transfer Resistance

The analysis presented in this section is also applicable to the description of solid dissolution in a liquid in which dissolved molecules of the solid undergo a first-order reaction with a reagent in the liquid phase. In the absorption of a pure gas (i.e., no gas-phase mass-transfer resistance), the surface concentration of dissolved gas in the liquid will remain constant (equal to the solubility c_{sat} of the gas in the liquid), assuming that the gas-phase concentration of the species being absorbed in the liquid is held constant. The concentration of dissolved gas in the bulk liquid can, nevertheless, be a function of the process time t.

In the surface renewal model, at any time t since the start of absorption, the surface of the liquid in which the gas is being absorbed is visualized as being populated by a mosaic of chaotic eddies or liquid elements having different ages, which range from zero to t. During its dwell time at the gas–liquid interface, a liquid element is assumed to be isolated from the bulk liquid. The unsteady-state diffusion of dissolved gas in a surface element, assumed to be of infinite depth, during its exposure time θ at the interface can be described by the diffusion–reaction equation, that is,

$$\frac{\partial c}{\partial \theta} = D\frac{\partial^2 c}{\partial x^2} - kc \tag{5.1}$$

with boundary and initial conditions given by

$$c = c_b\,(\theta = 0)\,e^{-k\theta} \text{ for } \theta > 0 \text{ at } x = \infty \tag{5.2}$$

$$c = c_{sat} \text{ for } \theta > 0 \text{ at } x = 0 \tag{5.3}$$

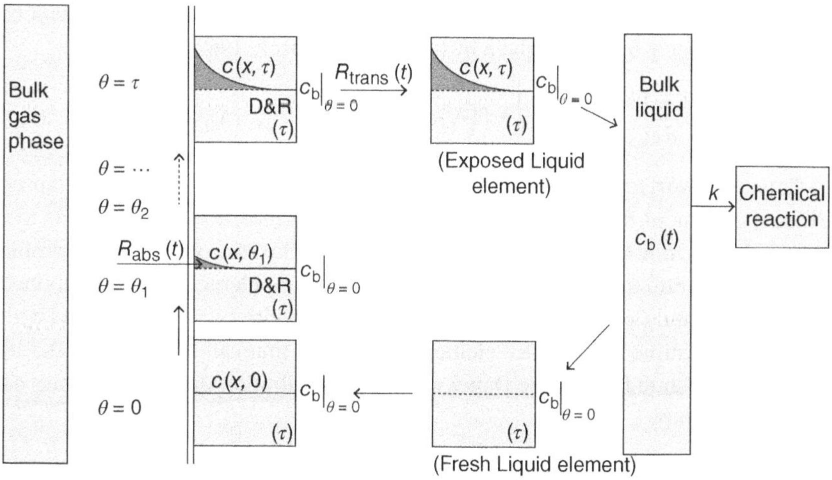

Figure 5.1 Schematic of the surface renewal model. The letters D and R denote diffusion and reaction, respectively. Source: Huang and Chatterjee (2023) / with permission of Elsevier.

and

$$c = c_b(\theta = 0) \text{ at } \theta = 0 \text{ for } 0 \leq x \leq \infty \tag{5.4}$$

Here, c and c_b are the concentrations of dissolved gas at location x (measured from the gas–liquid interface) in the surface element at time θ and in the bulk liquid, respectively, k is the first-order rate constant, and D is the diffusion coefficient of the dissolved gas in the liquid.

Figure 5.1 shows a schematic of the surface renewal model.

The boundary condition expressed by eq. 5.2, which specifies that the concentration of dissolved gas in the surface element decays in a first-order manner at $x = \infty$, has been previously used by Astarita and Beek (1962), Danckwerts (1970), and Lightfoot (1958) to analyze gas absorption accompanied by a first-order reaction with the penetration and surface renewal models.

The solution of the system represented by eqs. 5.1 to 5.4 is (Lightfoot, 1958; Astarita and Beek, 1962; Astarita, 1967)

$$\frac{c}{c_{\text{sat}}} = \frac{1}{2}\left[e^{x\sqrt{\frac{k}{D}}}\operatorname{erfc}\left(\frac{x}{2\sqrt{D\theta}} + \sqrt{k\theta}\right) + e^{-x\sqrt{\frac{k}{D}}}\operatorname{erfc}\left(\frac{x}{2\sqrt{D\theta}} - \sqrt{k\theta}\right)\right] + \frac{c_b(\theta = 0)}{c_{\text{sat}}}e^{-k\theta}\operatorname{erf}\left(\frac{x}{2\sqrt{D\theta}}\right) \tag{5.5}$$

The instantaneous rate of absorption $R_{\text{inst}}(\theta)$ at the gas–liquid interface can be obtained from eq. 5.5 and is given by (Astarita and Beek, 1962)

$$R_{\text{inst}}(\theta) = -D\frac{\partial c}{\partial x}\bigg|_{x=0} = \sqrt{kD}\left[\operatorname{erf}\left(\sqrt{k\theta}\right) + \frac{e^{-k\theta}}{\sqrt{\pi k\theta}}\right]c_{\text{sat}} - c_{\text{b}}\big|_{\theta=0}\sqrt{\frac{D}{\pi\theta}}e^{-k\theta} \tag{5.6}$$

Consider a surface element with an age of τ at time t. Since θ is the time elapsed since the element first arrived at the gas–liquid interface, $\theta = 0$ corresponds to a time of $t - \tau$, while $\theta = \tau$ corresponds to a time of t. Thus, the concentration profile $c(x, \tau, t)$ and rate of absorption $R_{\text{inst}}(\tau, t)$ in the liquid element may be obtained from eqs. 5.5 and 5.6, respectively, with $\theta = 0$ replaced with $t - \tau$ and θ replaced with τ. We now assume that surface elements have ages that can be characterized by the unsteady-state form of the Danckwerts age-distribution function (Chapter 4), which is given by

$$f_{\text{D}}(\tau, t) = \frac{Se^{-S\tau}}{1 - e^{-St}} \tag{5.7}$$

where S, the renewal frequency of liquid elements comprising the gas–liquid interface, is taken to be constant, that is, the level of turbulence or stirrer speed is assumed to be fixed.[1]

The transient rate of absorption $R_{\text{abs}}(t)$ of the gas can then be expressed as

$$R_{\text{abs}}(t) = \int_0^t R_{\text{inst}}(\tau, t)\, f_{\text{D}}(\tau, t)\, \mathrm{d}\tau \tag{5.8}$$

From eqs. 5.6 and 5.8, it follows that

$$R_{\text{abs}}(t) = \int_0^t \left[\frac{\sqrt{D}\,e^{-k\tau}}{\sqrt{\pi\tau}}\left(c_{\text{sat}} - c_{\text{b}}\big|_{t-\tau}\right) + c_{\text{sat}}\sqrt{kD}\operatorname{erf}\left(\sqrt{k\tau}\right)\right]\frac{Se^{-S\tau}}{1 - e^{-St}}\,\mathrm{d}\tau \tag{5.9}$$

which can be recast as

$$R_{\text{abs}}(t) = \frac{1}{1 - e^{-St}}\left[\begin{aligned}&\int_0^{St}\sqrt{\frac{DS}{\pi}}\cdot h\left(t - \frac{\alpha}{S}\right)\frac{e^{-\left(\frac{k+S}{S}\right)\alpha}}{\sqrt{\alpha}}\mathrm{d}\alpha\\ &+\int_0^{St} c_{\text{sat}}\sqrt{kD}\,e^{-\alpha}\operatorname{erf}\left(\sqrt{\frac{k\alpha}{S}}\right)\mathrm{d}\alpha\end{aligned}\right] \tag{5.10}$$

where

$$\alpha = S\tau \tag{5.11}$$

and

$$h(t) = c_{\text{sat}} - c_{\text{b}}(t) \tag{5.12}$$

1 In the absence of stirring, there is no surface renewal, and the situation reduces to one of transient diffusion and chemical reaction in a stagnant pool of liquid.

The Hatta number Ha is defined as $Ha = \sqrt{k/S}$. For $Ha \ll 1$ (i.e., $k \ll S$), eq. 5.10 reduces to the case of physical absorption, that is,

$$\underset{Ha \ll 1}{R_{\text{abs}}}(t) = k_{\text{L}} \left[c_{\text{sat}} - c_{\text{b}}(t)\right] \tag{5.13}$$

where k_{L} is the liquid-side mass-transfer coefficient given by the well-known expression of the surface renewal model, i.e.,

$$k_{\text{L}} = \sqrt{DS} \tag{5.14}$$

When $Ha \gg 1$ (i.e., $k \gg S$), eq. 5.10 reduces to the case of absorption with a fast first-order reaction, that is,

$$\underset{Ha \gg 1}{R_{\text{abs}}}(t) = \sqrt{kD}c_{\text{sat}} \tag{5.15}$$

It is to be noted that different authors define the condition for a fast first-order reaction differently. For example, according to van Elk (2001), for fast reactions, which are believed to occur predominantly near the gas–liquid interface, $Ha > 2$, whereas for slow reactions, which occur chiefly in the liquid bulk, $Ha < 0.2$. For a rapid reaction, the concentration of dissolved gas in the bulk liquid is expected to nearly fall to zero. Chatterjee (2010), who applied the film and surface renewal models to analyze a gas–liquid CFSTR operating at a steady state, found that the value of bulk-liquid concentration of dissolved gas was negligible compared to its interfacial concentration when $Ha \geq 0.5$. However, according to Shaikh and Al-Mubaiyedh (1998), the magnitude of Ha may not be the sole determining criterion, and additional factors like reactor type, operating conditions, and macromixing in the liquid phase (i.e., the influence of mass transfer) can also be important.

The convolution integral in eq. 5.10 can be evaluated by the following numerical procedure, which is based upon the integral mean value theorem. Let

$$I_{\text{A}} = \int_0^{St} h\left(t - \frac{\alpha}{S}\right) \frac{e^{-\left(\frac{k+S}{S}\right)\alpha}}{\sqrt{\alpha}} \text{d}\alpha \tag{5.16}$$

At time $t = t_1 > 0$, eq. 5.16 may be written as

$$I_{\text{A},1} \approx \frac{h(t_1) + h(0)}{2} \int_0^{St_1} \frac{e^{-\left(\frac{k+S}{S}\right)\alpha}}{\sqrt{\alpha}} \text{d}\alpha = \frac{h(t_1) + h(0)}{2} \left[F(St_1) - F(0)\right] \tag{5.17}$$

where the function $F(\beta)$ is defined by

$$F(\beta) = \sqrt{\frac{\pi S}{k+S}} \operatorname{erf}\left(\sqrt{\frac{k+S}{S}\beta}\right) \tag{5.18}$$

Similarly, at time $t = t_2 > t_1$, eq. 5.16 can be expressed as

$$I_{A,2} = \frac{h(t_2) + h(t_2 - t_1)}{2} \int_0^{St_1} \frac{e^{-\left(\frac{k+S}{S}\right)\alpha}}{\sqrt{\alpha}} d\alpha + \frac{h(t_2 - t_1) + h(0)}{2} \int_{St_1}^{St_2} \frac{e^{-\left(\frac{k+S}{S}\right)\alpha}}{\sqrt{\alpha}} d\alpha$$
$$= \frac{h(t_2) + h(t_2 - t_1)}{2} \left[F(St_1) - F(0)\right] + \frac{h(t_2 - t_1) + h(0)}{2} \left[F(St_2) - F(St_1)\right] \tag{5.19}$$

Therefore, by induction, it follows that at time $t = t_n > t_{n-1}$

$$I_{A,n} = \frac{1}{2} \sum_{j=1}^{n} \left[h(t_n - t_{j-1}) + h(t_n - t_j)\right] \left[F(St_j) - F(St_{j-1})\right] \tag{5.20}$$

where $t_0 = 0$. If the time levels are separated by a uniform increment Δt, that is, $t_1 = \Delta t$, $t_2 = t_1 + \Delta t$, $t_3 = t_2 + \Delta t$, …, $t_n = t_{n-1} + \Delta t$, it can be shown that eq. 5.20 simplifies to

$$I_{A,n} = \frac{1}{2} \sum_{j=1}^{n} \left[h(t_{n-j+1}) + h(t_{n-j})\right] \left[F(St_j) - F(St_{j-1})\right] \tag{5.21}$$

The rate of absorption $R_{abs}(t)$ can then be expressed as

$$R_{abs}(t) = \frac{\sqrt{DS}}{1 - e^{-St}} \cdot \frac{I_A}{\sqrt{\pi}} + P(t) \tag{5.22}$$

where the function $P(t)$ is given by

$$P(t) = \frac{c_{sat}\sqrt{kD}}{1 - e^{-St}} \left[\sqrt{\frac{k}{k+S}} \operatorname{erf}\left(\sqrt{(k+S)t}\right) - \operatorname{erf}\left(\sqrt{kt}\right) e^{-St}\right] \tag{5.23}$$

In order to obtain the net rate of transfer of dissolved gas to the bulk liquid due to surface renewal, we follow the procedure previously used by Chatterjee (2010). Letting

$$v(x, \theta) = c(x, \theta) e^{k\theta} - c_b(\theta = 0) \tag{5.24}$$

transforms eqs. 5.1 and 5.4 into the following system:

$$\frac{\partial v}{\partial \theta} = D \frac{\partial^2 v}{\partial x^2} \tag{5.25}$$

$$v = 0 \text{ for } \theta > 0 \text{ at } x = \infty \tag{5.26}$$

$$v = c_{sat} e^{k\theta} - c_b(\theta = 0) \text{ for } \theta > 0 \text{ at } x = 0 \tag{5.27}$$

and

$$v = 0 \text{ at } \theta = 0 \text{ for } 0 \le x \le \infty \tag{5.28}$$

In the above equations, v represents the liquid-phase concentration of a hypothetical gas that undergoes purely physical absorption, which has the value of zero in the bulk liquid. The rate of transfer of this hypothetical gas $R_{\text{trans,hyp}}(t)$ due to surface renewal from the interface to the bulk liquid is

$$R_{\text{trans,hyp}}(t) = S \int_0^\infty \int_0^t v(x, \theta) \frac{Se^{-S\theta}}{1 - e^{-St}} \, d\theta \, dx \tag{5.29}$$

The transient net rate of transfer $R_{\text{trans}}(t)$ of the real gas to the bulk liquid due to surface renewal is then given by (see eq. 5.24)

$$R_{\text{trans}}(t) = S \int_0^\infty \int_0^t e^{-k\theta} v(x, \theta) \frac{Se^{-S\theta}}{1 - e^{-St}} d\theta dx \tag{5.30}$$

Replacing $\theta = 0$ with $t - \tau$ and θ with τ, we obtain from eq. 5.30

$$R_{\text{trans}}(t) = \int_0^t e^{-(k+S)\tau} \left(\int_0^\infty v(x, \tau, t) \frac{S^2}{1 - e^{-St}} dx \right) d\tau \tag{5.31}$$

where $v(x, \tau, t)$ is defined by (see eq. 5.24)

$$v(x, \tau, t) = c(x, \tau, t) e^{k\tau} - c_b \big|_{t-\tau} \tag{5.32}$$

Using eqs. 5.5, 5.31, and 5.32, a long algebraic manipulation procedure finally yields

$$R_{\text{trans}}(t) = \frac{S^2}{1 - e^{-St}} \int_0^t \left[c_{\text{sat}} \sqrt{\frac{D}{k}} e^{-S\tau} \operatorname{erf}\left(\sqrt{k\tau} \right) - 2\sqrt{\frac{D}{\pi}} c_b \big|_{t-\tau} \sqrt{\tau} e^{-(k+S)\tau} \right] d\tau \tag{5.33}$$

which can be expressed as

$$R_{\text{trans}}(t) = \frac{2\sqrt{DS}}{1 - e^{-St}} \cdot \frac{I_B}{\sqrt{\pi}} + \frac{S}{k} P(t) \tag{5.34}$$

where

$$I_B = -\int_0^{St} c_b \left(t - \frac{\alpha}{S} \right) \sqrt{\alpha} e^{-\left(\frac{k+S}{S}\right)\alpha} d\alpha \tag{5.35}$$

with $P(t)$ given by eq. 5.23.

The convolution integral I_B can be evaluated by following the same procedure as that used previously for evaluating I_A (see eqs. 5.16–5.21). It can be shown that its numerical approximation at time level n, assuming the time levels to be separated by a uniform increment, is given by

$$I_{B,n} = \frac{1}{2} \sum_{j=1}^{n} \left[c_b \left(t_{n-j+1} \right) + c_b \left(t_{n-j} \right) \right] \left[G \left(St_j \right) - G \left(St_{j-1} \right) \right] \tag{5.36}$$

where the function $G(\beta)$ is defined by

$$G(\beta) = \frac{S}{k+S}\sqrt{\beta}\, e^{-\left(\frac{k+S}{S}\right)\beta} - \frac{\sqrt{\pi}}{2}\left(\frac{S}{k+S}\right)^{\frac{3}{2}} \operatorname{erf}\left(\sqrt{\frac{k+S}{S}\beta}\right) \tag{5.37}$$

In an ideal CFSTR, that is, one in which the bulk liquid is assumed to be perfectly mixed, c_b will eventually attain a constant value as $t \to \infty$, that is, when a steady state is reached. In this situation, eqs. 5.9 and 5.33 approach the following limits:

$$R_{abs}(t \to \infty) = \frac{S\sqrt{D}\left(c_{sat} - c_b\right) + k\sqrt{D}c_{sat}}{\sqrt{k+S}} \tag{5.38}$$

and

$$R_{trans}(t \to \infty) = \frac{\sqrt{DS}\left(\sqrt{S}\, c_{sat} - \frac{S^{\frac{3}{2}}\, c_b}{k+S}\right)}{\sqrt{k+S}} \tag{5.39}$$

The fraction f of absorbed gas that is transferred from the gas–liquid interface to the bulk liquid at steady state is given by

$$f(t \to \infty) = \frac{R_{trans}(t \to \infty)}{R_{abs}(t \to \infty)} = \frac{S\left(S\sqrt{D}c_{sat} - S\sqrt{D}c_b + k\sqrt{D}c_{sat}\right)}{(k+S)\left(S\sqrt{D}c_{sat} - S\sqrt{D}c_b + k\sqrt{D}c_{sat}\right)}$$

$$= \frac{S}{k+S} = \frac{1}{1+Ha^2} \tag{5.40}$$

Thus, the value of f at a steady state is equal to the reciprocal of the square of the enhancement factor (see eq. 5.104). In the case of physical absorption, $Ha = 0$ and $f(t \to \infty) = 1$. As Ha increases (i.e., for high values of the reaction rate constant k and low rates of surface renewal S), more of the absorbed gas will be consumed by a chemical reaction in the surface elements, and less of it will be available for transport to the bulk liquid due to the surface renewal mechanism.

5.2.2 Finite Gas-Phase Mass-Transfer Resistance

For this case, the governing equations are eqs. 5.2–5.4 with the interfacial boundary condition given by

$$k_G\left(y_b - H\, c|_{x=0}\right) = -D\left.\frac{\partial c}{\partial x}\right|_{x=0} \quad \text{for } \theta > 0 \ at\ x = 0 \tag{5.41}$$

where the bulk gas-phase concentration y_b of the solute gas is assumed to be held constant, H is Henry's constant, and k_G is the gas-phase mass-transfer coefficient.

Taking Laplace transforms (LTs) of eqs. 5.1, 5.2, 5.4, and 5.41 with respect to the variable θ gives the following image function (i.e., LT) $F(x, z)$ of $c(x, \theta)$:

$$\begin{aligned}
F(x,z) &= \mathcal{L}[c(x,\theta)] \\
&= \frac{\sqrt{D}}{\sqrt{D}h_1+\sqrt{k+z}}\left(\frac{h_2}{z}-\frac{c_b(\theta=0)\,h_1}{k+z}\right)e^{-\frac{x}{\sqrt{D}}\sqrt{k+z}}+\frac{c_b(\theta=0)}{k+z} \\
&= k\frac{h_2}{h_1}.\frac{1}{z}\frac{\sqrt{D}h_1e^{-\frac{x}{\sqrt{D}}\sqrt{k+z}}}{(k+z)\left(\sqrt{D}h_1+\sqrt{k+z}\right)}+\left(\frac{h_2}{h_1}-c_b(\theta=0)\right) \\
&\quad\times\frac{\sqrt{D}h_1e^{-\frac{x}{\sqrt{D}}\sqrt{k+z}}}{(k+z)\left(\sqrt{D}h_1+\sqrt{k+z}\right)}+\frac{c_b(\theta=0)}{k+z}
\end{aligned} \tag{5.42}$$

where $\mathcal{L}$ and z represent the LT operator and parameter, respectively, whereas h_1 and h_2 are defined by

$$h_1=\frac{k_G H}{D} \tag{5.43}$$

and

$$h_2=\frac{k_G y_b}{D} \tag{5.44}$$

The inversion of eq. 5.42 yields

$$\begin{aligned}
c(x,\theta) &= \frac{ky_b}{H}\left[-K_1e^{h_1x+Dh_1^2\theta-k\theta}\,\mathrm{erfc}\left(h_1\sqrt{D\theta}+\frac{x}{2\sqrt{D\theta}}\right)\right. \\
&\quad+K_2^-e^{-x\sqrt{\frac{k}{D}}}\,\mathrm{erfc}\left(\frac{x}{2\sqrt{D\theta}}-\sqrt{k\theta}\right) \\
&\quad\left.+K_2^+e^{x\sqrt{\frac{k}{D}}}\,\mathrm{erfc}\left(\frac{x}{2\sqrt{D\theta}}+\sqrt{k\theta}\right)\right] \\
&\quad-c_b(\theta=0)\left[-e^{h_1x+Dh_1^2\theta-k\theta}\,\mathrm{erfc}\left(h_1\sqrt{D\theta}+\frac{x}{2\sqrt{D\theta}}\right)\right. \\
&\quad\left.+\mathrm{erfc}\left(\frac{x}{2\sqrt{D\theta}}\right)e^{-k\theta}\right]+c_b(\theta=0)\,e^{-k\theta}
\end{aligned} \tag{5.45}$$

where K_1, K_2^-, and K_2^+ are defined by

$$K_1=\frac{1}{k}+\frac{1}{Dh_1^2-k} \tag{5.46}$$

$$K_2^-=\frac{1}{2}\left(\frac{1}{k}-\frac{1}{h_1\sqrt{Dk}+k}\right) \tag{5.47}$$

and

$$K_2^+ = \frac{1}{2}\left(\frac{1}{k} + \frac{1}{h_1\sqrt{Dk} - k}\right) \tag{5.48}$$

Differentiating eq. 5.45 with respect to θ and evaluating the resulting expression at $x = 0$ (i.e., the gas–liquid interface) give the instantaneous rate of absorption $R_{\text{inst}}(\theta)$ in a surface element as

$$\begin{aligned} R_{\text{inst}}(\theta) &= -D\left.\frac{\partial c(x,\theta)}{\partial x}\right|_{x=0} \\ &= \frac{kDy_b}{H}\left[K_1 h_1 e^{Dh_1^2\theta - k\theta}\,\text{erfc}\left(h_1\sqrt{D\theta}\right) - K_1\frac{e^{-k\theta}}{\sqrt{\pi D\theta}}\right. \\ &\quad + K_2^-\left(\sqrt{\frac{k}{D}}\text{erfc}\left(-\sqrt{k\theta}\right) + \frac{e^{-k\theta}}{\sqrt{\pi D\theta}}\right) \\ &\quad \left. - K_2^+\left(\sqrt{\frac{k}{D}}\text{erfc}\left(\sqrt{k\theta}\right) - \frac{e^{-k\theta}}{\sqrt{\pi D\theta}}\right)\right] \end{aligned} \tag{5.49}$$

where the argument of the complementary error function has been set to a negative form for conciseness, that is, $\text{erfc}\left(-\sqrt{k\theta}\right) = 2 - \text{erfc}\left(\sqrt{k\theta}\right)$.

Following a similar procedure as that used for the case of zero gas-phase mass-transfer resistance (Section 5.2.1), it may be shown that the transient (age-averaged) rate of gas absorption $R_{\text{abs}}(t)$ is given by

$$\begin{aligned} R_{\text{abs}}(t) &= \frac{DSk}{1 - e^{-St}}\frac{y_b}{H}\left\{K_1 h_1 f_1(t) - K_1\frac{\text{erf}\left(\sqrt{(k+S)t}\right)}{\sqrt{D(k+S)}}\right. \\ &\quad + K_2^-\left[\sqrt{\frac{k}{D}}f_2(t) + \frac{\text{erf}\left(\sqrt{(k+S)t}\right)}{\sqrt{D(k+S)}}\right] \\ &\quad \left. - K_2^+\left[\sqrt{\frac{k}{D}}f_3(t) - \frac{\text{erf}\left(\sqrt{(k+S)t}\right)}{\sqrt{D(k+S)}}\right]\right\} \\ &\quad - \frac{DSh_1}{1 - e^{-St}}\int_0^t c_b(t-\tau)\,e^{(Dh_1^2 - k - S)\tau}\,\text{erfc}\left(h_1\sqrt{D\tau}\right)\,\mathrm{d}\tau \end{aligned} \tag{5.50}$$

where the three functions $f_1(t)$, $f_2(t)$, and $f_3(t)$ are defined by

$$f_1(t) = \frac{1}{Dh_1^2 - k - S}\left[\operatorname{erfc}\left(h_1\sqrt{Dt}\right)e^{(Dh_1^2-k-S)t} + \frac{h_1\sqrt{D}}{\sqrt{(k+S)}}\operatorname{erf}\left(\sqrt{(k+S)t}\right) - 1\right] \tag{5.51}$$

$$f_2(t) = -\frac{2e^{-St}}{S} + \frac{1}{S}\operatorname{erfc}\left(\sqrt{kt}\right)e^{-St} + \frac{1}{S}\sqrt{\frac{k}{k+S}}\operatorname{erf}\left(\sqrt{(k+S)t}\right) + \frac{1}{S} \tag{5.52}$$

and

$$f_3(t) = -\frac{1}{S}\operatorname{erfc}\left(\sqrt{kt}\right)e^{-St} - \frac{1}{S}\sqrt{\frac{k}{k+S}}\operatorname{erf}\left(\sqrt{(k+S)t}\right) + \frac{1}{S} \tag{5.53}$$

The transient (age-averaged) net rate of transfer $R_{\text{trans}}(t)$ of dissolved gas to the bulk liquid can be obtained from eqs. 5.31 and 5.32 with $c(x, \tau, t)$ being given by eq. 5.45 in which $\theta = 0$ is to be replaced with $t - \tau$ and θ with τ, respectively, that is,

$$\begin{aligned} R_{\text{trans}}(t) = {} & \frac{S^2}{1-e^{-St}}\frac{ky_b}{H}\left\{-\frac{K_1}{h_1}\left[\frac{1-e^{-(k+S)t}}{k+S} - f_1(t)\right]\right. \\ & \left. - K_2^-\sqrt{\frac{D}{k}}\left[\frac{1-e^{-(k+S)t}}{k+S} - f_2(t)\right] + K_2^+\sqrt{\frac{D}{k}}\left[\frac{1-e^{-(k+S)t}}{k+S} - f_3(t)\right]\right\} \\ & + \frac{S^2}{1-e^{-St}}\int_0^t c_b(t-\tau)\left[\frac{1}{h_1}\left(e^{-(k+S)\tau} - e^{(Dh_1^2-k-S)\tau}\operatorname{erfc}\left(h_1\sqrt{D\tau}\right)\right)\right. \\ & \left. - 2\sqrt{\frac{D\tau}{\pi}}e^{-(k+S)\tau}\right]d\tau \end{aligned} \tag{5.54}$$

Similar to what was seen in Chapter 4 (Section 4.3), a striking feature of eqs. 5.50 and 5.54 is that the rates of absorption and dissolved-gas transfer involve convolution integrals that contain the bulk-liquid concentration c_b of the solute gas, that is, the values of these rates at any moment of time t are governed by the entire history of this concentration since the beginning of the absorption process at $t = 0$. The same is true for the zero gas-phase mass-transfer resistance case analyzed earlier – see eqs. 5.10 and 5.33 in Section 5.2.1.

Applying the numerical procedure outlined earlier in Section 5.2.1 for evaluating convolution integrals to eqs. 5.50 and 5.54 yields the following approximate equations for calculating the rates of absorption and dissolved-gas transfer at time level n:

$$R_{\text{abs}}(t_n) = f_{\text{abs}}(t_n) - \frac{DSh_1}{1-e^{-St_n}} \cdot II_{A,n} \tag{5.55}$$

and

$$R_{\text{trans}}(t_n) = f_{\text{trans}}(t_n) + \frac{S^2}{1-e^{-St_n}} \cdot II_{B,n} \tag{5.56}$$

where $f_{abs}(t_n)$ and $f_{trans}(t_n)$ are the parts of eqs. 5.50 and 5.54 that do not involve convolution integrals, whereas $II_{A,n}$ and $II_{B,n}$ are the numerical approximations of the convolution integrals in these two equations, respectively, that is,

$$f_{abs}(t) = \frac{DSk}{1-e^{-St}} \frac{y_b}{H} \left\{ K_1 h_1 f_1(t) - K_1 \frac{\text{erf}\left(\sqrt{(k+S)t}\right)}{\sqrt{D(k+S)}} + K_2^- \left[\sqrt{\frac{k}{D}} f_2(t) + \frac{\text{erf}\left(\sqrt{(k+S)t}\right)}{\sqrt{D(k+S)}} \right] - K_2^+ \left[\sqrt{\frac{k}{D}} f_3(t) - \frac{\text{erf}\left(\sqrt{(k+S)t}\right)}{\sqrt{D(k+S)}} \right] \right\} \tag{5.57}$$

$$f_{trans}(t) = \frac{S^2 k}{1-e^{-St}} \frac{y_b}{H} \left\{ -\frac{K_1}{h_1} \left[\frac{1-e^{-(k+S)t}}{k+S} - f_1(t) \right] - K_2^- \sqrt{\frac{D}{k}} \left[\frac{1-e^{-(k+S)t}}{k+S} - f_2(t) \right] + K_2^+ \sqrt{\frac{D}{k}} \left[\frac{1-e^{-(k+S)t}}{k+S} - f_3(t) \right] \right\} \tag{5.58}$$

$$II_{A,n} = \frac{1}{2} \sum_{j=1}^{n} \left[c_b\left(t_{n-j+1}\right) + c_b\left(t_{n-j}\right) \right] \left[f_1\left(t_j\right) - f_1\left(t_{j-1}\right) \right] \tag{5.59}$$

and

$$II_{B,n} = \frac{1}{2} \sum_{j=1}^{n} \left[c_b\left(t_{n-j+1}\right) + c_b\left(t_{n-j}\right) \right] \left[g_1\left(t_j\right) - g_1\left(t_{j-1}\right) \right] \tag{5.60}$$

In these equations, $f_1(t)$, $f_2(t)$, and $f_3(t)$ are given by eqs. 5.51–5.53, respectively, and $g_1(t)$ is defined by

$$g_1(t) = \frac{1}{h_1} \cdot \frac{1-e^{-(k+S)t}}{k+S} - \frac{f_1(t)}{h_1} + 2\sqrt{\frac{Dt}{\pi}} \frac{e^{-(k+S)t}}{k+S} - \sqrt{D}(k+S)^{-\frac{3}{2}} \,\text{erf}\left(\sqrt{(k+S)t}\right) \tag{5.61}$$

In an ideal CFSTR, as a steady state is approached (i.e., $t \to \infty$), $R_{abs}(t)$ and $R_{trans}(t)$ will tend toward the following two limits:

$$R_{\text{abs}}(t \to \infty) = \frac{Sky_{\text{b}}}{H}\left[\frac{K_1\left(-Dh_1 + \sqrt{D(k+S)}\right)}{Dh_1^2 - k - S} + \frac{K_2^- \sqrt{D}}{S}\left(\sqrt{k+S} + \sqrt{k}\right)\right.$$
$$\left. + K_2^+ \sqrt{D}\frac{k + S - \sqrt{k(k+S)}}{S\sqrt{k+S}}\right] - \frac{c_{\text{b}} D S h_1}{Dh_1^2 - k - S} \cdot \frac{h_1\sqrt{D} - \sqrt{k+S}}{\sqrt{k+S}} \tag{5.62}$$

and

$$R_{\text{trans}}(t \to \infty) = \frac{Sky_{\text{b}}}{H}\frac{S}{k+S}\left[\frac{K_1\left(-Dh_1 + \sqrt{D(k+S)}\right)}{Dh_1^2 - k - S}\right.$$
$$\left. + \frac{K_2^- \sqrt{D}}{S}\left(\sqrt{k+S} + \sqrt{k}\right) + K_2^+ \sqrt{D}\frac{k + S - \sqrt{k(k+S)}}{S\sqrt{k+S}}\right]$$
$$- \frac{S}{k+S} \cdot \frac{c_{\text{b}} D S h_1}{Dh_1^2 - k - S} \cdot \frac{h_1\sqrt{D} - \sqrt{k+S}}{\sqrt{k+S}} \tag{5.63}$$

From eqs. 5.62 and 5.63, it follows that the fraction f of absorbed gas that is transferred to the bulk liquid from the gas–liquid interface at steady state is given by

$$f(t \to \infty) = \frac{R_{\text{trans}}(t \to \infty)}{R_{\text{abs}}(t \to \infty)} = \frac{S}{k+S} = \frac{1}{1 + Ha^2} \tag{5.64}$$

The result was previously derived by Chatterjee (2010), which is the same as eq. 5.40 for the zero gas-phase mass-transfer resistance case.

5.3 Bulk Mass-Transfer Model

Consider the unsteady-state batch operation of a stirred-tank gas–liquid reactor (i.e., no inflow or outflow of liquid). The mass balance for dissolved gas in the bulk liquid is given by

$$\frac{\mathrm{d}c_{\text{b}}(t)}{\mathrm{d}t} = aR_{\text{trans}}(t) - kc_{\text{b}}(t) \tag{5.65}$$

with

$$c_{\text{b}} = c_{\text{b,init}} \text{ at } t = 0 \tag{5.66}$$

In eq. 5.66, $c_{\text{b,init}}$ [i.e., $c_{\text{b}}(0)$] is the initial dissolved-gas concentration of the bulk liquid (assumed to be zero in this work), whereas a is the gas–liquid interfacial contact area per unit volume of liquid. For gas–liquid mass transfer in agitated vessels, a value for a, if not known, can be estimated from correlations available in the literature (Perry *et al.*, 1984; Stegeman *et al.*, 1995).

5.3.1 Zero Gas-Phase Mass-Transfer Resistance

Substituting eq. 5.34 into eq. 5.65 yields the following integro-differential equation that describes the variation of the bulk-liquid dissolved-gas concentration with process time:

$$\frac{dc_b(t)}{dt} = a\frac{2\sqrt{DS}}{1-e^{-St}} \cdot \frac{I_B}{\sqrt{\pi}} + a\frac{S}{k}P(t) - kc_b(t) \tag{5.67}$$

Representing the time derivative in eq. 5.67 by a backward-difference approximation over the first time interval $[0, t_1]$ gives

$$\frac{c_b(t_1) - c_b(0)}{\Delta t} = a\frac{2\sqrt{DS}}{1-e^{-St_1}} \cdot \frac{I_{B,1}}{\sqrt{\pi}} + a\frac{S}{k}P(t_1) - kc_b(t_1) \tag{5.68}$$

where $t_1 = \Delta t$, and $I_{B,1}$ is to be obtained from eq. 5.36. Solving eq. 5.68 for $c_b(t_1)$ gives

$$c_b(t_1) = \frac{1+q\Delta t}{1-q\Delta t+k\Delta t} \cdot c_{b,init} + \frac{S}{k} \cdot \frac{a\,\Delta t}{1-q\Delta t+k\Delta t} \cdot P(t_1) \tag{5.69}$$

Here, q is defined by

$$q = a\sqrt{\frac{DS}{\pi}}\frac{1}{1-e^{-S\Delta t}}[G(S\Delta t) - G(0)] \tag{5.70}$$

where the function G is given by eq. 5.37.

After obtaining the value of c_b at the first time level, for all subsequent time levels, that is, $t_2 = t_1 + \Delta t, \ldots, t_n = t_{n-1} + \Delta t$, the following procedure is adopted for calculating $c_b(t_n)$, given the values of c_b at all earlier time levels. Representing the time derivative in eq. 5.67 by a centered-difference approximation over the time interval $[t_{n-2}, t_n]$ yields

$$\frac{c_b(t_n) - c_b(t_{n-2})}{2\Delta t} = a\frac{2\sqrt{DS}}{1-e^{-St_{n-1}}} \cdot \frac{I_{B,n-1}}{\sqrt{\pi}} + a\frac{S}{k}P(t_{n-1}) - kc_b(t_{n-1}) \quad \text{for } n \geq 2 \tag{5.71}$$

which upon algebraic manipulation results in

$$c_b(t_n) = c_b(t_{n-2}) + 2a\Delta t\frac{S}{k}P(t_{n-1}) + \frac{4a\sqrt{DS}\Delta t}{1-e^{-St_{n-1}}} \cdot \frac{I_{B,n-1}}{\sqrt{\pi}} - 2k\Delta t c_b(t_{n-1}) \quad \text{for } n \geq 2 \tag{5.72}$$

In eq. 5.72, $P(t_{n-1})$ and $I_{B,n-1}$ are to be obtained from eqs. 5.23 and 5.36, respectively.

5.3.2 Finite Gas-Phase Mass-Transfer Resistance

The mass balance for dissolved gas in the bulk liquid is given by eq. 5.65 with R_{trans} to be obtained from eq. 5.56. The procedure for calculating $c_{\text{b}}(t)$ is similar to that described earlier for the case of zero gas-phase mass-transfer resistance. Using a backward-difference approximation over the first time interval $[0, t_1]$, the expression for $c_{\text{b}}(t_1)$ is found to be

$$c_{\text{b}}\left(t_1\right) = \frac{1 + q'\Delta t}{1 - q'\Delta t + k\Delta t} \cdot c_{\text{b,init}} + \frac{a\,\Delta t}{1 - q\Delta t + k\Delta t} \cdot f_{\text{trans}}\left(\Delta t\right) \tag{5.73}$$

where q' is defined by

$$q' = \frac{aS^2}{2\left(1 - e^{-S\Delta t}\right)}\left[g_1\left(\Delta t\right) - g_1\left(0\right)\right] \tag{5.74}$$

and $f_{\text{trans}}(\Delta t)$ and $g_1(\Delta t)$ are to be obtained from eqs. 5.58 and 5.61, respectively.

For all subsequent time levels (i.e., $n \geq 2$), the following equation can be used to calculate $c_{\text{b}}(t_n)$:

$$c_{\text{b}}\left(t_n\right) = c_{\text{b}}\left(t_{n-2}\right) + 2a\Delta t f_{\text{trans}}\left(t_{n-1}\right) + \frac{2a\Delta t S^2}{1 - e^{-St_{n-1}}} \cdot II_{\text{B},n-1} - 2k\Delta t c_{\text{b}}\left(t_{n-1}\right) \tag{5.75}$$

where $f_{\text{trans}}(t_{n-1})$ and $II_{\text{B},n-1}$ are given by eqs. 5.58 and 5.60, respectively (with n replaced by $n - 1$).

5.4 Analytical Expressions for the Rates of Absorption and Dissolved-Gas Transfer Using the Laplace Transform Method

As discussed in Chapter 4 (Section 4.5), explicit mathematical formulas can be obtained for the rates of gas absorption and dissolved-gas transfer, $R_{\text{abs}}(t)$ and $R_{\text{trans}}(t)$, respectively, by using the LT technique and numerical inversion by the Gaver–Stehfest algorithm (Abate and Whitt, 2006). For the benefit of the reader, we again recapitulate it here.

5.4.1 Gaver–Stehfest Algorithm

Consider a function $\phi(t)$ whose LT is $\Phi(z)$, where z is the LT parameter. Given $\Phi(z)$, $\phi(t)$ can be calculated by the formula

$$\phi\left(t\right) = \frac{\ln 2}{t}\sum_{i=1}^{2M} w_i \cdot \Phi\left(\frac{i \ln 2}{t}\right) \tag{5.76}$$

where the weight w_i is obtained from

$$w_i = \frac{(-1)^{M+i}}{M!} \sum_{j=\lfloor \frac{i+1}{2} \rfloor}^{\min\{i,M\}} j^{M+1} C_j^M \, C_j^{2j} \, C_{i-j}^j \tag{5.77}$$

with $\lfloor y \rfloor$ being the greatest integer less than or equal to y, and C_l^m the binomial coefficient defined by

$$C_l^m = \binom{m}{l} = \frac{m!}{l!\,(m-l)!} \tag{5.78}$$

The significance of M has been explained in the previous chapter.

5.4.2 Zero Gas-Phase Mass-Transfer Resistance

For $St > 4, 1 - e^{-St} > 0.9817 \approx 1$, and, to a particularly good approximation, eq. 5.33 can be written as

$$R_{\text{trans}}(t) \approx S^2 \int_0^t \left[c_{\text{sat}} \sqrt{\frac{D}{k}}\, e^{-S\tau} \operatorname{erf}\left(\sqrt{k\tau}\right) - 2\sqrt{\frac{D}{\pi}}\, c_b\big|_{t-\tau} \sqrt{\tau} e^{-(k+S)\tau} \right] d\tau \tag{5.79}$$

Substituting eq. 5.79 into eq. 5.65 and taking the LT of both sides of the resulting equation yields

$$\mathcal{L}\left[\frac{dc_b(t)}{dt}\right] = aS^2 \cdot \mathcal{L}\left[\int_0^t \left(c_{\text{sat}} \sqrt{\frac{D}{k}}\, e^{-S\tau} \operatorname{erf}\left(\sqrt{k\tau}\right) - 2\sqrt{\frac{D}{\pi}}\, c_b\big|_{t-\tau} \sqrt{\tau} e^{-(k+S)\tau} \right) d\tau \right] - \mathcal{L}\left[kc_b(t)\right] \tag{5.80}$$

where $\mathcal{L}$ is the LT operator. Applying the derivative and convolution theorems of Laplace transformation to eq. 5.80 gives

$$\left[zF_b(z) - c_b(0)\right] = aS\sqrt{D}c_{\text{sat}} \left[\frac{1}{z} - \frac{1}{z+S}\right] \frac{1}{\sqrt{z+k+S}} - aS^2\sqrt{D}F_b(z)\frac{1}{(z+k+S)^{\frac{3}{2}}} - k\sqrt{D}F_b(z) \tag{5.81}$$

where z is the LT parameter and $F_b(z)$ is the LT of $c_b(t)$. Setting $c_b(0) = 0$ and solving eq. 5.81 for $F_b(z)$ yield the result

$$F_b(z) = \mathcal{L}\left[c_b(t)\right] = \frac{aS\sqrt{D}c_{\text{sat}}\left[\frac{1}{z} - \frac{1}{z+S}\right]\frac{1}{\sqrt{z+k+S}}}{z + \frac{S^2\sqrt{D}a}{(z+k+S)^{\frac{3}{2}}} + k} \tag{5.82}$$

From eq. 5.76, it follows that

$$c_{\mathrm{b}}(t) = \frac{\ln 2}{t} \sum_{i=1}^{2M} w_i \cdot F_{\mathrm{b}}\left(\frac{i \ln 2}{t}\right) \tag{5.83}$$

where the weight w_i and function $F_{\mathrm{b}}(z)$ are given by eqs. 5.77 and 5.82, respectively.

In order to obtain the rates of gas absorption and dissolved-gas transfer, the method of nonlinear least squares can be used to fit numerical values of $c_{\mathrm{b}}(t)$ obtained from eqs. 5.83 and 5.82 with the following expression suggested by the pseudo-steady-state (PSS) model, which will be presented later in Section 5.5.1 (see eq. 5.106):

$$c_{\mathrm{b}}(t) = A c_{\mathrm{sat}} \left[1 - e^{-Bt}\right] \tag{5.84}$$

where A and B are model parameters whose initial estimates are given by

$$A = \frac{1}{\frac{kE}{ak_{\mathrm{L}}} + \frac{1}{E^2}} \tag{5.85}$$

and

$$B = k + \frac{ak_{\mathrm{L}}}{E^3} \tag{5.86}$$

In eqs. 5.85 and 5.86, k_{L} is the liquid-side mass-transfer coefficient defined by eq. 5.14 and E is the enhancement factor (see eq. 5.104).

Substituting eq. 5.84 into eqs. 5.9 and 5.33 yields the following explicit expressions for the gas absorption and dissolved-gas transfer rates:

$$\begin{aligned} R_{\mathrm{abs}}(t) = \frac{1}{1 - e^{-St}} \Bigg\{ & (1 - A) c_{\mathrm{sat}} S \sqrt{\frac{D}{k + S}} \operatorname{erf}\left(\sqrt{(k + S)t}\right) \\ & + A c_{\mathrm{sat}} S \sqrt{\frac{D}{k + S - B}} e^{-Bt} \operatorname{erf}\left(\sqrt{(k + S - B)t}\right) \\ & + c_{\mathrm{sat}} \sqrt{kD} \left[-\operatorname{erf}\left(\sqrt{kt}\right) e^{-St} + \sqrt{\frac{k}{k + S}} \operatorname{erf}\left(\sqrt{(k + S)t}\right) \right] \Bigg\} \end{aligned} \tag{5.87}$$

and

$$\begin{aligned} R_{\mathrm{trans}}(t) = {} & \frac{c_{\mathrm{sat}} S}{1 - e^{-St}} \sqrt{\frac{D}{k}} \left[-\operatorname{erf}\left(\sqrt{kt}\right) e^{-St} + \sqrt{\frac{k}{k + S}} \operatorname{erf}\left(\sqrt{(k + S)t}\right) \right] \\ & - \frac{2 c_{\mathrm{sat}} S^2 A}{1 - e^{-St}} \sqrt{\frac{D}{\pi}} \left[-\frac{\sqrt{t} e^{-(k+S)t}}{k + S} \right. \\ & + \frac{1}{2}(k + S)^{-\frac{3}{2}} \sqrt{\pi} \operatorname{erf}\left(\sqrt{(k + S)t}\right) + \frac{\sqrt{t} e^{-(k+S)t}}{k + S - B} \\ & \left. - \frac{1}{2}(k + S - B)^{-\frac{3}{2}} \sqrt{\pi} e^{-Bt} \operatorname{erf}\left(\sqrt{(k + S - B)t}\right) \right] \end{aligned} \tag{5.88}$$

It is seen from these formulas that the rates of gas absorption and dissolved-gas transfer depend upon the following six parameters: c_{sat}, D, S, k, A, and B.

5.4.3 Finite Gas-Phase Mass-Transfer Resistance

Adopting a similar procedure with eqs. 5.54 and 5.65 as that followed for the case of zero gas-phase mass-transfer resistance (Section 5.4.2), it may be shown that

$$F_b(z) = \mathcal{L}\left[c_b(t)\right] = \frac{a\mathcal{U}(z)}{-aS^2\mathcal{R}(z) + k + z} \tag{5.89}$$

where the image functions $\mathcal{U}(z)$ and $\mathcal{R}(z)$ are given by

$$\mathcal{U}(z) = \frac{S^2 k y_b}{H}\left\{-\frac{K_1}{h_1}\left[\mathcal{W}(z) - L_1(z)\right] - K_2^-\sqrt{\frac{D}{k}}\left[\mathcal{W}(z) - L_2(z)\right] + K_2^+\sqrt{\frac{D}{k}}\left[\mathcal{W}(z) - L_3(z)\right]\right\} \tag{5.90}$$

and

$$\mathcal{R}(z) = \frac{1}{h_1(z+k+S)} - \frac{1}{h_1\left(z - Dh_1^2 + k + S\right)} + \frac{\sqrt{D}}{\left(z - Dh_1^2 + k + S\right)\sqrt{z+k+S}} - \frac{\sqrt{D}}{(z+k+S)^{\frac{3}{2}}} \tag{5.91}$$

Here, $\mathcal{W}(z)$, $L_1(z)$, $L_2(z)$, and $L_3(z)$ are defined by

$$\mathcal{W}(z) = \frac{1}{k+S}\left(\frac{1}{z} - \frac{1}{z+k+S}\right) \tag{5.92}$$

$$L_1(z) = \frac{1}{Dh_1^2 - k - S}\left[\frac{1}{\left(z - Dh_1^2 + k + S\right)} - \frac{h_1\sqrt{D}}{\left(z - Dh_1^2 + k + S\right)\sqrt{z+k+S}} + \frac{h_1\sqrt{D}}{z\sqrt{z+k+S}} - \frac{1}{z}\right] \tag{5.93}$$

$$L_2(z) = -\frac{2}{S(z+S)} + \frac{1}{S}\left[\frac{1}{z+S} - \frac{\sqrt{k}}{(z+S)\sqrt{z+k+S}}\right] + \frac{\sqrt{k}}{(zS)\sqrt{z+k+S}} + \frac{1}{zS} \tag{5.94}$$

and

$$L_3(z) = -\frac{1}{S}\left[\frac{1}{z+S} - \frac{\sqrt{k}}{(z+S)\sqrt{z+k+S}}\right] - \frac{\sqrt{k}}{(zS)\sqrt{z+k+S}} + \frac{1}{zS} \tag{5.95}$$

Numerical values of $c_b(t)$ can be obtained from eq. 5.83 with the function $F_b(z)$ given by eq. 5.89. These values can then be fitted with eq. 5.84 (with c_{sat} replaced with y_b/H) using nonlinear regression to determine the model parameters A and B, whose initial estimates are to be obtained from (see eq. 5.110)

$$A = \frac{aE^2K_{OL}}{aK_{OL} + kE^4} \tag{5.96}$$

and

$$B = k + \frac{aK_{OL}}{E^4} \tag{5.97}$$

where K_{OL} is an overall mass-transfer coefficient.

Substituting eq. 5.84 (with c_{sat} replaced with y_b/H) into eqs. 5.50 and 5.54 leads to the following explicit expressions for the gas absorption and dissolved-gas transfer rates:

$$R_{abs}(t) = f_{abs}(t) - \frac{ADSh_1}{1 - e^{-St}}\frac{y_b}{H}f_1(t) + \frac{ADSh_1 e^{-Bt}}{1 - e^{-St}}\frac{y_b}{H}f_4(t) \tag{5.98}$$

and

$$R_{trans}(t) = f_{trans}(t) + \frac{AS^2}{1 - e^{-St}}\frac{y_b}{H}g_1(t) - \frac{AS^2 e^{-Bt}}{1 - e^{-St}}\frac{y_b}{H}f_5(t) \tag{5.99}$$

Here, $f_{abs}(t)$ and $f_{trans}(t)$ are defined by eqs. 5.57 and 5.58, respectively, $f_1(t)$ and $g_1(t)$ are defined by eqs. 5.51 and 5.61, respectively, and $f_4(t)$ and $f_5(t)$ are given by

$$f_4(t) = \frac{1}{Dh_1^2 - k - S + B}\left[\operatorname{erfc}\left(h_1\sqrt{Dt}\right)e^{(Dh_1^2 - k - S + B)t} + \frac{h_1\sqrt{D}}{\sqrt{(k + S - B)}}\operatorname{erf}\left(\sqrt{(k + S - B)t}\right) - 1\right] \tag{5.100}$$

and

$$f_5(t) = \frac{1 - e^{-(k+S-B)t}}{h_1(k + S - B)} - \frac{1}{h_1}f_4(t) + 2\sqrt{\frac{Dt}{\pi}}\frac{e^{-(k+S-B)t}}{(k + S - B)} - \sqrt{D}(k + S - B)^{-\frac{3}{2}}\operatorname{erf}\left(\sqrt{(k + S - B)t}\right) \tag{5.101}$$

The rates of gas absorption and dissolved-gas transfer depend on the following seven parameters: y_b, H, D, S, k, A, and B.

5.5 Pseudo-Steady-State (PSS) Model

5.5.1 Zero Gas-Phase Mass-Transfer Resistance

The PSS model is defined by eqs. 5.65 and 5.66, with R_{trans} given by (see eq. 5.40)

$$R_{trans} = \frac{R_{abs}}{1 + Ha^2} \tag{5.102}$$

and R_{abs} by Chatterjee (2010)

$$R_{abs} = Ek_L\left(c_{sat} - \frac{c_b}{E^2}\right) \tag{5.103}$$

Here, k_L is given by eq. 5.14, and E is the enhancement factor defined by

$$E = \sqrt{1 + Ha^2} \tag{5.104}$$

and, as mentioned previously, Ha is the Hatta number, i.e.,

$$Ha = \sqrt{\frac{k}{S}} = \frac{\sqrt{kD}}{k_L} \tag{5.105}$$

The solution for $c_b(t)$, obtained by solving eq. 5.65 subject to eq. 5.66 (assuming $c_{b,init} = 0$), reads as

$$c_b(t) = \frac{c_{sat}}{\frac{kE}{ak_L} + \frac{1}{E^2}}\left[1 - e^{-\left(k + \frac{ak_L}{E^3}\right)t}\right] \tag{5.106}$$

Thus, for fast reactions and high liquid-side mass-transfer resistance, the steady-state value of the dissolved-gas bulk-liquid concentration $c_b(t \to \infty)$ can be significantly lower that its solubility c_{sat}, which is in accordance with physical intuition.

For fast reactions, $c_b \approx 0$, and it may be shown that eq. 5.103 can be reformulated as

$$(R_{abs}a)^2 = (k_L a)^2 (c_{sat})^2 + (ac_{sat})^2 Dk \tag{5.107}$$

The quantity $R_{abs}a$ is the rate of absorption per unit volume of packing (packed tower) or unit volume of liquid (stirred-tank gas–liquid reactor). Thus, by plotting the square of this quantity against the reaction rate constant k using experimental data, the gas–liquid interfacial contact area a and the liquid-side mass-transfer coefficient k_L can be estimated simultaneously from the slope and intercept of the plot. Figure 3 in Danckwerts (1965) shows such experimental plots for the absorption of CO_2 into a buffer solution in a column packed with 0.5-inch Raschig rings at different superficial liquid flowrates. In the experiments, k was varied by adding small quantities of arsenite to the buffer solution. Figure 2 in Cents *et al.* (2001), who studied the absorption of CO_2 into a 0.5 M K_2CO_3/0.5 M $KHCO_3$ solution in the absence and presence of a dispersed liquid phase in a stirred gas–liquid reactor, is another example. In their experiments, the magnitude of k was changed by adding small amounts of NaOCl, which served as a catalyst.

5.5.2 Finite Gas-Phase Mass-Transfer Resistance

For this case, the governing equations are eqs. 5.65, 5.66, and 5.102, with R_{abs} to be obtained from (Chatterjee, 2010)

$$R_{\text{abs}} = K_{\text{OL}} \left(\frac{y_{\text{b}}}{H} - \frac{c_{\text{b}}}{E^2} \right) \tag{5.108}$$

where K_{OL} is an overall mass-transfer coefficient defined by

$$\frac{1}{K_{\text{OL}}} = \frac{1}{Hk_{\text{G}}} + \frac{1}{Ek_{\text{L}}} \tag{5.109}$$

The solution for $c_{\text{b}}(t)$, assuming $c_{\text{b,init}} = 0$, is given by

$$c_{\text{b}}(t) = \frac{aE^2 K_{\text{OL}}}{aK_{\text{OL}} + kE^4} \frac{y_{\text{b}}}{H} \left[1 - e^{-\left(k + \frac{aK_{\text{OL}}}{E^4}\right)t} \right] \tag{5.110}$$

5.6 Absorption of Oxygen in Water Containing a Liquid-Phase Reagent (e.g., Na_2SO_3)

The different variants of the surface renewal model developed in this chapter are illustrated by applying them to the absorption of oxygen in water contained in a stirred batch cell whose surface is exposed to the gas, a schematic of which is shown in Figure 5.2. The DO is assumed to undergo a first-order chemical reaction with a liquid-phase reagent like Na_2SO_3, which is often used to remove DO in clean-water oxygen-transfer efficiency testing of aerators in wastewater treatment (Tchobanoglous and Burton, 1991).

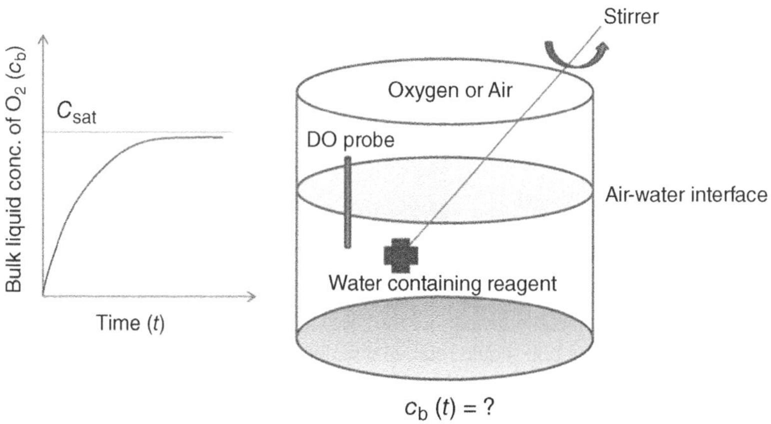

Figure 5.2 Schematic of a stirred batch cell. DO denotes dissolved oxygen. Source: Huang and Chatterjee (2023) / with permission of Elsevier.

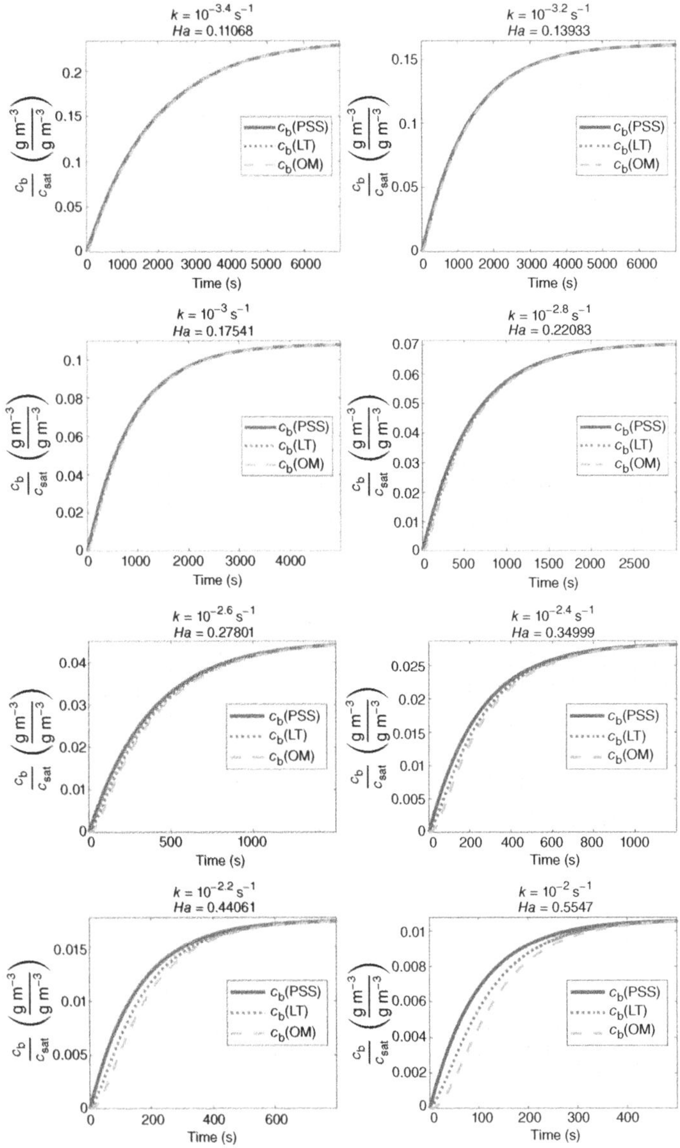

Figure 5.3 Variation of the bulk-liquid relative dissolved-gas concentration with process time in the absorption of pure oxygen into water contained in a stirred flask at 25 °C and atmospheric pressure for the case of zero gas-phase mass-transfer resistance and relatively fast first-order reaction as predicted by the surface renewal OM, LT, and PSS models. Parameter values used in the calculations, which were performed with the equations developed in Sections 5.2.1, 5.3.1, 5.4.2, and 5.5.1, are shown in Table 4.1. LT: Laplace transform; OM: original method; PSS: pseudo-steady-state. Source: Huang and Chatterjee (2023) / with permission of Elsevier.

The values of all parameters used in the calculations, except for the value of the reaction rate constant k, are taken from Chapter 4 (or see Huang and Chatterjee, 2023). To examine the effect of the chemical reaction on the absorption process, values of k varying over several orders of magnitude are assumed. In order to conserve space, only a brief overview of the results obtained by Huang and Chatterjee (2023) is presented in this section; the reader is referred to their publication for more details.

Figure 5.3 depicts the predicted fractional or relative DO concentration in the bulk liquid (i.e., c_b/c_{sat}) as a function of time when pure oxygen is absorbed in a stirred flask of water (i.e., zero gas-phase mass-transfer resistance) at 1 atm and 25 °C. The PSS model (Section 5.5.1) and the two versions of the surface renewal model [the original method (OM) – Sections 5.2.1 and 5.3.1 and the LT method – Section 5.4.2] are used in the predictions with the reaction rate constant k varying between $10^{-3.4}$ and 10^{-2} s^{-1}. The values of the parameters used in the calculations are reported in Table 4.1 (Chapter 4) and also in Table 1 of Huang and Chatterjee (2023). No correction was made to the value of c_{sat} to account for ionic effects. It is seen that, in general, the DO level rises rapidly at first, and thereafter, increasing slowly as a steady state is approached. The ultimate or steady-state DO concentration drops from 22.86% to 1.05% (OM and LT) of the saturation DO value of 39.04 g m^{-3} (prevailing at the gas–liquid interface), as k increases from $10^{-3.4}$ to 10^{-2} s^{-1}, that is, as the Hatta number Ha varies from 0.111 to 0.555. Thus, for $Ha \geq 0.5$, the bulk-liquid concentration of dissolved gas is negligible as compared to its interfacial concentration, a result which, as mentioned earlier, was also arrived at by Chatterjee (2010) for both film and surface renewal models in the case of a gas–liquid CFSTR operating at steady state.

Figure 5.3 also shows the excellent agreement between the PSS model and the two versions (OM and LT) of the surface renewal model for $Ha < 0.221$, that is, when the steady-state bulk concentration of DO is above 7% of its saturation value. However, as the velocity of the chemical reaction increases (i.e., $Ha > 0.278$), this concentration reduces drastically (i.e., it is only a few percent of c_{sat}), and some discrepancies between the models can be observed during the initial few minutes after the beginning of absorption. But even under such conditions of fast reaction, the fraction f of absorbed gas transferred to the bulk liquid is still significant. Thus, steady-state values of f (eq. 5.40) are 0.988 and 0.765 for $Ha = 0.111$ and 0.555 with the bulk DO concentration dropping to 23.48% and 1.06% (PSS model) of its saturation value, respectively, which implies that a significant portion of the absorbed gas is consumed by reaction in the bulk liquid. It is commonly accepted that if the enhancement factor $E \approx 1$, physical absorption predominates, and the reaction has a negligible effect (Danckwerts, 1970). The values of E calculated by eq. 5.104 are 1.01 and 1.14 for $Ha = 0.111$ and 0.555, respectively. Thus, in spite of E being close to 1, the reaction in the bulk liquid can significantly depress the dissolved-gas bulk-liquid concentration. According to van Elk *et al.* (2000), for slow reactions ($Ha < 0.2$), the influence of the reaction occurring in the liquid bulk is significant, and the enhancement factor, being equal to 1 for such reactions, provides no additional information.

There is a noticeable feature of the concentration-time curves of the surface renewal model (both OM and LT) in Figure 5.3 during the initial moments of absorption, which is especially evident for values of k greater than $10^{-2.6}$ s^{-1} (i.e., $Ha > 0.278$). These curves have a sigmoidal shape in contrast to the strictly convex nature of the PSS model with respect to the time axis.

Huang and Chatterjee (2023) also studied the behavior of the relative DO concentration as a function of process time for smaller values of the reaction rate constant k (i.e., slow reaction) varying from 10^{-7} to 10^{-4} s^{-1} (Ha lying in the range 0.002 to 0.056) and found excellent agreement between the PSS and the rigorous surface renewal models (OM and LT). The steady-state DO concentration of the bulk liquid ranged from almost 100% ($k = 10^{-7}$ s^{-1}) to about 54% ($k = 10^{-4}$ s^{-1}) of the saturation DO concentration. The case for $k = 10^{-7}$ s^{-1} ($Ha = 0.002$) was virtually identical to that for physical absorption (k or $Ha = 0$), which was discussed in Chapter 4 (see also Huang and Chatterjee, 2021).

Figure 5.4 shows the rates of gas absorption and dissolved-gas transfer to the bulk liquid as predicted by the surface renewal model (OM and LT) for a fast first-order reaction ($k = 10^{-3.4}$ to 10^{-2} s^{-1} or $Ha = 0.111$ to 0.555) for short times ($t = 0$ to 300 s) after the commencement of absorption when there is no gas-phase mass-transfer resistance. The following are observed in this figure: First, there is excellent agreement between the OM and LT versions of the surface renewal model. Second, the gas absorption rate decays in an exponential-like fashion with process time, whereas, in contrast, the dissolved-gas transfer rate increases smoothly from a value of zero and runs parallel to the absorption rate curve from about 150 s onward. Third, as the steady state is approached, a constant gap emerges between the two curves, whose magnitude is equal to the rate of consumption of dissolved gas by a chemical reaction in the surface elements. This gap, which is difficult to discern for small values of k, widens as the reaction velocity increases. This can be compared to the case of physical gas absorption, in which the two rates merge into a single curve after an initial period of time has elapsed, which then either declines with process time if the volume of the liquid into which the absorption occurs is finite (Chapter 4; Huang and Chatterjee, 2021) or remains invariant if the liquid volume is large (Chapter 3; Horvath and Chatterjee, 2018).

Figure 5.4 also shows the dynamic behavior of f – the ratio of the dissolved-gas transfer and gas absorption rates. With the passage of time, this ratio ascends from a value of zero and subsequently levels out after 200–250 s have elapsed. As indicated earlier, the steady-state value of f (eq. 5.40) declines from 0.988 to 0.765 as Ha

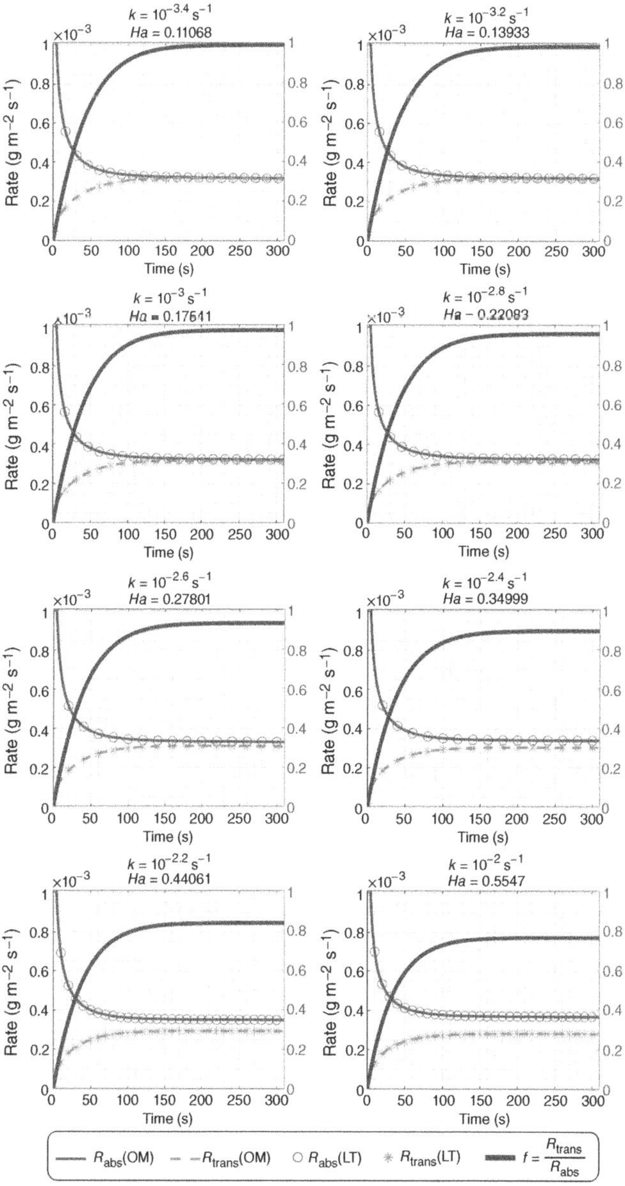

Figure 5.4 Evolution of the rates of gas absorption and dissolved-gas transfer in the absorption of pure oxygen into water contained in a stirred flask at 25 °C and atmospheric pressure for the case of zero gas-phase mass-transfer resistance and relatively fast first-order reaction as predicted by the surface renewal model using two methods of calculation: OM = original method (Sections 5.2.1 and 5.3.1) and LT = Laplace transform (eqs. 5.87 and 5.88 in Section 5.4.2). The other parameter values used in the calculations are shown in Table 4.1. Source: Huang and Chatterjee (2023) / with permission of Elsevier.

changes from 0.111 to 0.555 – a fivefold variation in Ha. In the former case, most of the reaction occurs in the bulk liquid, whereas, in the latter, the reaction occurring in the surface elements becomes more important. Overall, in both cases, a substantial amount of dissolved gas is consumed by chemical reaction, which was also seen in Figure 5.3.

The variation of the relative bulk-liquid dissolved-gas concentration $[c_b/(y_b/H)]$ with process time for the case of oxygen absorption from air (i.e., finite gas-phase mass-transfer resistance) in a stirred flask of water at 1 atm and 25 °C was examined by Huang and Chatterjee (2023) using parameter values given in Table 2 of their paper (see also Table 4.2 in Chapter 4). In their calculations, the gas-phase mass-transfer coefficient k_G was assumed to vary over six orders of magnitude covering the range of negligible ($k_G = 3 \times 10^{-3}$ m s^{-1}) to significant ($k_G = 3 \times 10^{-9}$ m s^{-1}) gas-phase mass-transfer resistance. Equations presented for the rigorous surface renewal model (OM and LT versions) and the PSS model in Sections 5.2.2, 5.3.2, 5.4.3, and 5.5.2 were used to generate the results. The value of the reaction rate constant k was kept fixed at 10^{-3} s^{-1} (i.e., $Ha = 0.175$ and $E = 1.015$). As the gas-phase mass-transfer coefficient k_G decreased from 3×10^{-3} m s^{-1} to 3×10^{-9} m s^{-1}, the steady-state value of the bulk-liquid relative DO concentration $c_b/(y_b/H)$ dropped from about 10.7% to 0.2%, which can be contrasted to the case of physical absorption examined in Chapter 4 where, for the same change in the magnitude of k_G, the decrease was from 100% to 6% – see Figure 4.3 in Chapter 4 or Figure 3 in Huang and Chatterjee (2021). There was very good agreement between the surface renewal (OT and LT) and PSS models with regard to the prediction of the bulk-liquid DO level, with some discrepancy existing between them at short times, which, however, lessened with the progress of time. Also, over this enormous six-order range of decrease in k_G, the liquid-phase mass-transfer resistance $[K_{OL}/(Ek_L)]$ reduced from 99.99% to 1.18%, while the gas-phase resistance $[K_{OL}/(Hk_G)]$ increased from 0.01% to 98.82% of the total mass-transfer resistance, that is, the major portion of the mass-transfer resistance shifted to the gas side, which caused a great suppression of the DO level and its rate of increase with time.

For the case of finite gas-phase mass-transfer resistance, Huang and Chatterjee (2023) found that the dynamic behavior of R_{abs}, R_{trans}, and f was qualitatively similar to that shown in Figure 5.4 for the zero gas-phase mass-transfer resistance case, with the steady-state absorption and transfer rates being an order of magnitude smaller as can be observed in Figure 7 of their paper, which shows excellent agreement between the OM and LT versions of the surface-renewal model.

Huang and Chatterjee (2023) performed a detailed comparison of the LT and PSS models since the latter model is much easier to use. Both zero and finite gas-phase mass-transfer resistance cases, presented in Sections 5.4 and 5.5, were considered. Very good agreement was found between the models except for fast ($Ha \geq 0.555$)

and very slow ($Ha \leq 0.002$) reactions. However, the PSS model has a major drawback, i.e., it cannot predict the increase in the dissolved-gas transfer rate R_{trans} as a function of process time t. As can be inferred from eq. 5.103, the rate of gas absorption R_{abs} will decrease in a hyperbolic manner with process time. However, in the PSS model, the ratio $f = R_{trans}/R_{abs}$ is solely a function of the Hatta number Ha (see eq. 5.102), which also holds for the finite gas-phase mass-transfer resistance case, i.e., it is constant. Hence, R_{trans} will also follow the same trend as R_{abs} with the progress of time, which is not correct. In order to resolve this deficiency, an amalgam of the PSS and LT models for performing unsteady-state gas absorption calculations is proposed: (1) Fit experimental data of transient bulk-liquid dissolved-gas concentration with eq. 5.84. This regression will yield values of the constants A and B from which the magnitudes of k and k_L or K_{OL} can be obtained from eqs. 5.85 and 5.86 (zero gas-phase mass-transfer resistance case) or eqs. 5.96 and 5.97 (finite gas-phase mass-transfer resistance case) together with eq. 5.105, provided that the gas–liquid interfacial contact area a is known or can be estimated from an appropriate correlation. (2) Determine the surface renewal frequency S from either eqs. (5.14) (zero gas-phase mass-transfer resistance case) or eqs. 5.14 and 5.109 (finite gas-phase mass-transfer resistance case). (3) Calculate R_{abs} and R_{trans} using eqs. 5.87 and 5.88 (zero gas-phase mass-transfer resistance case) or eqs. 5.98 and 5.99 (finite gas-phase mass-transfer resistance case).

It was seen in the case of physical absorption (Chapter 4) that an increase in the surface renewal rate S dramatically reduces the initial period of separation between the gas-absorption and dissolved-gas transfer rate curves. To study the effect of a heightened S when the dissolved gas undergoes a first-order reaction in the liquid phase and to test the surface renewal model further, calculations are performed for the absorption of oxygen in a stirred flask of water for no gas-phase mass-transfer resistance but at a faster stirrer speed of 1025 RPM, which corresponds to a value of $k_L = 7.64$ cm h^{-1} or 2.122×10^{-5} m s^{-1} in the experiments of Hutchinson and Sherwood (1937). Using the value of the diffusion coefficient D of oxygen in water shown in Table 4.1, eq. 5.14 yields an estimate of 0.2108 s^{-1} for S. All other parameter values used in the calculations are the same as those given in this table. Results of the calculations are presented in Figures 5.5 and 5.6, which are discussed next.

Figure 5.5 shows the relative bulk-liquid DO concentration as a function of time for "slow" (171 RPM) and "fast" (1025 RPM) stirring as predicted by the rigorous surface renewal (OM and LT) and PSS models for different values of the reaction rate constant k. It is assumed that the value of the gas–liquid interfacial mass-transfer area a does not change with a change in stirring speed, i.e., the exposed liquid surface in the flask is unbroken. The nature of the bulk-liquid DO concentration–time curves is the same as that seen earlier in Figure 5.3. It is also seen that at a fixed value of k, an increased stirrer speed (i.e., a greater value of S

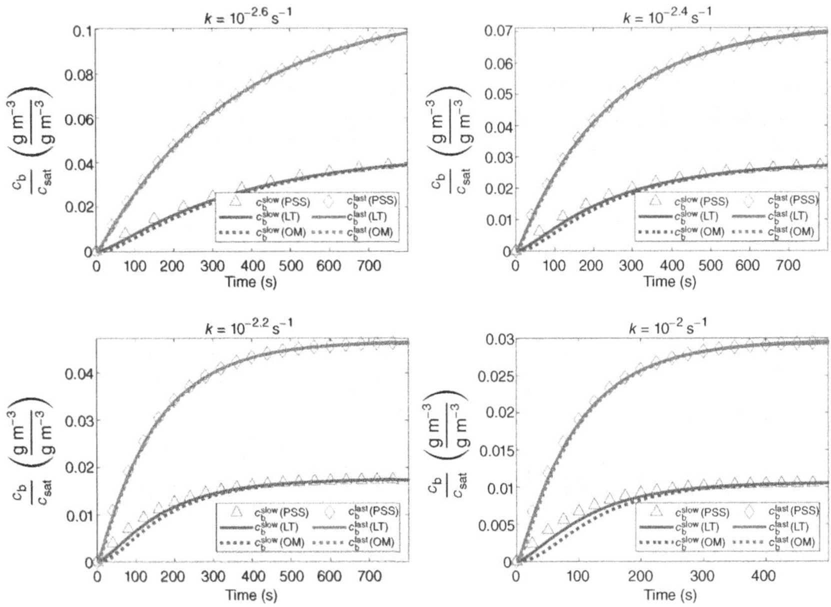

Figure 5.5 Influence of slow ($S = 0.0325\ \mathrm{s}^{-1}$) and fast ($S = 0.2108\ \mathrm{s}^{-1}$) stirring rates on the behavior of the transient bulk-liquid relative dissolved-gas concentration in the absorption of pure oxygen into water contained in a stirred flask at 25 °C and atmospheric pressure for the case of zero gas-phase mass-transfer resistance and relatively fast reaction as predicted by the surface renewal OM, LT, and PSS models. The values of other parameters used in the calculations, which were performed with the equations developed in Sections 5.2.1, 5.3.1, 5.4.2, and 5.5.1, are shown in Table 4.1. LT: Laplace transform; OM: original method; PSS: pseudo-steady-state. Source: Huang and Chatterjee (2023) / with permission of Elsevier.

or lower value of Ha) leads to a considerable enhancement of the bulk-liquid DO level due to a greater rate of dissolved-gas transfer from the gas–liquid interface. The difference between the models is somewhat larger at a lower stirrer speed and for fast reaction, especially during the first few minutes of absorption; however, the agreement between them is quite good overall.

Figure 5.6 exhibits the rates of absorption and dissolved-gas transfer calculated using the OM and LT methods (presented in Sections 5.2.1, 5.3.1, and 5.4.2) over the first 300 s after the start of absorption at the two different stirrer speeds. As the stirrer speed rises from 171 to 1025 RPM, the surface renewal frequency S increases from 0.0325 to 0.2108 Hz. Thus, a sixfold rise in stirrer speed, which increases S by a factor of 6.5 causing thereby a great increase in the ratio f of the rate of dissolved-gas transfer to that of absorption, reduces the initial period of separation

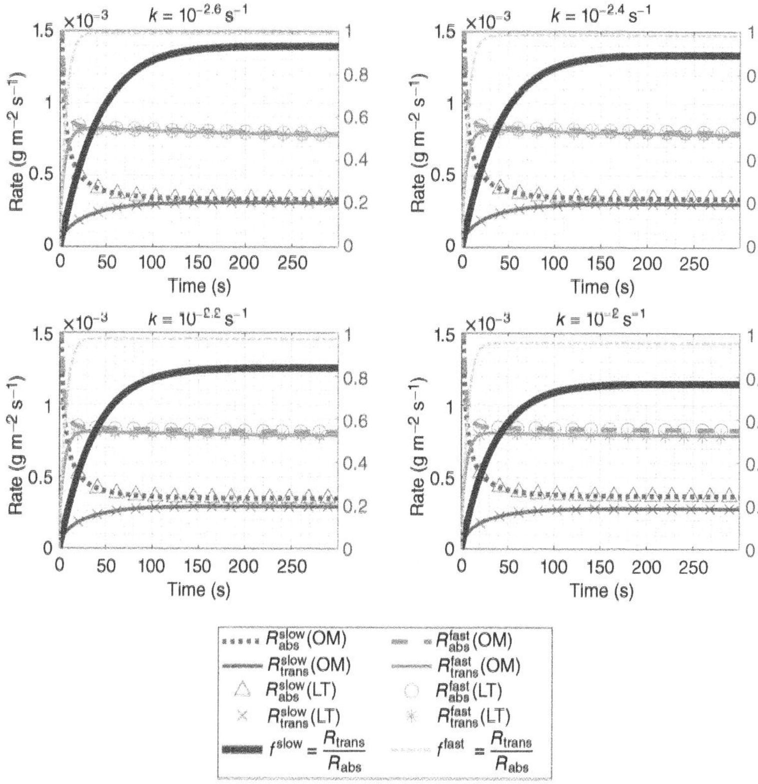

Figure 5.6 Influence of slow ($S = 0.0325\ s^{-1}$) and fast ($S = 0.2108\ s^{-1}$) stirring rates on the evolution of the rates of absorption and dissolved-gas transfer in the absorption of pure oxygen into water contained in a stirred flask at 25 °C and atmospheric pressure for the case of zero gas-phase mass-transfer resistance and relatively fast reaction as predicted by the surface renewal OM and LT methods. The values of other parameters used in the calculations, which were performed with the equations developed in Sections 5.2.1, 5.3.1, and 5.4.2, are given in Table 4.1. LT: Laplace transform; OM: original method. Source: Huang and Chatterjee (2023) / with permission of Elsevier.

between the gas-absorption and dissolved-gas transfer rate curves from about 150 s (2.5 min) to 30 s (0.5 min) – a reduction of 80%.

As was indicated earlier, the two values of the surface-renewal frequency used in this chapter ($S = 0.0325$ and 0.2108 Hz) were estimated from eq. 5.14 and the experimental values of k_L that were reported by Hutchinson and Sherwood (1937) for the transient absorption of pure oxygen in a stirred flask containing water (stirrer speeds of 171 RPM and 1025 RPM) whose surface was exposed to the gas. The interfacial gas–liquid contact area per unit volume of liquid was $a = 14.7\ m^{-1}$ (Table 4.1). Richards *et al.* (1964) investigated the absorption of pure CO_2 into carbonate–bicarbonate buffer solutions (containing arsenite as a catalyst) in a 4-inch (internal diameter) column packed with 0.5-inch Raschig rings. Dissolved CO_2 underwent a first-order reaction in these solutions. They reported values of a (interpreted here as the gas–liquid interfacial area per unit volume of packing) and S lying in the ranges 82–144.5 m^{-1} and 0.96–2.91 Hz, respectively, as the liquid flow rate changed from 0.072 to 0.362 $cm^3\ cm^{-2}\ s^{-1}$. de Waal and Beek (1967) studied the absorption of oxygen in sodium sulfite solutions in a column (diameter = 1 ft. and height = 10 ft) packed with 1-inch Raschig rings. Assuming a first-order reaction between dissolved O_2 and Na_2SO_3, they estimated values of a = 55–95.4 m^{-1} and S = 8–30 Hz as the liquid and gas mass velocities ranged over 10,000–80,000 and 500–3500 $kg\ m^{-2}\ h^{-1}$, respectively. Thus, the values of S and a used for the batch cell absorber in this chapter are considerably smaller than those reported by these investigators.

Notation

a	gas–liquid interfacial contact area per unit volume of liquid (m^{-1})
A	defined by eq. 5.85 or eq. 5.96
B	defined by eq. 5.86 or 5.97
c	concentration of solute gas at location x in a liquid element at time θ ($kmol\ m^{-3}$ or $g\ m^{-3}$)
c_b	concentration of solute gas in the bulk liquid ($kmol\ m^{-3}$ or $g\ m^{-3}$)
$c_{b,init}$	initial concentration of solute gas in the bulk liquid ($kmol\ m^{-3}$ or $g\ m^{-3}$)
c_{sat}	solubility of solute gas in the liquid or equal to y_b/H ($kmol\ m^{-3}$ or $g\ m^{-3}$)
C_1^m	binomial coefficient; defined by eq. 5.78
D	diffusion coefficient of solute gas in the liquid ($m^2\ s^{-1}$)
E	enhancement factor; defined by eq. 5.104
f	fraction of absorbed gas transferred from the gas–liquid interface to the bulk liquid
f_{abs}	defined by eq. 5.57 ($kmol\ m^{-2}\ s^{-1}$ or $g\ m^{-2}\ s^{-1}$)
f_{trans}	defined by eq. 5.58 ($kmol\ m^{-2}\ s^{-1}$ or $g\ m^{-2}\ s^{-1}$)
f_1, f_2, f_3, f_4, and f_5	defined by eqs. 5.51–5.53, 5.100, and 5.101, respectively (s or m s)
$f_D(\tau, t)$	Danckwerts surface age distribution function given by eq. 5.7 (s^{-1})

$F(\beta)$	defined by eq. 5.18
$F_b(z)$	LT of $c_b(t)$
g_1	defined by eq. 5.61 (m s)
$G(\beta)$	defined by eq. 5.37
h	defined by eq. 5.12 (kmol m^{-3} or g m^{-3})
H	Henry's constant
h_1 and h_2	defined by eqs. 5.43 and 5.44, respectively (m^{-1} and kmol m^{-4} or g m^{-4})
Ha	Hatta number (equal to $\sqrt{k/S}$)
i	index of summation
I_A	integral defined by eq. 5.16 (kmol m^{-3} or g m^{-3})
$I_{A,1}$, $I_{A,2}$, and $I_{A,n}$	values of I_A at t equal to t_1, t_2, and t_n, respectively (kmol m^{-3} or g m^{-3})
I_B	integral defined by eq. 5.35 (kmol m^{-3} or g m^{-3})
$I_{B,1}$, $I_{B,n-1}$, and $I_{B,n}$	values of I_B at t equal to t_1, t_{n-1}, and t_n, respectively (kmol m^{-3} or g m^{-3})
$II_{A,n}$ and $II_{B,n}$	values of II_A and II_B at t equal to t_n given by eqs. 5.59 and 5.60, respectively (kmol s m^{-3} or g s m^{-3} and kmol s m^{-2} or g s m^{-2})
$II_{B,n-1}$	value of II_B at t equal to t_{n-1} (kmol s m^{-2} or g s m^{-2})
j	index of summation
k	first-order rate constant (s^{-1})
k_G	gas-phase mass-transfer coefficient (m s^{-1})
k_L	liquid-phase mass-transfer coefficient (m s^{-1})
K_{OL}	overall mass-transfer coefficient defined by eq. 5.109 (m s^{-1})
K_1, K_2^-, and K_2^+	defined by eqs. (5.46–5.48), respectively (s)
$\mathcal{L}$	LT operator
$L_1(z)$, $L_2(z)$, and $L_3(z)$	image functions defined by eqs. (5.93–5.95), respectively
m, M, l	integer
$P(t)$	defined by eq. 5.23 (kmol m^{-2} s^{-1} or g m^{-2} s^{-1})
q	defined by eq. 5.70 (s^{-1})
q'	defined by eq. 5.74 (s^{-1})
R_{abs}	age-averaged rate of absorption of solute gas (kmol m^{-2} s^{-1} or g m^{-2} s^{-1})
R_{inst}	instantaneous rate of absorption of solute gas in a liquid element (kmol m^{-2} s^{-1} or g m^{-2} s^{-1})
R_{trans}	age-averaged net rate of transfer of dissolved gas to the bulk liquid due to surface renewal (kmol m^{-2} s^{-1} or g m^{-2} s^{-1})

$R_{trans,hyp}$	age-averaged net rate of transfer of hypothetical dissolved gas to the bulk liquid due to surface renewal (kmol m^{-2} s^{-1} or g m^{-2} s^{-1})
$\mathcal{R}(z)$	image function defined by eq. 5.91
S	frequency or rate of renewal of liquid elements at the gas–liquid interface (Hz or s^{-1})
t	process time (s)
t_1 and t_2	values of t at time levels 1 and 2, respectively (s)
t_{n-2}, t_{n-1}, and t_n	values of t at time levels $n-2$, $n-1$, and n, respectively (s)
$\mathcal{U}(z)$	image function defined by eq. 5.90
$v(x, \theta)$	liquid-phase concentration of a hypothetical gas undergoing purely physical absorption having zero concentration in the bulk liquid; defined by eq. 5.24 (kmol m^{-3} or g m^{-3})
w_i	weight; defined by eq. 5.77
$\mathcal{W}(z)$	image function defined by eq. 5.92
x	distance into a liquid element measured from the gas–liquid interface (m)
y_b	concentration of solute gas in the bulk gas phase (kmol m^{-3} or g m^{-3})
z	LT parameter

Greek Letters

α	defined by eq. 5.11
β	argument in eqs. 5.18 and 5.37
Δt	uniform time increment (s)
θ	time of exposure of a liquid element at the gas–liquid interface (s)
τ	age of a liquid element at the gas–liquid interface at process time t (s)
$\phi(t)$	function of t
$\Phi(z)$	LT (or image function) of $\phi(t)$

Acronyms

CFSTR	continuous-flow stirred-tank reactor
DO	dissolved oxygen
LT	Laplace transform
OM	original method
PSS	pseudo-steady-state

Homework Exercises

i) Verify that eq. 5.10 reduces to eq. 5.13 when $Ha \ll 1$ (physical absorption) and to eq. 5.15 when $Ha \gg 1$ (absorption with fast reaction).

ii) Show that eq. 5.1 is converted into eq. 5.25 by using the transformation given by eq. 5.24.

iii) Show that for an ideal CFSTR as steady state is approached, eqs. 5.9 and 5.33 reduce to eqs. 5.38 and 5.39, respectively.

iv) Derive eqs. 5.106 and 5.110 of the PSS model.

v) Pure hydrogen at atmospheric pressure and 25 °C is absorbed in 85 g of water contained in a stirred flask. Dissolved hydrogen undergoes a first-order chemical reaction in the water and the rate constant $k = 0.1\ s^{-1}$. The water, which does not contain any initial dissolved hydrogen, has an upper surface area of $34\ cm^2$ that is exposed to the gas. The value of the liquid-side mass-transfer coefficient k_L is 5.93×10^{-5}, and the diffusion coefficient D and solubility of hydrogen c_{sat} in water are $6 \times 10^{-9}\ m^2\ s^{-1}$ and $1.6\ g\ m^{-3}$, respectively. Using the equations developed in Sections 5.4.2 and 5.5.1, plot the (1) rates of absorption and dissolved-gas transfer; (2) fraction of absorbed gas transferred to the bulk liquid; and (3) dissolved hydrogen concentration in the bulk liquid as functions of process time. Approximately, how long will it take for the absorption process to attain a steady state?

vi) Consider a steady-state gas–liquid CFSTR in which a dissolved gas undergoes a first-order reaction with a reagent present in the liquid. The gas solubility in the liquid is c_{sat}, and the gas–liquid interfacial contact area is a. The liquid volumetric flowrate Q into and out of the reactor and the volume of liquid V_{liq} in the reactor are constant. The concentration of dissolved gas in the inlet liquid stream is zero, and the gas-phase mass-transfer resistance is negligible. Develop equations for the rate of gas absorption and concentration of dissolved gas in the liquid stream leaving the reactor.

References

Abate, J. and Whitt, W. (2006). "A unified framework for numerically inverting Laplace transforms," *INFORMS Journal on Computing*, 18(4), pp. 408–421. Available at: http://dx.doi.org/10.1287/ijoc.1050.0137. See also http://www.columbia.edu/~ww2040/JoC.pdf.

Astarita, G. (1967). *Mass transfer with chemical reaction*. Amsterdam: Elsevier.

Astarita, G. and Beek, W.J. (1962). "Effects of liquid phase mixing in a packed tower on first-order chemical absorption," *Chemical Engineering Science*, 17(9), pp. 665–674. Available at: http://dx.doi.org/10.1016/0009-2509(62)85023-4.

Cents, A.H.G., Brilman, D.W.F. and Versteeg, G.F. (2001). "Gas absorption in an agitated gas–liquid–liquid system," *Chemical Engineering Science*, 56(3), pp. 1075–1083. Available at: http://dx.doi.org/10.1016/S0009-2509(00)00324-9.

Chatterjee, S.G. (2010). "A theoretical analysis of dissolved-gas transfer to the bulk liquid in gas absorption with a first-order reaction," *Indian Chemical Engineer*, 51(3), pp. 177–193. Available at: http://dx.doi.org/10.1080/00194500903315476.

Chen, T.-I. and Barron, C.H. (1972). "Some aspects of the homogeneous kinetics of sulfite oxidation," *Industrial & Engineering Chemistry Fundamentals*, 11(4), pp. 466–470. Available at: http://dx.doi.org/10.1021/i160044a006.

Danckwerts, P.V. (1950). "Absorption by simultaneous diffusion and chemical reaction," *Transactions of the Faraday Society*, 46, p. 300. Available at: http://dx.doi.org/10.1039/tf9504600300.

Danckwerts, P.V. (1951). "Significance of liquid-film coefficients in gas absorption," *Industrial & Engineering Chemistry*, 43(6), pp. 1460–1467. Available at: http://dx.doi.org/10.1021/ie50498a055.

Danckwerts, P.V. (1965). "The absorption of gases in liquids," *Pure and Applied Chemistry*, 10(4), pp. 625–642. Available at: http://dx.doi.org/10.1351/pac196510040625.

Danckwerts, P.V. (1970). *Gas-liquid reactions*. New York: McGraw-Hill.

Danckwerts, P.V., Kennedy, A.M. and Roberts, D. (1963). "Kinetics of CO2 absorption in alkaline solutions—II," *Chemical Engineering Science*, 18(2), pp. 63–72. Available at: http://dx.doi.org/10.1016/0009-2509(63)80015-9.

de Waal, K.J.A. and Beek, W.J. (1967). "A comparison between chemical absorption with rapid first-order reactions and physical absorption in one packed column," *Chemical Engineering Science*, 22(4), pp. 585–593. Available at: http://dx.doi.org/10.1016/0009-2509(67)80041-1.

DeCoursey, W.J. (1974). "Absorption with chemical reaction: development of a new relation for the Danckwerts model," *Chemical Engineering Science*, 29(9), pp. 1867–1872. Available at: http://dx.doi.org/10.1016/0009-2509(74)85003-7.

Hartridge, H. and Roughton, F.J.W. (1923). "A method of measuring the velocity of very rapid chemical reactions," *Proceedings of the Royal Society of London. Series A, Containing Papers of a Mathematical and Physical Character*, 104(726), pp. 376–394. Available at: http://dx.doi.org/10.1098/rspa.1923.0116.

Horvath, I.R. and Chatterjee, S.G. (2018). "A surface renewal model for unsteady-state mass transfer using the generalized Danckwerts age distribution function," *Royal Society Open Science*, 5(5), 172423. Available at: http://dx.doi.org/10.1098/rsos.172423.

Huang, H. and Chatterjee, S.G. (2021). "Transient physical gas absorption in a stirred liquid using the surface renewal model of mass transfer," *Chemical Engineering Science*, 234, 116449. Available at: http://dx.doi.org/10.1016/j.ces.2021.116449.

Huang, H. and Chatterjee, S.G. (2023). "A rigorous surface renewal model for transient gas absorption with first-order chemical reaction in a stirred liquid," *Chemical Engineering Science*, 268, 118393. Available at: http://dx.doi.org/10.1016/j.ces.2022.118393.

Hutchinson, M.H. and Sherwood, T.K. (1937). "Liquid film in gas absorption," *Industrial & Engineering Chemistry*, 29(7), pp. 836–840. Available at: http://dx.doi.org/10.1021/ie50331a023.

Kayihan, F. and Sandall, O.C. (1974). "Gas absorption with first-order reaction in turbulent liquid films," *AIChE Journal*, 20(2), pp. 402–404. Available at: http://dx.doi.org/10.1002/aic.690200233.

Kucka, L., Richter, J., Kenig, E.Y. and Górak, A. (2003). "Determination of gas–liquid reaction kinetics with a stirred cell reactor," *Separation and Purification Technology*, 31(2), pp. 163–175. Available at: http://dx.doi.org/10.1016/S1383-5866(02)00179-X.

Lamb, W.B., Springer, T.G. and Pigford, R.L. (1969). "An interface impedance bridge," *Industrial & Engineering Chemistry Fundamentals*, 8(4), pp. 823–827. Available at: http://dx.doi.org/10.1021/i160032a039.

Lesage, F., Midoux, N., Latifi, A. and M. (2002). "Momentum transfer in a fixed-bed reactor described by the surface renewal model," *Chemical Engineering Science*, 57(24), pp. 5115–5122. Available at: http://dx.doi.org/10.1016/S0009-2509(02)00426-8.

Lightfoot, E.N. (1958). "Steady state absorption of a sparingly soluble gas in an agitated tank with simultaneous irreversible first-order reaction," *AIChE Journal*, 4(4), pp. 499–500. Available at: http://dx.doi.org/10.1002/aic.690040425.

Lightfoot, E.N. (1962). "Gas absorption with simultaneous irreversible first-order reaction. II. Comparison of results for the film and penetration models," *AIChE Journal*, 8(5), pp. 710–712. Available at: http://dx.doi.org/10.1002/aic.690080531.

Onda, K., Sada, E. and Takeuchi, H. (1968). "Gas absorption with chemical reaction in packed columns," *Journal of Chemical Engineering of Japan*, 1(1), pp. 62–66. Available at: http://dx.doi.org/10.1252/jcej.1.62.

Perry, R.H., Green, D.W. and Maloney, J.O. (eds.) (1984). *Perry's chemical engineers' handbook*. Sixth edition. New York: McGraw-Hill.

Reith, T. and Beek, W.J. (1973). "The oxidation of aqueous sodium sulphite solutions," *Chemical Engineering Science*, 28(6), pp. 1331–1339. Available at: http://dx.doi.org/10.1016/0009-2509(73)80084-3.

Richards, G.M., Ratcliff, G.A. and Danckwerts, P.V. (1964). "Kinetics of CO2 absorption—III," *Chemical Engineering Science*, 19(5), pp. 325–328. Available at: http://dx.doi.org/10.1016/0009-2509(64)80001-4.

Shaikh, A.A. and Al-Mubaiyedh, U.A. (1998). "On the bulk-liquid reaction in isothermal reactive gas absorption," *Chemical Engineering Journal*, 70(1), pp. 65–70. Available at: http://dx.doi.org/10.1016/S1385-8947(98)00077-1.

Sivaji, K. and Murty, G.S.R.N. (1982). "Kinetics of sulfite oxidation reaction," *Industrial & Engineering Chemistry Fundamentals*, 21(4), pp. 344–352. Available at: http://dx.doi.org/10.1021/i100008a005.

Srivastava, R.D., Mcmillan, A.F. and Harris, I.J. (1968). "The kinetics of oxidation of sodium sulphite," *The Canadian Journal of Chemical Engineering*, 46(3), pp. 181–184. Available at: http://dx.doi.org/10.1002/cjce.5450460308.

Stegeman, D., Ket, P.J., v. d. Kolk, H.A., Bolk, J.W., Knop, P.A. and Westerterp, K.R. (1995). 'Interfacial Area and Gas Holdup in an Agitated Gas-Liquid Reactor under Pressure', *Industrial & Engineering Chemistry Research*, 34(1), pp. 59–71. Available at: http://dx.doi.org/10.1021/ie00040a003.

Tchobanoglous, G. and Burton, F.L. (1991). *Wastewater engineering: treatment, disposal, and reuse*. Third edition. New York: McGraw-Hill.

Tontiwachwuthikul, P., Meisen, A. and Lim, C.J. (1992). "CO_2 absorption by NaOH, monoethanolamine and 2-amino-2-methyl-1-propanol solutions in a packed column," *Chemical Engineering Science*, 47(2), pp. 381–390. Available at: http://dx.doi.org/10.1016/0009-2509(92)80028-B.

van Elk, E.P. (2001). Gas-liquid reactions – influence of liquid bulk and mass transfer on process performance. The Netherlands: Thesis, Twente University. Available at: https://ris.utwente.nl/ws/portalfiles/portal/6074250/t0000030.pdf (Accessed: 15 October 2024).

van Elk, E.P., Borman, P.C., Kuipers, J.A.M. and Versteeg, G.F. (2000). "Modelling of gas–liquid reactors — implementation of the penetration model in dynamic modelling of gas–liquid processes with the presence of a liquid bulk," *Chemical Engineering Journal*, 76(3), pp. 223–237. Available at: http://dx.doi.org/10.1016/S1385-8947(99)00162-X.

Wilkinson, P.M., Doldersum, B., Cramers, P.H.M.R. and Van Dierendonck, L.L. (1993). "The kinetics of uncatalyzed sodium sulfite oxidation," *Chemical Engineering Science*, 48(5), pp. 933–941. Available at: http://dx.doi.org/10.1016/0009-2509(93)80331-J.

6

Gas Absorption with First-Order Reaction in a Packed Tower

6.1 Introduction

A packed column is a common device used in chemical process operations such as absorption, stripping, and distillation. Ideally, the purpose of a packed column is to improve contact between a gas phase and a liquid phase to facilitate mass transfer of a component or components between the two phases. The column is usually filled with randomly dumped or structured packing made of metal or plastic, which fills the empty space of the column and provides added surface area for gas–liquid contact. The process design of such columns (e.g., determination of the volume of packing required for a given removal efficiency of a component in the inlet gas stream) will involve both interfacial and bulk-liquid mass balances for the dissolved gas, liquid-phase reagents, and products of chemical reactions.

As discussed in Chapter 1, two well-known theoretical descriptions of the mass-transfer process at the gas–liquid interface are the film and surface renewal models. The film model, proposed by Lewis and Whitman (1924), pictures stagnant gas and liquid films at the interface with the dissolved gas crossing the liquid film, whose thickness depends upon the prevailing hydrodynamics, by steady-state molecular diffusion. In contrast, the surface renewal model, proposed by Danckwerts (1951), visualizes the gas–liquid interface to be continuously replenished by fresh liquid elements arriving at the interface from the bulk liquid. The rate of replenishment of the liquid elements comprising the interface is characterized by the surface renewal frequency S, which depends upon the state of interfacial turbulence.

Because of its simplicity (i.e., steady-state interfacial mass transport described by ordinary differential equations), the film model has usually been the choice for the design and analysis of plate/packed absorption and distillation columns, e.g., in commercial process simulators and the computer model for gas absorption with

Applications of the Surface Renewal Model of Mass Transfer, First Edition.
Hengshuo Huang and Siddharth G. Chatterjee.

Companion website: www.wiley.com/go/huang/surfacerenewalmodel

chemical reaction in packed columns provided by De Leye and Froment (1986). However, as discussed in Chapter 1, the more complex surface renewal model in which the interfacial mass transfer is represented by partial differential equations, is considered to be more realistic than the film model because it predicts that the liquid-side mass-transfer coefficient k_L is proportional to the square root of the diffusion coefficient D of the dissolved gas in the liquid – a result that has often been confirmed experimentally. This evidence is in sharp contrast to the prediction of the film model according to which $k_L \propto D$, and which therefore raises the question of whether it is scientifically appropriate to use this model to describe the turbulent mass transfer at the gas–liquid interface and for studying both single component and multicomponent absorption, which has often been the practice, based solely on the criteria of simplicity and practical utility.

The preceding discussion lays the groundwork for using the surface renewal model for analysis and design of gas–liquid contacting equipment. Chapters 4 and 5 (or see Huang and Chatterjee, 2021, 2023) presented rigorous surface renewal models (i.e., which distinguish between the rate of absorption at the gas–liquid interface, R_{abs}, and the rate of dissolved gas-transfer to the bulk liquid, R_{trans}) to describe physical gas absorption and gas absorption with a first-order chemical reaction in an unsteady-state stirred batch cell. However, to our knowledge, there is no rigorous surface renewal model available for chemical gas absorption in a packed tower. We note that a partial solution for gas absorption with a first-order chemical reaction in a packed column, assuming only liquid-side resistance to mass transfer, was attempted by Astarita (1967), who, besides not elaborating on the type of the interfacial mass-transfer model used, also neglected to account for the consumption of dissolved gas by reaction in the interfacial region (i.e., did not distinguish between R_{abs} and R_{trans}) – see eq. 7.1 in his textbook (Astarita, 1967).

It is well known that both film and penetration models, the latter being a special case of the surface renewal model, give close numerical values for the enhancement factor for many types of gas–liquid reactions (Danckwerts, 1970). Brian *et al.* (1961) solved the Higbie penetration model for a second order gas–liquid reaction numerically and found good agreement (within 16%) of the enhancement factor with that calculated by the film model (van Krevelen and Hoftijzer, 1948). Predictions made by an approximate expression for the enhancement factor for a second-order gas–liquid reaction using the surface renewal model that was derived by DeCoursey (1974) agreed reasonably well (i.e., having a maximum deviation of 14%) with the numerical calculations of Brian *et al.* (1961). However, a quantitative equivalence between the film and penetration/surface renewal models has yet to be established for chemical gas absorption in a packed tower. It should be emphasized that a qualitative equivalence between them is impossible since they are based on radically different mass-transfer mechanisms occurring at the gas–liquid interface as discussed previously.

The chief objectives of this chapter are the following: (1) To develop and compare the theoretical predictions of rigorous surface renewal and film models, which account for both gas- and liquid-side mass-transfer resistances, and also distinguish between R_{abs} and R_{trans} for gas absorption accompanied by a first-order chemical reaction in a packed column; (2) to investigate the numerical equivalency of the two models; and (3) to extend the well-known transfer unit concept, commonly used to describe physical gas absorption in a packed column, to the case of absorption with reaction – a beginning was made in this direction by Astarita (1967). Despite the considerable literature on gas absorption in liquids, all of this has yet to be established, to the best of our knowledge.

The work described in this chapter is also an example of using the surface renewal model to analyze chemical gas absorption in a packed tower in a comprehensive and integrated fashion. Illustrative calculations using the surface renewal and film models are presented for three sample gas–liquid systems with widely different solubility and mass-transfer parameters.

6.2 Surface Renewal Model

Using the surface renewal model of mass transfer to describe the absorption of a gas in a liquid where it is consumed by a first-order reaction, Chatterjee (2010) derived the following relation for the rate of absorption R_{abs} for steady-state conditions:

$$R_{abs} = K_{OG}\left(y_b - \frac{Hc_b}{E^2}\right) = K_{OL}\left(\frac{y_b}{H} - \frac{c_b}{E^2}\right) \tag{6.1}$$

where y_b is the concentration of the solute gas in the bulk gas phase, c_b is the concentration of solute gas in the bulk liquid phase, H is Henry's constant, and E is what has been conventionally called the enhancement factor. K_{OL} and K_{OG} are overall liquid-phase and gas-phase mass-transfer coefficients which include the effect of chemical reaction; these are given by

$$\frac{1}{K_{OL}} = \frac{1}{Hk_G} + \frac{1}{Ek_L} \tag{6.2}$$

$$\frac{1}{K_{OG}} = \frac{1}{k_G} + \frac{H}{Ek_L} \tag{6.3}$$

$$E = \sqrt{1 + Ha^2} \tag{6.4}$$

$$Ha = \sqrt{\frac{k}{S}} = \frac{\sqrt{kD}}{k_L} \tag{6.5}$$

and

$$k_L = \sqrt{DS} \tag{6.6}$$

Here, k_G and k_L are the gas- and liquid-side physical mass-transfer coefficients, Ha is the Hatta number, D is the binary diffusion coefficient of the solute gas in the liquid solvent in which the absorption is occurring (also referred to as D_L later in this manuscript), S is the rate of surface renewal, and k is the first-order reaction rate constant. Eqs. 6.2 to 6.6 are well known in the literature whereas eq. 6.1, which may not be as familiar, reduces to an equation presented by Danckwerts (1970) for the surface renewal model for the case of no gas-phase mass-transfer resistance. It is clearly observed in this equation that the chemical reaction, represented by the factor E, affects both the overall mass-transfer coefficient given by eqs. 6.2 or 6.3 and the overall driving force for absorption. We further note that $K_{OL} = HK_{OG}$, i.e., they are not linearly independent quantities.

The fraction f of absorbed gas that is transferred at steady state from the gas–liquid interface to the bulk liquid due to the surface renewal mechanism is given by eq. 5.64 in Chapter 5 (or see Chatterjee, 2010), i.e.,

$$f = \frac{R_{trans}}{R_{abs}} = \frac{1}{1 + Ha^2} = \frac{1}{E^2} \tag{6.7}$$

For physical absorption, $Ha = 0$ and $f = 1$ whereas for $Ha \to \infty$ (i.e., fast chemical reaction), $f \to 0$. For $Ha > 3$, less than 10% of the absorbed gas is transferred to the bulk liquid.

The gas-phase mass balance for the solute gas in a packed tower, a schematic of which is shown in Figure 6.1, can be expressed as

$$G\frac{dy_b}{dz} = R_{abs}aA_c = K_{OG}\left(y_b - \frac{Hc_b}{E^2}\right)aA_c \tag{6.8}$$

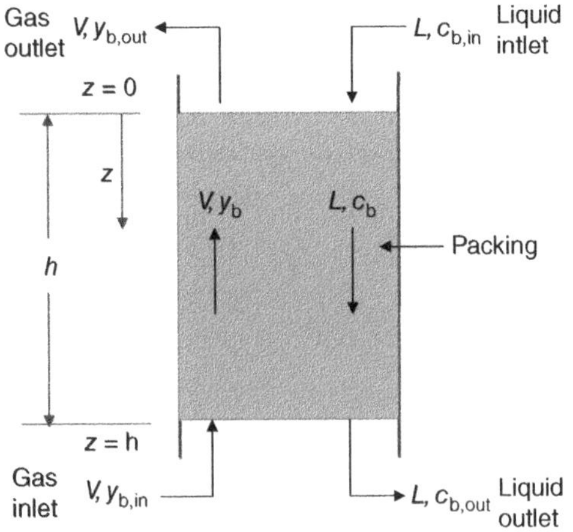

Figure 6.1 Schematic of a packed absorption tower.

where a is the gas–liquid interfacial area of contact per unit volume of packing, A_c is the area of cross-section of the tower, G is the volumetric flow rate of bulk gas in the tower, which is assumed to be constant (i.e., dilute system), and z is the downward distance measured from the top of the packing.

The mass balance for the solute gas in the liquid phase is given by

$$L\frac{dc_b}{dz} = R_{trans}a\,A_c - k\,c_b\phi A_c = f\,K_{OL}\left(\frac{y_b}{H} - \frac{c_b}{E^2}\right)a\,A_c - k\,c_b\phi A_c \tag{6.9}$$

where ϕ is the fractional liquid holdup in the column and L the volumetric flow rate of liquid in the tower, which is also assumed to be constant.

It should be noted that R_{abs} appears in the gas-phase mass balance (eq. 6.8) whereas R_{trans} does so in the liquid-phase mass balance (eq. 6.9). It is to be reiterated that this has often not been the practice in the literature where R_{trans} is tacitly assumed to be the same as R_{abs}, which as evident from eq. 6.7, is not theoretically correct except in the case of physical absorption (i.e., when $Ha = 0$).

We define the following quantities:

$$H_{OG} = \frac{\frac{G}{A_c}}{K_{OG}a} \tag{6.10}$$

$$H_{OL} = \frac{\frac{L}{A_c}}{K_{OL}a} \tag{6.11}$$

and

$$k^* = \frac{k\,\phi}{K_{OL}a} \tag{6.12}$$

Here, H_{OG} and H_{OL} are the heights of the overall gas- and liquid-phase mass-transfer units, respectively, which include the effect of chemical reaction, and k^* is a dimensionless rate constant.

Eqs. 6.8 and 6.9 then become

$$H_{OG}\frac{dy_b}{dz} = y_b - \frac{Hc_b}{E^2} \tag{6.13}$$

and

$$H_{OL}\frac{dc_b}{dz} = f\left(\frac{y_b}{H} - \frac{c_b}{E^2}\right) - k^*c_b \tag{6.14}$$

We introduce the following dimensionless quantities:

$$z^* = \frac{z}{h},\quad y_b^* = \frac{y_b}{y_{b,in}},\quad c_b^* = \frac{Hc_b}{y_{b,in}}$$

where h is the height of packing in the column and $y_{b,in}$ is the inlet concentration of the solute gas in the gas phase at the bottom of the packing (i.e., at $z = h$). Eqs. 6.13

and 6.14 are then transformed into

$$\frac{dy_b^*}{dz^*} = N_{OG}y_b^* - \frac{N_{OG}}{E^2}c_b^* \tag{6.15}$$

where

$$N_{OG} = \frac{h}{H_{OG}} \tag{6.16}$$

and

$$\frac{dc_b^*}{dz^*} = fN_{OL}y_b^* - N_{OL}\left(\frac{f}{E^2} + k^*\right)c_b^* \tag{6.17}$$

with

$$N_{OL} = \frac{h}{H_{OL}} \tag{6.18}$$

In these equations, N_{OG} and N_{OL} are the numbers of the overall gas-phase and liquid-phase mass-transfer units, respectively, which include the effect of chemical reaction.

The solution of eqs. 6.15 and 6.17 is given by

$$y_b^*(z^*) = c_1 e^{\lambda_A z^*} + c_2 e^{\lambda_B z^*} \tag{6.19}$$

and

$$c_b^*(z^*) = c_1 D_A e^{\lambda_A z^*} + c_2 D_B e^{\lambda_B z^*} \tag{6.20}$$

where

$$\lambda_{A,B} = \frac{N_{OG} - N_{OL}\left(k^* + \frac{f}{E^2}\right) \pm \sqrt{\left[N_{OG} + N_{OL}\left(k^* + \frac{f}{E^2}\right)\right]^2 - \frac{4fN_{OG}N_{OL}}{E^2}}}{2} \tag{6.21}$$

with

$$D_A = E^2\left(1 - \frac{\lambda_A}{N_{OG}}\right) \tag{6.22}$$

and

$$D_B = E^2\left(1 - \frac{\lambda_B}{N_{OG}}\right) \tag{6.23}$$

By analysis of the discriminant of eq. 6.21, it can be shown the criterion for it to be equal to or greater than zero, i.e., for λ_A and λ_B to be real numbers, is $\left(\frac{L}{HG} - 1\right)^2 \geq 0$, which is always true.

In eqs. 6.19 and 6.20, c_1 and c_2 are constants that have to be determined from the boundary conditions at the top ($z = 0$, i.e., $z^* = 0$) and bottom ($z = h$, i.e., $z^* = 1$) of the column. At $z^* = 0$, $c_b^* = c_{b,in}^* = Hc_{b,in}/y_{b,in}$ where $c_{b,in}$ is the solute gas concentration in the entering liquid at the top of the column. Also, at $z^* = 1$, $y_b^* = y_{b,in}^* = y_{b,in}/y_{b,in} = 1$. Upon using these values in eqs. 6.19 and 6.20, it can be shown that

$$c_1 = \frac{c_{b,in}^* - D_B y_{b,out}^*}{D_A - D_B} \tag{6.24}$$

and

$$c_2 = \frac{D_A y_{b,out}^* - c_{b,in}^*}{D_A - D_B} \tag{6.25}$$

where

$$y_{b,out}^* = \frac{y_{b,out}}{y_{b,in}} = \frac{(D_A - D_B) - c_{b,in}^*\left(e^{\lambda_A} - e^{\lambda_B}\right)}{D_A e^{\lambda_B} - D_B e^{\lambda_A}} \tag{6.26}$$

Here, $y_{b,out}$ is the solute gas concentration in the gas phase at the top of the column. Substituting eqs. 6.22 and 6.23 into eq. 6.26 yields

$$y_{b,out}^* = \frac{\frac{E^2(\lambda_B - \lambda_A)}{N_{OG}} - c_{b,in}^*\left(e^{\lambda_A} - e^{\lambda_B}\right)}{E^2\left[\left(1 - \frac{\lambda_A}{N_{OG}}\right)e^{\lambda_B} - \left(1 - \frac{\lambda_B}{N_{OG}}\right)e^{\lambda_A}\right]} \tag{6.27}$$

The dimensionless concentration of the solute gas in the liquid leaving at the bottom of the tower, which can be obtained by setting $z^* = 1$ in eq. 6.20, is given by

$$c_{b,out}^* = \frac{Hc_{b,out}}{y_{b,in}} = c_1 D_A e^{\lambda_A} + c_2 D_B e^{\lambda_B} \tag{6.28}$$

where c_1 and c_2 are given by eqs. 6.24 and 6.25, respectively.

If $c_{b,in}^* = 0$, (i.e., no solute gas is present in the inlet liquid), these equations simplify to

$$y_{b,out}^* = \frac{\lambda_B - \lambda_A}{(N_{OG} - \lambda_A)e^{\lambda_B} - (N_{OG} - \lambda_B)e^{\lambda_A}} \tag{6.29}$$

and

$$\begin{aligned} c_{b,out}^* &= \frac{D_A D_B y_{b,out}^*}{(D_A - D_B)}\left(e^{\lambda_B} - e^{\lambda_A}\right) \\ &= \frac{E^2 (N_{OG} - \lambda_A)(N_{OG} - \lambda_B)}{N_{OG}}\left[\frac{e^{\lambda_B} - e^{\lambda_A}}{(N_{OG} - \lambda_A)e^{\lambda_B} - (N_{OG} - \lambda_B)e^{\lambda_A}}\right] \end{aligned} \tag{6.30}$$

From eqs. 6.2, 6.4–6.6, and 6.12, it can be shown that

$$k^* = \left(\frac{k_L \phi}{Da}\right) Ha^2 \left(\frac{EZ + 1}{E}\right) \tag{6.31}$$

where the non-dimensional quantity Z is defined by

$$Z = \frac{k_L}{Hk_G} \tag{6.32}$$

Eqs. 6.2 and 6.3, which state the law of addition of resistances for gas absorption in a liquid with a first-order chemical reaction, can be expressed as

$$H_{OL} = \frac{H_L^{PA}}{E} + \left(\frac{L}{HG}\right) H_G^{PA} \tag{6.33}$$

and

$$H_{OG} = H_G^{PA} + \left(\frac{HG}{L}\right) \frac{H_L^{PA}}{E} \tag{6.34}$$

In eqs. 6.33 and 6.34, "PA" stands for physical absorption (discussed later in Section 6.2.1), i.e., H_L^{PA} and H_G^{PA} are the heights of the liquid- and gas-phase mass-transfer units for physical absorption, which are given by

$$H_L^{PA} = \frac{\frac{L}{A_c}}{k_L a} \tag{6.35}$$

and

$$H_G^{PA} = \frac{\frac{G}{A_c}}{k_G a} \tag{6.36}$$

Eqs. 6.33 and 6.34 indicate that H_{OG} and H_{OL} are related by

$$H_{OG} = \left(\frac{HG}{L}\right) H_{OL} \tag{6.37}$$

i.e., they are not linearly independent. It follows from eqs. 6.16, 6.18, and 6.37 that

$$N_{OG} = \left(\frac{L}{HG}\right) N_{OL} \tag{6.38}$$

The general calculation procedure for determining the gas- and liquid-phase concentration profiles in the absorption column is as follows:

1) Specify the type of packing and values for the height h of packing, tower cross-sectional area A_c (calculated from the column diameter), liquid volumetric flow rate L (obtained by dividing the operating liquid mass flow rate L_M by the liquid density ρ_L), gas volumetric flow rate G (obtained by dividing the operating gas mass flow rate G_M by the gas density ρ_G), inlet liquid-phase concentration of dissolved gas $c_{b,in}$ (at the top of the column), inlet gas-phase concentration of solute gas $y_{b,in}$ (at the bottom of the column), Henry's constant H, and reaction rate constant k. Instead of specifying L_M and G_M directly, the operating liquid and vapor mass velocities can be specified from which the liquid and vapor mass flow rates in the column can be calculated by multiplying the respective mass velocity with A_c, which was the procedure adopted in this work.

2) Provide values for the gas-side physical mass-transfer coefficient k_G, liquid-side physical mass-transfer coefficient k_L, fractional liquid holdup ϕ, diffusion coefficient D of the dissolved gas in the liquid solvent, and gas–liquid interfacial contact area a from available laboratory or pilot-plant data or estimate their values from suitable empirical correlations, as done in this work.
3) Calculate the heights of the liquid- and gas-phase mass-transfer units for physical absorption, i.e., H_L^{PA} and H_G^{PA} from eqs. 6.35 and 6.36, respectively.
4) Calculate the Hatta number Ha by eq. 6.5 and the enhancement factor E by eq. 6.4.
5) Calculate the heights of the overall liquid- and gas-phase mass-transfer units for chemical absorption, i.e., H_{OL} and H_{OG} by eqs. 6.33, 6.34 or 6.37, respectively.
6) Determine the number of overall liquid- and gas-phase mass-transfer units for chemical absorption, i.e., N_{OL} and N_{OG} from eqs. 6.18 and 6.16 or 6.38, respectively.
7) Calculate the dimensionless parameter Z and dimensionless rate constant k^* using eqs. 6.32 and 6.31, respectively. Also, calculate $c_{b,in}^* = Hc_{b,in}/y_{b,in}$, i.e., the dimensionless dissolved gas concentration in the liquid entering at the top of the column.

At the end of the last step of the calculation procedure, the values of the following five dimensionless quantities will be known: N_{OL}, N_{OG}, k^*, Ha, and $c_{b,in}^*$. The dimensionless gas- and liquid-phase concentration profiles of the solute gas can then be obtained from eqs. 6.7 and 6.19 to 6.26. For $c_{b,in}^* \neq 0$, the outlet solute concentrations of the gas and liquid leaving at the top and bottom of the tower are given by eqs. 6.27 and 6.28, respectively. If $c_{b,in}^* = 0$, these can be calculated from eqs. 6.29 and 6.30, respectively.

From eqs. 6.2, 6.3, and 6.32, it may be shown that

$$\frac{H_{OL}}{H_{OL}^{PA}} = \frac{H_{OG}}{H_{OG}^{PA}} = \frac{1}{E}\left(\frac{Hk_G + Ek_L}{Hk_G + k_L}\right) = \frac{1}{E}\left(\frac{EZ+1}{Z+1}\right) \tag{6.39}$$

where H_{OL}^{PA} and H_{OG}^{PA} are the heights of the overall liquid- and gas-phase mass-transfer units for physical absorption (see eqs. 6.44 and 6.45). Eq. 6.39 clearly indicates that the chemical reaction reduces the heights of the overall transfer units compared to those for physical absorption since $E > 1$.

Some special cases are examined next.

6.2.1 Physical Absorption

In this case $Ha = 0$, which implies that $k^* = 0, f = 1$, and $E = 1$. Eq. 6.21 then gives

$$\lambda_A = N_{OG}^{PA} - N_{OL}^{PA} \tag{6.40}$$

and

$$\lambda_B = 0 \tag{6.41}$$

where N_{OL}^{PA} and N_{OG}^{PA} are the numbers of the overall liquid- and gas-phase mass-transfer units for physical absorption, respectively. They are given by

$$N_{OL}^{PA} = \frac{h}{H_{OL}^{PA}} \tag{6.42}$$

and

$$N_{OG}^{PA} = \frac{h}{H_{OG}^{PA}} \tag{6.43}$$

with H_{OL}^{PA} and H_{OG}^{PA} being obtained by setting $E = 1$ in eqs. 6.33 and 6.34, respectively, i.e.,

$$H_{OL}^{PA} = H_L^{PA} + \left(\frac{L}{HG}\right) H_G^{PA} \tag{6.44}$$

and

$$H_{OG}^{PA} = H_G^{PA} + \left(\frac{HG}{L}\right) H_L^{PA} \tag{6.45}$$

where H_L^{PA} and H_G^{PA} are to be obtained from eqs. 6.35 and 6.36, respectively.

These are well-known equations available in standard chemical engineering textbooks.

6.2.2 Liquid-Phase Control

In this case, the dominant resistance to mass transfer lies in the liquid phase. Thus, if $EZ \ll 1$, then $Z \ll 1$ (i.e., $k_L \ll Hk_G$) since $E \geq 1$, and eq. 6.39 gives

$$H_{OL} \approx \frac{H_{OL}^{PA}}{E} \approx \frac{H_L^{PA}}{E} \tag{6.46}$$

and

$$H_{OG} \approx \frac{H_{OG}^{PA}}{E} \approx \frac{H_G^{PA}}{ZE} \tag{6.47}$$

Thus, as mentioned before, the chemical reaction (i.e., $E > 1$) reduces the heights of the overall liquid- and gas-phase mass-transfer units compared to those for physical absorption ($E = 1$).

Also

$$N_{OG} = \frac{h}{H_{OG}} \approx \frac{h}{\frac{H_{OG}^{PA}}{E}} \approx \frac{h}{H_G^{PA}} ZE \tag{6.48}$$

and

$$N_{OL} = \frac{h}{H_{OL}} \approx \frac{h}{\frac{H_{OL}^{PA}}{E}} \approx \frac{h}{H_L^{PA}}E \tag{6.49}$$

Further, for $EZ \ll 1$, i.e., $Z \ll 1$, eq. 6.31 gives

$$k^* \approx \left(\frac{k_L \phi}{Da}\right) \frac{Ha^2}{E} \tag{6.50}$$

where E is to be obtained from eq. 6.4. It should be noted that the dimensional group $k_L \phi /(Da)$ is a physical quantity only, i.e., it is independent of the chemical reaction.

The procedure for calculating the gas- and liquid-phase concentration profiles is the same as given earlier in Section 6.2 except that H_{OL}, H_{OG}, N_{OG}, N_{OL}, and k^* are to be obtained from eqs. 6.46 to 6.50.

6.2.3 Gas-Phase Control

For this situation, the dominant resistance to mass transfer lies in the gas phase. Thus, $Z \gg 1$ (i.e., $k_L \gg Hk_G$) and since $E \geq 1$, eq. 6.39 yields

$$H_{OG} \approx H_{OG}^{PA} \approx H_G^{PA} \tag{6.51}$$

and

$$H_{OL} \approx H_{OL}^{PA} \approx H_L^{PA} Z \tag{6.52}$$

which reveals that the chemical reaction has almost no effect on the heights of the overall mass-transfer units.

The number of overall gas- and liquid-phase transfer units are given by

$$N_{OG} \approx \frac{h}{H_{OG}^{PA}} \approx \frac{h}{H_G^{PA}} \tag{6.53}$$

and

$$N_{OL} \approx \frac{h}{H_{OL}^{PA}} \approx \frac{h}{H_L^{PA} Z} \tag{6.54}$$

For $Z \gg 1$, eq. 6.31 reduces to

$$k^* \approx \left(\frac{k_L \phi}{Da}\right) Ha^2 Z \tag{6.55}$$

The calculation procedure is identical to that for the case of liquid-phase control with the modification that eqs. 6.51 to 6.55 replace eqs. 6.46 to 6.50.

6.3 Film Model

Danckwerts (1970) presented the following equations for R_{abs} and R_{trans} for the film model:

$$R_{abs} = Ek_L \left(c_i - \frac{c_b}{\cosh Ha} \right) \tag{6.56}$$

and

$$R_{trans} = Ek_L \left(\frac{c_i}{\cosh Ha} - c_b \right) \tag{6.57}$$

where the enhancement factor E is given by

$$E = \frac{Ha}{\tanh Ha} \tag{6.58}$$

and

$$Ha = \frac{\sqrt{kD}}{\frac{D}{\delta}} = \frac{\sqrt{kD}}{k_L} \tag{6.59}$$

with δ being the thickness of the liquid film at the gas–liquid interface. In the above, c_i is the liquid-side interfacial concentration of dissolved gas, which can be obtained from a mass balance for solute gas at the gas–liquid interface, i.e.,

$$R_{abs} = k_G \left(y_b - Hc_i \right) = Ek_L \left(c_i - \frac{c_b}{\cosh Ha} \right) \tag{6.60}$$

which yields

$$c_i = \frac{k_G y_b + \frac{Ek_L c_b}{\cosh Ha}}{Hk_G + Ek_L} \tag{6.61}$$

Eliminating c_i from eqs. 6.56 and 6.57 by using eq. 6.61 and performing some algebraic manipulation, gives

$$R_{abs} = K_{OG} \left(y_b - \frac{Hc_b}{E_1^2} \right) = K_{OL} \left(\frac{y_b}{H} - \frac{c_b}{E_1^2} \right) \tag{6.62}$$

and

$$R_{trans} = f_1 K_{OL} \left(\frac{y_b}{H} - \frac{c_b}{E_2^2} \right) \tag{6.63}$$

where

$$E_1 = \sqrt{\cosh Ha} \tag{6.64}$$

$$f_1 = \frac{1}{\cosh Ha} = \frac{1}{E_1^2} \tag{6.65}$$

and

$$E_2 = \frac{1}{\sqrt{\cosh IIa\,(1 + ZIIa \tanh IIa)}} \tag{6.66}$$

where K_{OL}, K_{OG}, and Z are to be obtained from eqs. 6.2, 6.3, and 6.32, respectively. Eqs. 6.62 and 6.63 of the film model are the analogs of eqs. 6.1 and 6.7 of the surface renewal model.

The gas- and liquid-phase mass balances for the solute gas are expressed by

$$G\,\frac{dy_b}{dz} = R_{abs} a\,A_c = K_{OG}\left(y_b - \frac{Hc_b}{E_1^2}\right) a\,A_c \tag{6.67}$$

and

$$L\,\frac{dc_b}{dz} = R_{trans} a\,A_c - k\,c_b \phi\,A_c = f_1 K_{OL}\left(\frac{y_b}{H} - \frac{c_b}{E_2^2}\right) a\,A_c - k\,c_b \phi\,A_c \tag{6.68}$$

The non-dimensional forms of eqs. 6.67 and 6.68 can be obtained in a similar fashion as done in Section 6.2 for the surface renewal model; these are given by

$$\frac{dy_b^*}{dz^*} = N_{OG} y_b^* - \frac{N_{OG}}{E_1^2} c_b^* \tag{6.69}$$

and

$$\frac{dc_b^*}{dz^*} = f_1 N_{OL} y_b^* - N_{OL}\left(\frac{f_1}{E_2^2} + k^*\right) c_b^* \tag{6.70}$$

which can be compared with eqs. 6.15 and 6.17 of the surface renewal model. In eqs. 6.69 and 6.70, the number of overall gas- and liquid-phase transfer units, i.e., N_{OG} and N_{OL}, are given by eqs. 6.16 and 6.18, respectively, with the heights of the overall gas- and liquid-phase transfer units, i.e., H_{OG} and H_{OL}, being defined by eqs. 6.10 and 6.11, respectively. Also, the dimensionless reaction rate constant k^* is given by 6.31.

The solution of the system of eqs. 6.69 and 6.70, which will yield the gas- and liquid-phase concentration profiles of the solute gas, is given by eqs. 6.19 and 6.20 with

$$\lambda_{A,B} = \frac{N_{OG} - N_{OL}\left(k^* + \frac{f_1}{E_2^2}\right) \pm \sqrt{\left[N_{OG} + N_{OL}\left(k^* + \frac{f_1}{E_2^2}\right)\right]^2 - \frac{4 f_1 N_{OG} N_{OL}}{E_1^2}}}{2} \tag{6.71}$$

The same criterion, as that mentioned earlier for the surface renewal model, will hold and λ_A and λ_B given by eq. 6.71 will always be real.

The constants c_1 and c_2 are given by eqs. 6.24 and 6.25, respectively, with D_A and D_B to be obtained from eqs. 6.22 and 6.23, respectively, in which E has to be replaced by E_1 as defined by eq. 6.64.

The outlet gas- and liquid-phase concentrations of the solute gas can be obtained from eqs. 6.27 (with E replaced by E_1) and 6.28, respectively. In the special case of $c^*_{b,in} = 0$ (i.e., no solute gas present in the inlet liquid), they are given by eqs. 6.29 and 6.30 (with E replaced by E_1).

Similar to the case of the surface renewal model as seen earlier in Section 6.2, the five dimensionless parameters which govern the film model are N_{OG}, N_{OL}, k^*, Ha, and $c^*_{b,in}$. The calculation procedure to obtain the gas- and liquid-phase concentration profiles along the tower is the same as that for the surface renewal model, with the enhancement factor E being obtained from eq. 6.58.

6.3.1 Physical Absorption

The relevant equations for this case are the same as those for the surface renewal model, which were presented in Section 6.2.1 earlier.

6.3.2 Liquid-Phase Control

In this case, $Z \ll 1$ (i.e., $k_L \ll Hk_G$) and the equations in Section 6.2.2, derived previously for the surface renewal model, also hold for the film-model with E (enhancement factor) for the film model being obtained from eq. 6.58 and k^* given by (see eq. 6.31)

$$k^* \approx \left(\frac{k_L \phi}{Da}\right) Ha \tanh Ha \tag{6.72}$$

For $Z \ll 1$, eq. 6.66 reduces to

$$E_2 \approx \frac{1}{\sqrt{\cosh Ha}} \tag{6.73}$$

6.3.3 Gas-Phase Control

When the dominant resistance to mass transfer lies in the gas phase, i.e., when $Z \gg 1$ or $k_L \gg Hk_G$, the relevant equations for the film model are the same as those derived earlier for the surface renewal model in Section 6.2.3.

For $Z \gg 1$, eq. 6.66 simplifies to

$$E_2 \approx \frac{1}{\sqrt{Z Ha \sinh Ha}} \tag{6.74}$$

6.4 Mass-Transfer and Hydraulic Correlations

Mass-transfer correlations are needed in order to estimate a (gas–liquid interfacial area of contact, assumed to be equal to the wetted surface of the packing), k_L (liquid-side mass-transfer coefficient), k_G (gas-side mass-transfer coefficient), and ϕ (fractional liquid holdup in the column). For the first three parameters, the empirical equations proposed by Onda *et al.* (1968a) for columns filled with random packings can be used whereas the correlation of Fu *et al.* (2020) can be employed to estimate the liquid holdup.

The correlation for a is given by:

$$\frac{a}{a_t} = 1 - \exp\left[-1.45\left(\frac{\sigma_c}{\sigma_L}\right)^{0.75} N_{Re}^{0.1} N_{Fr}^{-0.05} N_{We}^{0.2}\right] \tag{6.75}$$

where σ_c is the critical surface tension of the packing material, σ_L is the surface tension of the liquid, a_t is the packing surface area, while

$$N_{Re} = \frac{\frac{L_M}{A_c}}{a_t \mu_L} \quad \text{(Reynolds number)} \tag{6.76}$$

$$N_{Fr} = \frac{\left(\frac{L_M}{A_c}\right)^2 a_t}{\rho_L^2 g} \quad \text{(Froude Number)} \tag{6.77}$$

and

$$N_{We} = \frac{\left(\frac{L_M}{A_c}\right)^2}{\rho_L \sigma_L a_t} \quad \text{(Weber Number)} \tag{6.78}$$

Here, μ_L and ρ_L are the viscosity and density of the liquid, respectively, and g is the acceleration due to gravity.

The correlation for k_L is given by:

$$k_L\left(\frac{\rho_L}{\mu_L g}\right)^{\frac{1}{3}} = 0.0051\left(\frac{\frac{L_M}{A_c}}{a\mu_L}\right)^{\frac{2}{3}}\left(\frac{\mu_L}{\rho_L D_L}\right)^{-\frac{1}{2}}(a_t d_p)^{0.4} \tag{6.79}$$

where D_L is the diffusion coefficient of the dissolved gas in the liquid and d_p is the packing size.

The gas-phase mass-transfer coefficient k_G is obtained from the following equations:

$$\frac{k_g RT}{a_t D_G} = \alpha\left(\frac{\frac{G_M}{A_c}}{a_t \mu_G}\right)^{0.7}\left(\frac{\mu_G}{\rho_G D_G}\right)^{\frac{1}{3}}(a_t d_p)^{-2.0} \tag{6.80}$$

and

$$k_G = k_g \frac{P_{col} MW_G}{\rho_G} \tag{6.81}$$

where μ_G is the gas viscosity, ρ_G is the gas density, D_G is the gas-phase diffusion coefficient of the solute gas component in the carrier gas, P_{col} is the column pressure, and MW_G is the molecular weight of the gas. The constant α equals 5.23 for ring and saddle packings larger than about 12 mm (0.5 in) and 2.00 for smaller packings (Onda *et al.*, 1968b).

The fractional liquid holdup ϕ can be calculated using the following equation (Fu *et al.*, 2020):

$$\phi = \frac{0.56 {N_{Re}}^{-0.4} {N_{Fr}}^{0.47} {N_{Ka}}^{0.17}}{(1 - 0.32 \cos\theta)^{2.73} \varepsilon^{1.22}} \tag{6.82}$$

where ε is the fractional void volume of the packing, θ is the contact angle of the packing (= 12.7° for Raschig rings), and Ka is the Kapitza number defined by

$$Ka = \sigma_L \left(\frac{\rho_L}{{\mu_L}^4 g} \right)^{\frac{1}{3}} \tag{6.83}$$

The flooding gas velocity is estimated from the following correlation (Leva, 1953, 1954):

$$\log Y = -0.270 \log^2(X) - 1.076 \log(X) - 1.705 \tag{6.84}$$

where the flow parameters X and Y are defined by

$$X = \frac{L_s}{G_{fl}} \sqrt{\frac{\rho_L}{\rho_G}} \tag{6.85}$$

and

$$Y = \frac{G_{fl}^2 \rho_G {\rho_w}^2 a_t \mu_L^{0.2}}{g {\rho_L}^3 \varepsilon^3} \tag{6.86}$$

Here, ε is the porosity or voidage of the packing, ρ_W is the density of water, L_s is the operating superficial liquid velocity (m s^{-1}), and G_{fl} is the superficial flooding gas velocity (m s^{-1}). The percentage flood is defined by

$$\%\ \text{flood} = \frac{G_s}{G_{fl}} \times 100 \tag{6.87}$$

where G_s is the operating superficial gas velocity (m s^{-1}).

6.5 Illustrative Calculations

Calculations are performed for three sample gas–liquid systems shown in Table 6.1 assuming that the absorption is occurring in a column (diameter:

Table 6.1 Values of physical parameters for the sample gas–liquid systems.

Parameter	System 1 (air-acetone-water)	System 2 (air-SO_2-water)	System 3 (air-O_2-Na_2SO_3 soln.)
type of packing	1-in ceramic Raschig ring		
critical surface tension of packing (σ_c, mN m^{-1})	61	61	61
nominal packing size (d_p, m)	0.0254	0.0254	0.0254
packing surface area (a_t, m^2 m^{-3})	190	190	203[a)]
packing factor (F_p, m^{-1})	510	510	510
voidage (ε)	0.74	0.74	0.74
column temperature (°C)	25	20	30
column pressure (P_{col}, atm)	1	1	1
packed height (h, m)	5	5	3.048[a)]
column diameter (d, m)	0.5	0.5	0.3048[a)]
liquid mass velocity (L_M/A_c, kg m^{-2} s^{-1})	2.14	2.14	11.11 (Table 1 in[a)]; second case)
vapor mass velocity (G_M/A_c, kg m^{-2} s^{-1})	0.935	0.935	0.1389 (Table 1 in[a)]; second case)
liquid density (ρ_L, kg m^{-3})	997.045	998.204	995.65
vapor density (ρ_G, kg m^{-3})	1.184	1.206	1.164
liquid viscosity (μ_L, kg m^{-1} s^{-1})	8.937×10^{-4}	1.005×10^{-3}	0.7972×10^{-3}
vapor viscosity (μ_G, kg m^{-1} s^{-1})	1.837×10^{-5}	1.812×10^{-5}	1.86×10^{-5}
inlet liquid concentration of solute gas ($c_{b,in}$, kgmol m^{-3})	0	0	0
diffusivity of solute gas in liquid (D_L, m^2 s^{-1})	1.03×10^{-9}	1.49×10^{-9}	2.229×10^{-9}[a)]
diffusivity of solute gas in air (D_G, m^2 s^{-1})	1.27×10^{-5}	1.38×10^{-5}	2.08×10^{-5}[b)]
Henry's law constant (H)	1.54×10^{-3}	2.431×10^{-2}[d)]	33.159[c)]
surface tension of liquid (σ_L, mN m^{-1})	71.97	72.75	71.2

a) de Waal and Beek (1967).
b) Badino (2009).
c) Sander (2015),
d) The value of H for SO_2 absorption in water at an average temperature of 25°C for the data of Terraglio and Manganelli (1967) is 2.625×10^{-2}.

0.3–0.5 m) packed with ceramic Raschig rings (nominal size: 0.0254 m) up to a height of 3–5 m. It is assumed to operate at atmospheric pressure and in the temperature range of 20–30°C. Most of the physical properties for Systems 1 and 2 are obtained from Perry *et al.* (1984) whereas those for System 3 are for an experimental gas–liquid system to be described later. Two operating liquid and vapor mass velocities are used: 2.14 and 11.11 kg m^{-2} s^{-1} for the liquid, and 0.139 and 0.935 kg m^{-2} s^{-1} for the gas, respectively. The concentration of dissolved gas in the incoming liquid fed to the top of the column is taken as zero. It is to be noted that acetone undergoes physical absorption in water (System 1) whereas when SO_2 is absorbed in water (System 2), the total SO_2 in solution is distributed amongst the following three species: H_2SO_3 (sulfurous acid), HSO_3^- (bisulfite ion) and SO_3^{-2} (sulfite ion), with the transformations between these forms occurring almost instantaneously. Thus, Systems 1 and 2, in the context of this chapter, are hypothetical – they are chosen only to study the effect of different physical properties on the absorption process. System 3, on the other hand, is an actual gas–liquid system studied by de Waal and Beek (1967) who postulated that the oxidation of sodium sulfite solutions in the presence of cobaltous ions was first order in oxygen over a wide range of sulfite concentrations. However, they also mentioned, citing another investigation, that the reaction could be second order in oxygen. According to Reith and Beek (1973), the oxidation reaction is first-order only for pure oxygen at atmospheric pressure but second order when oxygen is absorbed from a gas mixture. The wide difference in values of the Henry's law constant H amongst the three systems in Table 6.1 is to be noted.

Table 6.2 reports values of some of the calculated parameters for the three gas–liquid systems shown in Table 6.1. Although the mass-transfer coefficients k_L and k_G are more or less of the same order of magnitude for the three systems, there is a great difference in the magnitude of Z. As one conceptually moves from System 1 to System 3, the extent of liquid-phase mass-transfer control increases dramatically along with a great decrease in gas solubility.

Values of physical parameters used in the calculations, except for the value of the Hatta number Ha, are taken from Table 6.1. To examine the effect of the chemical reaction on the absorption process, different values of Ha are assumed for Systems 1 and 2 and calculated for System 3 as will be explained later. The complete or full surface renewal and film models are used in the calculations illustrated in Sections 6.5.1 to 6.5.6.

6.5.1 System 1

The dimensionless gas- and liquid-phase concentration profiles for System 1, as predicted by the surface renewal and film models, are shown in Figure 6.2 as a

Table 6.2 Values of some calculated physical parameters for the sample gas–liquid systems.

Parameter	System 1 (air-acetone-water)	System 2 (air-SO_2-water)	System 3 (air-O_2-Na_2SO_3 soln.)
liquid-phase mass-transfer coefficient (k_L, m s^{-1})	6.55×10^{-5}	7.22×10^{-5}	2.34×10^{-4}
gas-phase mass-transfer coefficient (k_G, m s^{-1})	2.90×10^{-2}	3.06×10^{-2}	1.07×10^{-2}
gas–liquid interfacial area of contact (a, m^{-1})	78.3	77.0	76.0 (from de Waal and Beek, 1967)
fractional total liquid holdup (ϕ)	0.0415	0.0424	0.0785
height of liquid-phase mass-transfer unit for physical absorption (H_L^{PA}, m)	0.419	0.386	0.6263
height of gas-phase mass-transfer unit for physical absorption (H_G^{PA}, m)	0.348	0.329	0.146
Z [$k_L/(Hk_G)$]	1.466	0.097	6.595×10^{-4}
liquid volumetric flow rate (L, m^3 s^{-1})	4.214×10^{-4}	4.210×10^{-4}	5.655×10^{-6}
vapor volumetric flow rate (G, m^3 s^{-1})	1.551×10^{-1}	1.552×10^{-1}	6.046×10^{-5}
flood (%)	48.5	42.0	11.33

function of the dimensionless distance along the column. The value of Z for this system is 1.466 (Table 6.2) showing that both gas- and liquid-phase mass-transfer resistances need to be considered. Values of Ha ranging from 0 to 10 are used in the calculations. It is evident that the two models show excellent agreement over the range of Hatta numbers considered, with the gas- and liquid phase profiles exhibiting concavity with respect to the horizontal (i.e., z^*) axis. As Ha increases from 0 (physical absorption) to 0.1, the gas-phase concentration profile is barely affected although there is a slight decrease in the outlet gas concentration (at $z^* = 0$). However, the liquid-phase concentration profile of the dissolved gas is lowered with a rise in Ha with a much larger decrease in the outlet liquid concentration (at $z^* = 1$). Also, the gas- and liquid phase concentrations reach low values over a significant portion of the column as Ha increases, indicating inefficient operation. Thus, for $Ha = 0.1$, only about 50% of the column is utilized for mass transfer.

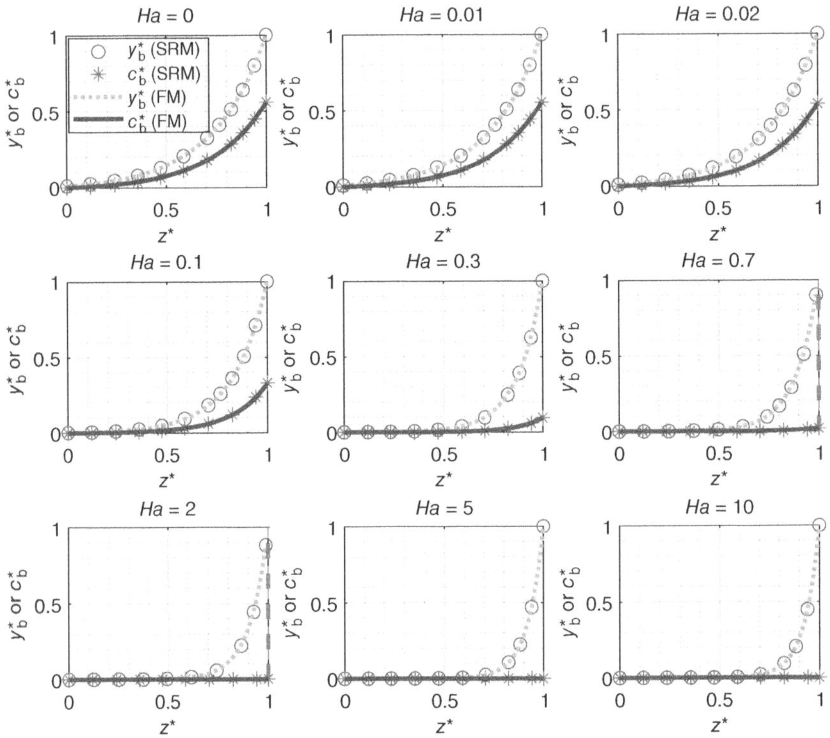

Figure 6.2 Dimensionless gas- and liquid-phase concentration profiles of the solute gas along the packed tower for System 1 as predicted by the surface renewal model (SRM) and film model (FM) as a function of the Hatta number *Ha* varying from 0 to 10. The operating conditions and physical properties of the system are reported in the first column of Table 6.1.

For fast reactions ($Ha > 0.7$), there is virtually no dissolved gas present in the liquid in the column – see Chapter 5 and Chatterjee (2010).

6.5.2 System 2

The values of H (2.431×10^{-2}) and Z (0.097) for System 2 are about 16 times greater and 15 times smaller, respectively, compared to System 1, which indicates that System 2 has low gas solubility and mass transfer is primarily controlled by the resistance in the liquid phase.

The dimensionless gas- and liquid-phase concentration profiles for System 2, as predicted by the surface renewal and film models, are shown in Figure 6.3 as a function of the dimensionless distance along the column with the Hatta number *Ha* varying from 0 to 10. In the case of physical absorption (i.e., $Ha = 0$), the outlet

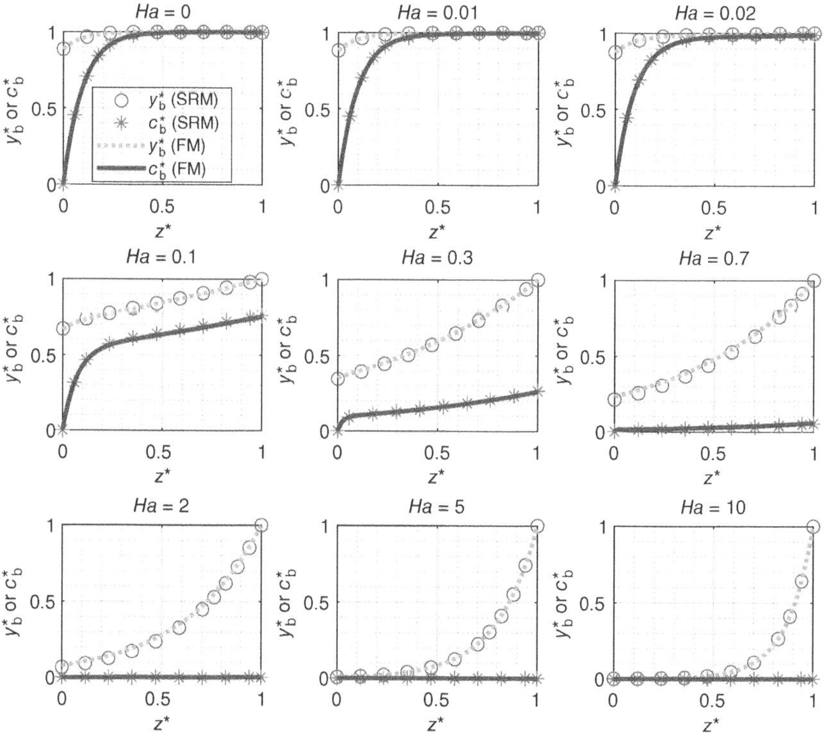

Figure 6.3 Dimensionless gas- and liquid-phase concentration profiles of the solute gas along the packed tower for System 2 as predicted by the surface renewal model (SRM) and film model (FM) as a function of the Hatta number *Ha* varying from 0 to 10. The operating conditions and physical properties of the system are reported in the second column of Table 6.1.

concentration of the solute in the gas phase at the top of the tower ($z^* = 0$) is 90% of its inlet value at the bottom of the tower ($z^* = 0$); i.e., only about 10% of the solute in the incoming gas is absorbed in the liquid. For $Ha > 0.1$, the chemical reaction starts having an effect thereby increasing the absorption of the solute gas in the liquid, i.e., increasing the effective mass-transfer driving force. For all values of *Ha* in Figure 6.3, similar to that observed for System 1 (Figure 6.2), the film and surface renewal models show excellent agreement for System 2 also. However, the shape of the concentration profiles for low values of *Ha* (0 to 0.1; slow reaction) is convex in sharp contrast to System 1. As *Ha* increases beyond 0.1, there is a lowering of the gas-phase concentration profile, which now transitions to a concave shape, with a drastic reduction in the concentration of dissolved gas in the liquid in the tower as *Ha* increases beyond 2 (fast reaction) and the solute concentration in the gas leaving the tower is virtually zero.

6.5.3 System 3

As discussed in Chapter 5, the absorption of oxygen (O_2) in sodium sulfite solution (Na_2SO_3) in the presence of cobaltous ions is accompanied by a chemical reaction which can be first order in dissolved O_2 and in which varying the concentration of cobaltous ions alters the magnitude of the reaction rate constant over a relatively wide range from about 40 s^{-1} to 5000 s^{-1} (de Waal and Beek, 1967). A packed column (0.3048 m inner diameter and 3.048 m packed height) operating at 1 atm and 30 °C was used in the experiments of these authors for different values of the reaction rate constant (by changing the concentration of cobaltous ions in the incoming liquid). Operating conditions and values of physical properties and reaction rate constant k are derived from their paper (de Waal and Beek, 1967) and are shown in the last column of Table 6.1 (except for the reaction rate constant, k).

The value of 2.08×10^{-5} m^2 s^{-1} for the binary diffusion coefficient of O_2 in air (D_G) at 30°C shown in Table 6.1 was obtained by correcting the value of 1.78×10^{-5} m^2 s^{-1} at STP (standard temperature and pressure) given by Badino (2009) with the "3/2" temperature correction factor from Chapman-Enskog theory. de Waal and Beek (1967) reported a value of 1.71×10^{-9} m^2 s^{-1} for the diffusivity of dissolved O_2 in aqueous Na_2SO_3 solution at 20 °C. This was corrected for viscosity and temperature using the Wilke-Chang correlation to obtain a value of $D_L = 2.229 \times 10^{-9}$ m^2 s^{-1} at 30°C. de Waal and Beek (1967) also reported four different values of the first-order rate constant k (32, 56, 68, and 2500 s^{-1}) and a surface renewal frequency S of 8 s^{-1} at liquid and vapor mass velocities of 40 000 and 500 kg m^{-2} h^{-1}, respectively; these were used to calculate values of the Hatta number by the equation $Ha = \sqrt{k/S}$, which were also used in the film model calculations so as to maintain consistency in the comparison between the two models.

The values of H (33.159; no correction was made for ionic effects) and Z (6.595×10^{-4}) reveal that mass transfer between gas and liquid is severely restricted in System 3. This almost complete absence of mass transfer explains the flat gas- and liquid phase concentration profiles in Figure 6.4 for values of Ha up to 2.915. It is only when $Ha = 17.678$ that the gas-phase concentration profile of the solute starts to bend downwards appreciably with the outlet concentration being approximately 80% of its inlet value, which indicates the occurrence of some mass transfer of the solute from gas to liquid. The concentration of solute in the liquid (i.e., dissolved-gas concentration) is vanishingly small at all values of Ha. Once again, as observed in Figure 6.4, there is excellent agreement between the film and surface renewal models.

According to the surface renewal model, the liquid-side physical mass-transfer coefficient is given by eq. 6.6. Substituting values of $D_L = 2.229 \times 10^{-9}$ m^2 s^{-1} and

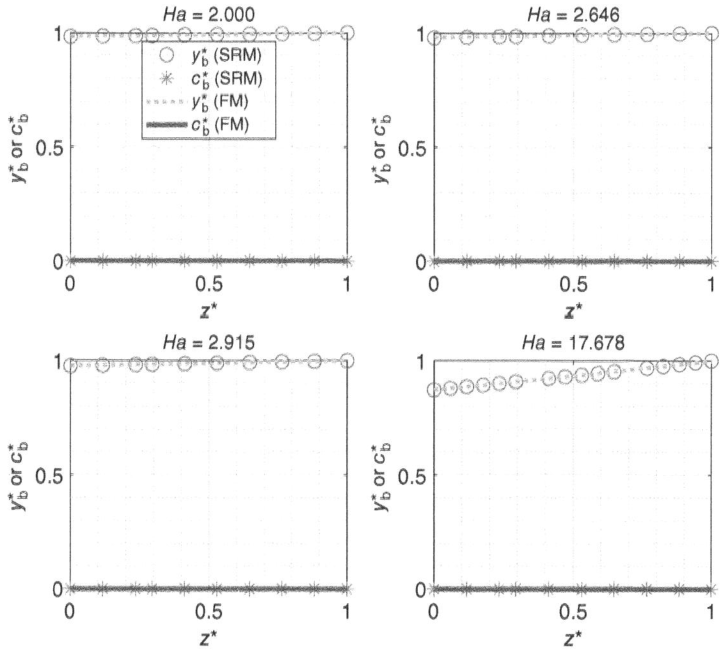

Figure 6.4 Dimensionless gas- and liquid-phase concentration profiles of the solute gas along the packed tower as predicted by the surface renewal model (SRM) and film model (FM) in the absorption of O_2 in Na_2SO_3 solution (System 3) for four different values of the Hatta number *Ha* (2.00, 2.646, 2.915, and 17.68) derived from the work of de Waal and Beek (1967). The operating conditions and physical properties of the system are reported in the third column of Table 6.1.

$S = 8\ s^{-1}$ into this equation yields $k_L = 1.34 \times 10^{-4}$ m s^{-1}, which is about 57% of the value shown in Table 6.2 for System 3, the latter being estimated from eq. 6.79. However, using the smaller value of k_L in the calculations has an insignificant effect on the results, e.g., the value of the dimensionless outlet gas concentration $y^*_{b,out}$ increases by 2.5% for $Ha = 17.678$.

6.5.4 Outlet Gas- and Liquid-Phase Concentrations as Functions of Hatta Number

Figure 6.5 shows the behavior of the dimensionless gas and liquid concentrations leaving the tower at the top and bottom, respectively, as a function of the Hatta number *Ha* (varying over seven orders of magnitude) for the three gas–liquid systems in Table 6.1. To compare the systems on an equivalent basis, the tower

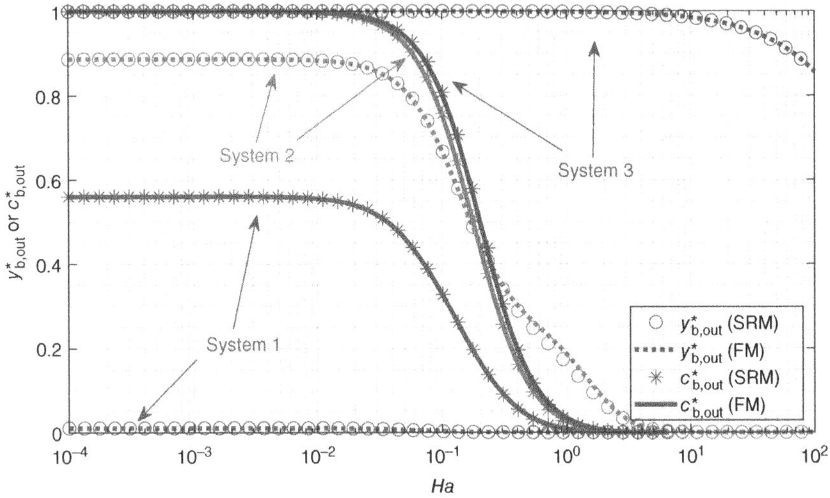

Figure 6.5 Variation of the dimensionless outlet gas and liquid concentrations with Hatta number for the three gas–liquid systems shown in Table 6.1.

operating conditions (i.e., gas and liquid mass velocities), tower diameter, and packed height for System 3 were assumed to be the same as those for the other two systems, with no changes being made to the physical properties and temperature (see Table 6.1). We note that this equivalence will also be assumed in Section 6.5.6. The following conclusions are reached from Figure 6.5:

1) There is excellent consistency between the film and surface renewal models.
2) For all three systems, the dissolved-gas concentration in the outlet liquid is negligible for $Ha > 1$.
3) For all three systems, there is very little change in the gas and liquid outlet concentrations for $Ha < 10^{-2}$.
4) There is a great difference between the three systems regarding the behavior of the outlet gas concentration. For System 1, it is virtually zero for all values of Ha whereas for System 2, it is around 0.9 for values of Ha up to 4×10^{-2} after which it declines and becomes negligible for $Ha > 5$. For System 3, which is the most solubility and mass-transfer limited among the three systems, the outlet gas concentration is almost 1 (signifying no mass transfer from gas to liquid) up to $Ha = 10$. Beyond this value, it begins a slow decline due to the initiation of mass transfer of the solute from gas to liquid because of the enhanced speed of the chemical reaction.
5) For System 2, concentrations of both phases descend synchronously as Ha increases beyond a certain point. However, for System 3, there is "retardance,"

i.e., the gas concentration declines only after the liquid concentration has been effectively reduced to zero due to a high value of *Ha*.

6) For all three systems, the outlet liquid concentration of the dissolved gas is negligible for $Ha > 1$, which roughly aligns with the conclusion reached earlier by Chatterjee (2010) in his analysis of a gas–liquid reactor using the surface renewal model.

6.5.5 Effect of Operational Gas and Liquid Mass Velocities on Gas Concentration Profile

Figure 6.6 shows plots of the dimensionless concentration profile of the gas as a function of gas or liquid mass velocity for System 2, which is liquid-phase controlled. The gas mass velocity G_M is varied from 0.2 kg m^{-2} s^{-1} to 1.2 kg m^{-2} s^{-1} using an increment of 0.2 kg m^{-2} s^{-1} (keeping the liquid mass velocity constant at 2.14 kg m^{-2} s^{-1}; see Table 6.1) whereas the liquid mass velocity L_M is changed from 1 kg m^{-2} s^{-1} to 11 kg m^{-2} s^{-1} using an increment of by 2 kg m^{-2} s^{-1} (keeping G_M constant at 0.935 kg m^{-2} s^{-1}; see Table 6.1). The value of the Hatta number *Ha* is assumed to be equal to 1.

As L_M is changed from 1 kg m^{-2} s^{-1} to 11 kg m^{-2} s^{-1}, % flood varies from 9 to 54% whereas when G_M is increased from 0.2 kg m^{-2} s^{-1} to 1.2 kg m^{-2} s^{-1}, % flood varies from 40 to 77%. It can be expected that as L_M is increased, mass transfer of the solute gas into the liquid will be promoted (Aroonwilas *et al.*, 2001; Park *et al.*, 2004; Yu *et al.*, 2011). This expectation is validated in Figure 6.6a where it is observed that the gas concentration profile in the tower is continuously lowered as L_M is increased, thereby leading to heightened absorption efficiency. The main cause for this is an increase in the interfacial gas–liquid contact area a (see eq. 6.75; Aroonwilas and Tontiwachwuthikul, 1997) and the liquid-phase mass-transfer coefficient k_L (see eq. 6.79).

In contrast, Figure 6.6b also shows that an increase in G_M leads to a rise in the gas concentration profile i.e., mass transfer between gas and liquid is diminished, which may be contrary to intuition. However, it can be observed from eqs. 6.75 to 6.79 that a and k_L are independent of the gas mass velocity. And although the gas-phase mass-transfer coefficient k_G increases with an increase in G_M, as shown by eqs. 6.80 and 6.81, System 2 is liquid-phase controlled as was seen earlier. A higher value of G_M will shorten the gas–liquid contact time in the tower, thereby causing a reduction in absorption efficiency (Park *et al.*, 2004). In industrial operations, a high gas flow rate can entrain liquid leading to a reduction in the liquid flow rate in the tower, which has a deleterious effect on the efficiency of mass transfer (Yan *et al.*, 2008).

As observed earlier, the film and surface renewal models show quite good agreement with each other in Figure 6.6.

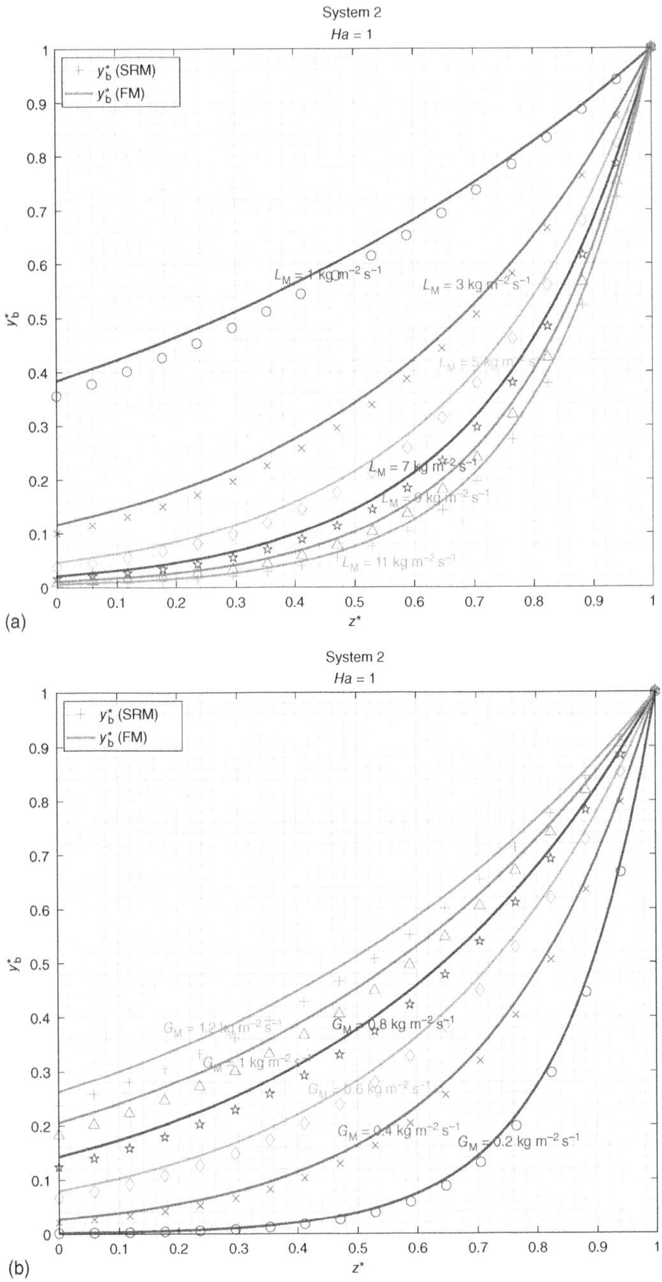

Figure 6.6 Variation of the gas concentration profile with liquid (a) and gas (b) mass velocities for System 2 ($Ha = 1$). Symbols and lines represent the surface renewal and film models, respectively.

6.5.6 Effect of Hatta Number on the Height and Number of Overall Gas- and Liquid-Phase Mass-Transfer Units

The predicted values of the heights of overall gas- and liquid-phase mass-transfer units (i.e., H_{OG} and H_{OL}) for the film and surface renewal models over a six-order of magnitude change in the Hatta number *Ha* are compared in Figure 6.7 for the three systems. In the comparison, the equivalence assumed in Section 6.5.4 is also invoked.

For $Ha < 0.1$ (i.e., slow reaction), E (enhancement factor) ≈ 1 and H_{OG} and H_{OL} change insignificantly, although the difference in values for the three systems are significant. For example, the H_{OG} values are 0.6, 3.8, and 3000 m for Systems 1, 2 and 3, respectively, the last being severely mass-transfer limited – see eq. 6.47 and the very small value of Z in Table 6.2. The values of H_{OL} for all three systems are approximately 1 m or smaller, and except for System 1, are much smaller than the corresponding H_{OG} values.

For $Ha > 1$ and Systems 2 and 3, H_{OG} and H_{OL} begin to decline dramatically with an increase in *Ha* indicating that the chemical reaction begins to have a significant

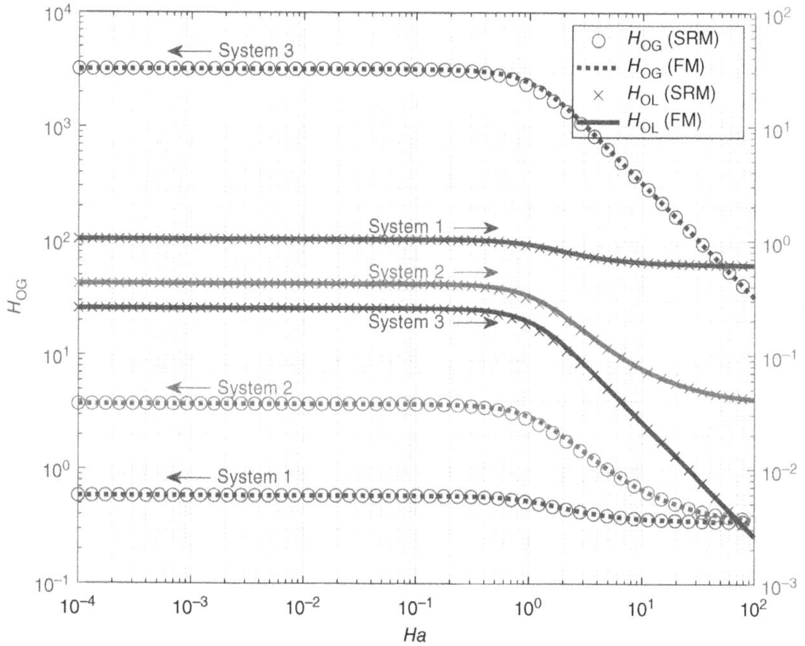

Figure 6.7 Variation of the height (m) of overall gas- and liquid-phase mass-transfer units with Hatta number for the film and surface renewal models.

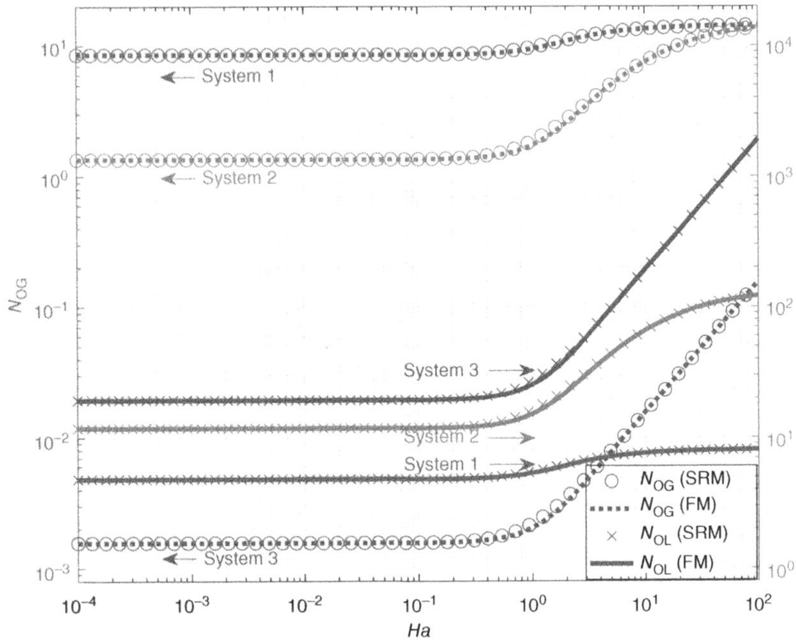

Figure 6.8 Variation of the number of overall gas- and liquid-phase mass-transfer units with Hatta number for the film and surface renewal models.

effect in aiding mass transfer. Thus, H_{OL} decreases from 0.4 to 0.04 m and from 0.28 to 0.0028 m for Systems 2 and 3, respectively, as *Ha* changes from 1 to 100.

The variation in the number of gas- and liquid-phase mass-transfer units (i.e., N_{OG} and N_{OL}) as a function of *Ha* varying from 10^{-4} to 10^{2} is shown Figure 6.8. This plot is an inverse mirror image of Figure 6.6 since the number of transfer units is inversely proportional to the height of a transfer unit because their (multiplicative) product is a fixed constant, i.e., equal to the height of the packed section.

The excellent agreement between the film and surface renewal models is once again evident in both Figure 6.7 and Figure 6.8. Also, the linear relation between H_{OG} and H_{OL}, as given by eq. 6.37, can be clearly observed in Figure 6.7 for all three systems.

In spite of the very good agreement between the film and surface renewal models for gas absorption with a first-order reaction in a packed tower at an outer or superficial level, e.g., gas- and liquid-phase concentration profiles, they differ at an inner or deeper level, e.g., as regards the fraction of absorbed gas transferred to the bulk liquid. Equation 6.7 of the surface renewal model predicts this ratio, which only depends upon the Hatta number, to be constant along the column.

In contrast, eqs. 6.62 to 6.66 of the film model show that the behavior of this ratio is more complex, and which can vary along the column.

The numerical equivalency between the film and surface renewal models for other types of gas–liquid systems and gas–liquid reactions (second order, reversible, etc.) is an open question, which remains to be investigated. The case of gas absorption with an instantaneous chemical reaction in a packed tower is examined in Chapter 7.

Finally, both film and surface renewal models ultimately need validation against experimental packed-column data. However, the theoretical lens to be used to analyze or correlate such data needs to be decided on beforehand.

Notation

a	gas–liquid interfacial contact area per unit volume of packing (m^{-1})
A_c	area of cross-section of the tower (m^2)
c_b	concentration of solute gas in the bulk liquid at location z in the tower (kgmol m^{-3})
$c_{b,in}$	concentration of dissolved gas in the inlet liquid at the top of the packed section (kgmol m^{-3})
$c_{b,out}$	concentration of dissolved gas in the outlet liquid at the bottom of the packed section (kgmol m^{-3})
$c^*_{b,in}$	dimensionless inlet liquid-phase concentration of dissolved gas at the top of the tower $\left(\frac{c_{b,in}}{Hy_{b,in}}\right)$
$c^*_{b,out}$	dimensionless outlet liquid-phase concentration of dissolved gas at the bottom of the tower $\left(\frac{Hc_{b,out}}{y_{b,in}}\right)$
c^*_b	dimensionless liquid-phase concentration of dissolved gas at location z^* in the tower $\left(\frac{Hc_b}{y_{b,in}}\right)$
c_1, c_2	defined by eqs. 6.24 and 6.25, respectively
c_i	liquid-side interfacial concentration of dissolved gas at location z in the tower (kgmol m^{-3})
d	column diameter (m)
D_A, D_B	defined by eqs. 6.22 and 6.23, respectively
D_L (or D), D_G	binary diffusion coefficients of solute gas in the liquid solvent and carrier gas (i.e., air), respectively ($m^2\ s^{-1}$)
d_p	nominal packing size (m)
E	enhancement factor
E_1, E_2	defined by eqs. 6.64 and 6.66, respectively

f	fraction of absorbed gas transported to the bulk liquid due to the surface renewal mechanism
f_1	defined by eq. 6.65
F_p	packing factor (m^{-1})
g	acceleration due to gravity ($m\,s^{-2}$)
G	volumetric flow rate of the bulk gas in the tower ($m^3\,s^{-1}$)
G_{fl}	gas superficial flooding velocity ($m\,s^{-1}$)
G_M	gas mass flow rate ($kg\,s^{-1}$)
G_s	operating superficial gas velocity ($m\,s^{-1}$)
h	height of packing in the column (m)
H	Henry's constant
H_G^{PA}	height of gas-phase mass-transfer unit for physical absorption (m)
H_L^{PA}	height of liquid-phase mass-transfer unit for physical absorption (m)
H_{OG}	height of overall gas-phase mass-transfer unit (m)
H_{OG}^{PA}	height of overall gas-phase mass-transfer unit for physical absorption (m)
H_{OL}	height of liquid-phase mass-transfer unit (m)
H_{OL}^{PA}	height of overall liquid-phase mass-transfer unit for physical absorption (m)
Ha	Hatta number $\left(\frac{\sqrt{kD}}{k_L}\right)$
k	first-order reaction rate constant (s^{-1})
k^*	dimensionless rate constant defined by eq. 6.12
k_G	gas-phase physical mass-transfer coefficient ($m\,s^{-1}$)
k_L	liquid-phase physical mass-transfer coefficient ($m\,s^{-1}$)
K_{OG}	overall gas-phase mass-transfer coefficient ($m\,s^{-1}$)
K_{OL}	overall liquid-phase mass-transfer coefficient ($m\,s^{-1}$)
L	volumetric flow rate of liquid in the tower ($m^3\,s^{-1}$)
L_M	liquid mass flow rate ($kg\,s^{-1}$)
L_s	operating superficial liquid velocity ($m\,s^{-1}$)
N_{OG}	number of overall gas-phase mass-transfer units
N_{OL}	number of overall liquid-phase mass-transfer units
P_{col}	column pressure (atm)
R_{abs}	rate of absorption of solute gas at a location z in the tower (kgmol $m^{-2}\,s^{-1}$)
R_{trans}	net rate of transfer of dissolved gas from the gas–liquid interface to the bulk liquid at a location z in the tower (kgmol $m^{-2}\,s^{-1}$)
S	surface renewal frequency (Hz or s^{-1})
X	flow parameter defined by eq. 6.85
Y	flow parameter defined by eq. 6.86

$y_{b,in}$ inlet concentration of solute gas in the gas entering at the bottom of the packing (kgmol m^{-3})

$y_{b,out}$ outlet concentration of solute gas in the gas leaving at the top of the packing (kgmol m^{-3})

$y^*_{b,out}$ dimensionless outlet gas-phase concentration $\left(\frac{y_{b,out}}{y_{b,in}}\right)$

y_b concentration of solute gas in the bulk gas phase at location z in the tower (kgmol m^{-3})

y^*_b dimensionless gas-phase concentration at location z^* in the tower $\left(\frac{y_b}{y_{b,in}}\right)$

z downward distance measured from the top of the packing (m)

Z dimensionless quantity defined by eq. 6.32

z^* dimensionless distance $= \frac{z}{h}$

Greek Letters

δ thickness of the liquid film at the gas–liquid interface in the film model (m)

λ_A, λ_B defined by eq. 6.21 for the surface renewal model or eq. 6.71 for the film model

ϕ fractional liquid holdup in the column

ρ_G and ρ_L vapor and liquid densities, respectively (kg m^{-3})

ρ_W density of water (kg m^{-3})

σ_c critical surface tension of packing (mN m^{-1})

σ_L surface tension of liquid (mN m^{-1})

Homework Exercises

i) Show that the solution of eqs. 6.15 and 6.17 of the surface renewal model is given by eqs. 6.19 and 6.20 where λ_A, λ_B, D_A, and D_B are defined by eqs. 6.21 to 6.23.

ii) Show that the solution of eqs. 6.69 and 6.70 of the film model is given by eqs. 6.19 and 6.20 with λ_A and λ_B defined by eq. 6.71.

iii) Assuming that $c_b \approx 0$ (e.g., for $Ha > 0.5$), develop an expression for the height of packing h required for 99% removal of a solute gas from an inlet gas stream using the surface renewal model. Prepare tables or plots showing the variation of h with Ha for Systems 1 and 2. The physical parameters of these systems are given in Tables 6.1 and 6.2.

iv) By substituting eq. 6.13 into 6.14 of the surface renewal model, derive an equation for $c_b(z)$ in the packed column. Explain the significance of your result.

v) Plot the absolute difference in predictions of y_b^* and c_b^* for System 2, which is liquid-phase controlled, between the full and approximate surface renewal models as a function of z^* for *Ha* varying from 0 to 10. Comment on your results.

Acknowledgment

We thank Matthew Cripps for performing exploratory calculations with the surface renewal model presented in this chapter.

References

Aroonwilas, A. and Tontiwachwuthikul, P. (1997). "High-efficiency structured packing for CO_2 separation using 2-amino-2-methyl-1-propanol (AMP)," *Separation and Purification Technology*, 12(1), pp. 67–79. Available at: https://doi.org/10.1016/S1383-5866(97)00037-3.

Aroonwilas, A., Tontiwachwuthikul, P. and Chakma, A. (2001). "Effects of operating and design parameters on CO_2 absorption in columns with structured packings," *Separation and Purification Technology*, 24(3), pp. 403–411. Available at: https://doi.org/10.1016/S1383-5866(01)00140-X.

Astarita, G. (1967). *Mass transfer with chemical reaction*. Amsterdam: Elsevier.

Badino, G. (2009). "The legend of carbon dioxide heaviness," *Journal of Cave and Karst Studies*, 71(1), pp. 100–107. Available at: https://legacy.caves.org/pub/journal/PDF/v71/cave-71-01-100.pdf.

Brian, P.L.T., Hurley, J.F. and Hasseltine, E.H. (1961). "Penetration theory for gas absorption accompanied by a second order chemical reaction," *AIChE Journal*, 7(2), pp. 226–231. Available at: https://doi.org/10.1002/aic.690070212.

Chatterjee, S.G. (2010). "A theoretical analysis of dissolved-gas transfer to the bulk liquid in gas absorption with a first-order reaction," *Indian Chemical Engineer*, 51(3), pp. 177–193. Available at: https://doi.org/10.1080/00194500903315476.

Danckwerts, P.V. (1951). "Significance of liquid-film coefficients in gas absorption," *Industrial & Engineering Chemistry*, 43(6), pp. 1460–1467. Available at: https://doi.org/10.1021/ie50498a055.

Danckwerts, P.V. (1970). *Gas-liquid reactions*. New York: McGraw-Hill.

DeCoursey, W.J. (1974). "Absorption with chemical reaction: development of a new relation for the Danckwerts model," *Chemical Engineering Science*, 29(9), pp. 1867–1872. Available at: https://doi.org/10.1016/0009-2509(74)85003-7.

Fu, Y., Bao, J., Singh, R., Wang, C. and Xu, Z. (2020). "Investigation of countercurrent flow profile and liquid holdup in random packed column with local CFD data," *Chemical Engineering Science*, 221, 115693. Available at: https://doi.org/10.1016/j.ces.2020.115693.

Huang, H. and Chatterjee, S.G. (2021). "Transient physical gas absorption in a stirred liquid using the surface renewal model of mass transfer," *Chemical Engineering Science*, 234, 116449. Available at: https://doi.org/10.1016/j.ces.2021.116449.

Huang, H. and Chatterjee, S.G. (2023). "A rigorous surface renewal model for transient gas absorption with first-order chemical reaction in a stirred liquid," *Chemical Engineering Science*, 268, 118393. Available at: https://doi.org/10.1016/j.ces.2022.118393.

van Krevelen, D.W. and Hoftijzer, P.J. (1948). "Kinetics of gas-liquid reactions part I. General theory," *Recueil des Travaux Chimiques des Pays-Bas*, 67(7), pp. 563–586.

Leva, M. (1953). *Tower packings and packed tower design*. United States Stoneware Company: Second.

Leva, M. (1954). "Flow through irrigated dumped packings-pressure drop, loading, flooding," *Chemical Engineering Progress Symposium Series*, 50(10), pp. 51–59.

Lewis, W.K. and Whitman, W.G. (1924). "Principles of gas absorption," *Industrial & Engineering Chemistry*, 16(12), pp. 1215–1220. Available at: https://doi.org/10.1021/ie50180a002.

De Leye, L. and Froment, G.F. (1986). "Rigorous simulation and design of columns for gas absorption and chemical reaction—I," *Computers & Chemical Engineering*, 10(5), pp. 493–504. Available at: https://doi.org/10.1016/0098-1354(86)85017-7.

Onda, K., Sada, E. and Takeuchi, H. (1968a). "Gas absorption with chemical reaction in packed columns," *Journal of Chemical Engineering of Japan*, 1(1), pp. 62–66. Available at: https://doi.org/10.1252/jcej.1.62.

Onda, K., Takeuchi, H. and Okumoto, Y. (1968b). "Mass transfer coefficients between gas and liquid phases in packed columns," *Journal of Chemical Engineering of Japan*, 1(1), pp. 56–62. Available at: https://doi.org/10.1252/jcej.1.56.

Park, S.-Y., Min, B.-M., Lee, J.-S., Nam, S.-C., Han, K.-H. and Hyun, J.-S. (2004). "Absorption characteristic of continuous CO_2 absorption process," *Prepr. Pap.-Am. Chem. Soc. Div. Fuel Chem.*, 49(1), pp. 249–250. Available at: https://api.semanticscholar.org/CorpusID:138352546.

Perry, R.H., Green, D.W. and Maloney, J.O. (eds.) (1984). *Perry's chemical engineers' handbook*. Sixth edition. New York: McGraw-Hill.

Reith, T. and Beek, W.J. (1973). "The oxidation of aqueous sodium sulphite solutions," *Chemical Engineering Science*, 28(6), pp. 1331–1339. Available at: https://doi.org/10.1016/0009-2509(73)80084-3.

Sander, R. (2015). "Compilation of Henry's law constants (version 4.0) for water as solvent," *Atmospheric Chemistry and Physics*, 15(8), pp. 4399–4981. Available at: https://doi.org/10.5194/acp-15-4399-2015.

Terraglio, F.P. and Manganelli, R.M. (1967). "The absorption of atmospheric sulfur dioxide by water solutions," *Journal of the Air Pollution Control Association*, 17(6), pp. 403–406. Available at: https://doi.org/10.1080/00022470.1967.10468999.

de Waal, K.J.A. and Beek, W.J. (1967). "A comparison between chemical absorption with rapid first-order reactions and physical absorption in one packed column," *Chemical Engineering Science*, 22(4), pp. 585–593. Available at: https://doi.org/10.1016/0009-2509(67)80041-1.

Yan, S., Fang, M., Zhang, W., Zhong, W., Luo, Z. and Cen, K. (2008). "Comparative analysis of CO_2 separation from flue gas by membrane gas absorption technology and chemical absorption technology in China," *Energy Conversion and Management*, 49(11), pp. 3188–3197. Available at: https://doi.org/10.1016/j.enconman.2008.05.027.

Yu, H., Morgan, S., Allport, A., Cottrell, A., Do, T., McGregor, J., Wardhaugh, L. and Feron, P. (2011). "Results from trialling aqueous NH_3 based post-combustion capture in a pilot plant at Munmorah power station: Absorption," *Chemical Engineering Research and Design*, 89(8), pp. 1204–1215. Available at: https://doi.org/10.1016/j.cherd.2011.02.036.

7

Gas Absorption with Instantaneous Reaction in a Packed Tower

7.1 Introduction

In many industrial operations, the occurrence of gas absorption accompanied by an instantaneous (i.e., extremely rapid) chemical reaction is of critical significance. A chemical reaction for which the rate of reaction is substantially higher compared to the rate of diffusion can be regarded as instantaneous, examples of which are the absorption of hydrogen sulfide in a solution of sodium hydroxide (i.e., proton transfer mechanism), absorption of ammonia in sulfuric acid, etc.

Figure 7.1 depicts a process of gas absorption accompanied by an instantaneous chemical reaction in a quiescent liquid in which a solute A in the gas phase is absorbed at the gas–liquid interface, diffuses in its dissolved form in the liquid, and then reacts instantly with a reagent B in the liquid phase. Because the two species A and B cannot coexist together, a reaction plane (or front) is generated parallel to the gas-liquid interface, which moves with time (under unsteady-state conditions) due to the incessant consumption of the species B in the liquid phase by reaction with dissolved A. Therefore, the process of gas absorption accompanied by an instantaneous chemical reaction is conceptually similar to the conventional moving-boundary problem – often called a Stefan problem. Figure 7.1 also shows the concentration profile of a product P generated by the reaction between A and B.

Simple hydrodynamic models of the gas–liquid interface are commonly used to analyze the phenomenon of gas absorption in a liquid, typical examples of which are the film, penetration, and surface renewal models of mass transfer. Danckwerts (1950) proposed a mathematical formulation for describing gas absorption with instantaneous chemical reaction and derived solutions of the governing diffusion equations under unsteady-state conditions, assuming no gas-phase mass-transfer resistance and a moving boundary condition in infinite liquid media. Sherwood and Pigford (1952) presented film (Hatta) and penetration (Higbie) models for gas

Applications of the Surface Renewal Model of Mass Transfer, First Edition.
Hengshuo Huang and Siddharth G. Chatterjee.

Companion website: www.wiley.com/go/huang/surfacerenewalmodel

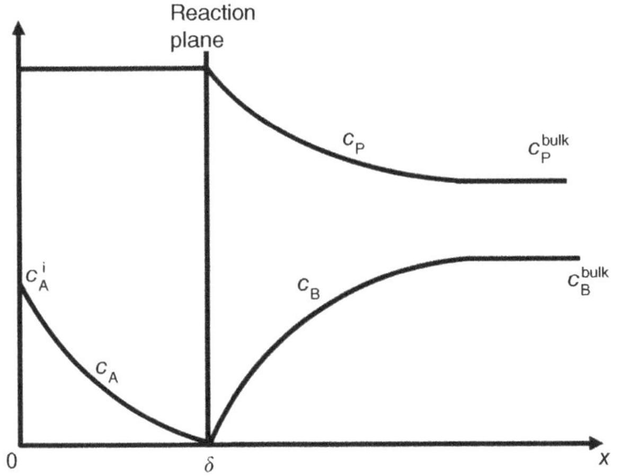

Figure 7.1 Liquid-phase interfacial concentration profiles in gas absorption with instantaneous chemical reaction in a quiescent liquid (A = solute gas, B = liquid-phase reagent, and P = product).

absorption accompanied by an instantaneous bimolecular chemical reaction. An excellent review of gas absorption with an instantaneous reaction in the earlier literature has been provided by Astarita (1967), who also discussed both film and penetration models. Using the film model, he analyzed gas absorption with an instantaneous reaction in a packed tower operating in both countercurrent and cocurrent fashion. Following the earlier work of Secor and Southworth (1961), he divided the tower into a "surface reaction" zone and an "inner reaction" zone. In the former, the reaction occurs at the gas–liquid interface, whereas in the latter, the reaction plane is located somewhere within the liquid. Secor and Southworth (1961) had included an additional "physical absorption" zone in their analysis of gas absorption with an instantaneous reaction in a packed tower operating in a countercurrent manner under steady-state conditions using the film model to describe interfacial mass transfer. However, it can be shown from their packed-tower model, which included both gas- and liquid-side mass-transfer resistances, that the resulting expression for the height of the packed section is indeterminate since it depends upon two unknown (or unspecified) concentrations of the solute gas within and at the exit of the tower. This is because they did not impose additional continuity conditions on the gas- and liquid-phase concentrations of the solute gas and reactant at the boundary between any two zones, as done in the present work. It should also be noted that neither they nor Astarita (1967) presented any numerical results using their models.

The phenomenon of simultaneous absorption of a mixture of two gases (hydrogen sulfide and carbon dioxide) in aqueous sodium hydroxide solutions was studied by Astarita and Gioia (1965), who obtained a solution based on the film model, assuming equal diffusivities of the reacting species and two reaction planes. A theoretical solution based on the penetration model was presented by Onda *et al.* (1972), who also assumed the presence of two reaction fronts in the liquid phase. Predicted reaction factors were in good agreement with experimental data obtained in a wetted-wall column. Earlier, Onda *et al.* (1971) had investigated the behavior of the reaction plane in the absorption of sulfur dioxide and hydrogen sulfide into jelly containing sodium sulfite or sodium hydroxide aqueous solutions. They derived analytical solutions for five different categories of instantaneous chemical reaction based on the penetration model and verified that two reaction planes existed in the reaction between sulfur dioxide and sodium hydroxide, which is a two-step consecutive instantaneous reaction. They also found that the location of the moving reaction plane was proportional to the square root of time, which they measured through the change of color tone of an indicator. Sada and Ameno (1973) presented a model for gas absorption accompanied by an instantaneous irreversible chemical reaction in a liquid of finite depth. They utilized the concept of negative concentration (to be used later in this chapter) in deriving their analytical solution and found that the reaction plane location was initially proportional to the square root of time, as predicted by the penetration model but moved in an accelerated manner at later times. Peng *et al.* (2020) developed a two-film model for predicting the removal efficiency of sulfur dioxide by absorbing it in ammonia solution in a packed column. In their analysis, they assumed the ammonia flue gas desulfurization reaction to be instantaneous, although it is known to be second order and only considered the instantaneous reaction zone of the tower, besides not including product formation. Their theoretical predictions agreed well with experimental data on the variation of desulfurization efficiency with pH, liquid–gas ratio, inlet SO_2 concentration, flue gas velocity, and temperature. Hikita *et al.* (1985) theoretically studied the problem of gas absorption accompanied by an instantaneous reaction in a laminar falling film of a pseudoplastic liquid. They presented plots of the reaction factor and distance of the reaction plane from the gas–liquid interface as functions of the Graetz number, concentration ratio, diffusivity ratio, and the power-law index of the liquid.

It was mentioned in Chapter 6 that both film and penetration models, the latter being a special case of the surface renewal model, predict close numerical values for the enhancement factor for many different types of gas–liquid reactions (Danckwerts, 1970). For example, Brian *et al.* (1961) found the enhancement factor for a second-order gas–liquid reaction calculated numerically using the Higbie penetration model to be in good agreement (i.e., within 16%) with the value calculated with the film model of van Krevelen and Hoftijzer (1948). Predictions made by an approximate expression for the enhancement factor for a second-order gas–liquid reaction using the surface renewal model that was derived by DeCoursey (1974) agreed fairly well (i.e., with a maximum deviation of 14%) with the numerical calculations of Brian *et al.* (1961). It should be noted that this expression contains the enhancement factor for an infinitely rapid chemical reaction. However, to our knowledge, no comparison between film and surface renewal models exists in the literature for the case of gas absorption with instantaneous reaction in a packed tower because there is no surface renewal model available for this purpose.

This chapter presents rigorous film and surface renewal models for gas absorption accompanied by an instantaneous chemical reaction in a packed tower and compares their predictions against each other for the cases of zero and finite gas-phase mass-transfer resistance. In contrast to Secor and Southworth (1961) and Astarita (1967), the tower, which is assumed to operate in a countercurrent manner, is conceived, from physical considerations, as being composed of the following two sections: (1) An upper "instantaneous reaction" region in which the concentration of the liquid-phase reagent B is nonzero and where the product P is created and (2) a lower "physical absorption" region where B has been completely depleted. Between these two regions, the gas- and liquid-phase compositions are assumed to be continuous. As will be seen later, this theoretical formulation leads to a system of ordinary differential equations (ODEs) for the film model and differential algebraic equations (DAEs) in the case of the surface renewal model. Illustrative calculations using both models are presented for the absorption of gaseous sulfur dioxide (SO_2) into an aqueous solution of sodium sulfite (Na_2SO_3) in a packed tower operating at steady state.

The case of instantaneous reaction between a dissolved gas and liquid-phase reagent treated in this chapter is a limiting case of a chemical reaction of generalized kinetics occurring between the same.

7.2 Theoretical Analysis

In the countercurrent packed tower conceptualized in this work, fresh liquid solvent (i.e., water) containing a reagent B is fed at the top of the tower and absorbs a solute A from the gas phase. In the liquid phase, dissolved A reacts instantaneously with B and produces a nonvolatile product P. As the liquid traverses downward through the packing, the concentration of B in the liquid continuously decreases due to reaction with A until, at a certain location in the tower, called the "critical location," it becomes equal to zero (i.e., B is completely depleted). Thereafter, that is, below the critical location, it follows that solute A will be absorbed in the liquid only via the mechanism of physical absorption. The region from the top of the tower to the critical location, that is, the zone of instantaneous reaction in which the product P is created in the liquid phase by reaction between dissolved A and reagent B, will be referred to as Region I, whereas the section of the tower below the critical location, that is, the zone of physical absorption, will be called Region II. We also assume that the gas–liquid system is dilute, that is, gas and liquid flow rates in the packed column are constant, and there is absorption of only a single species A (i.e., single-component absorption) from the gas phase into the absorbent liquid.

The overall (instantaneous) irreversible reaction between dissolved A and a liquid-phase reagent B producing a product P can be conceptualized as

$$A + bB \rightarrow pP \tag{7.1}$$

where b and p are stoichiometric coefficients of the liquid-phase reagent and product, respectively.

7.2.1 Film Model

Figure 7.2 shows steady-state concentration profiles of A, B, and P at the gas–liquid interfacial region in the liquid film for a situation where there is no gas-phase mass-transfer resistance. Note that the concentrations of A and B are zero at the reaction plane ($x = \beta$) since both cannot coexist based on the assumption of an instantaneous chemical reaction occurring between them.

Here, c_B^{bulk} is the molar concentration of B in the bulk liquid, c_A^i is the molar concentration of dissolved gas A at the gas–liquid interface on the liquid side, y_A^i

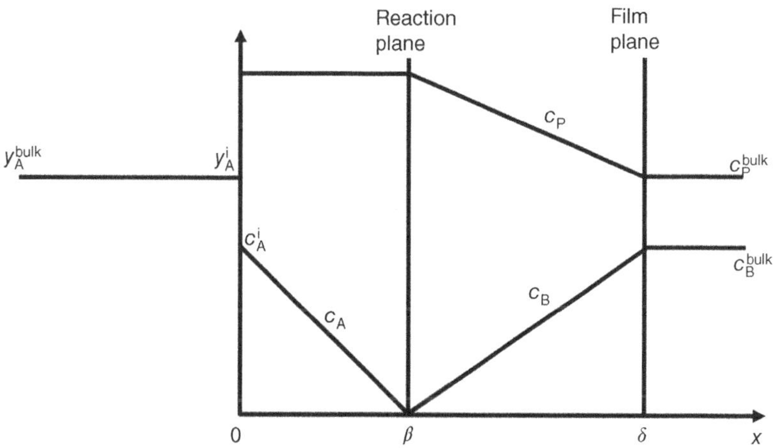

Figure 7.2 Concentration profiles in gas absorption with instantaneous chemical reaction (A = solute gas, B = liquid-phase reagent, and P = product) for the film model in the case of zero gas-phase mass-transfer resistance. β and δ are the location of the reaction front and the liquid film thickness, respectively.

and y_A^{bulk} are the molar concentrations of A at the gas–liquid interface on the gas side and in the bulk gas, respectively, whereas c_P^{bulk} is the molar concentration of P in the bulk liquid.

The width δ of the liquid film is connected to the liquid-phase physical mass-transfer coefficient k_L by the equation

$$k_L = \frac{D_A}{\delta} \tag{7.2}$$

where D_A is the binary diffusion coefficient of dissolved A in the liquid solvent. The interfacial concentrations y_A^i and c_A^i are related by

$$c_A^i = \frac{y_A^i}{H} \tag{7.3}$$

where H is Henry's constant. We note that when there is no gas-phase mass-transfer resistance, $y_A^i = y_A^{bulk}$.

The rate of absorption of solute A at the gas–liquid interface can be obtained from Fick's first law, that is,

$$\begin{aligned} R_A &= -D_A\left(\frac{dc_A}{dx}\right)_{x=0} \\ &= \frac{D_A c_A^i}{\beta} \end{aligned} \tag{7.4}$$

where $x = \beta$ is the location of the reaction plane.

Since b moles of B react with one mole of A and the reaction occurs at the reaction plane $x = \beta$, one has

$$-D_A\left(\frac{dc_A}{dx}\right)_{x=\beta} = \frac{D_B}{b}\left(\frac{dc_B}{dx}\right)_{x=\beta} \tag{7.5}$$

In a packed tower, the above process will occur everywhere except where reactant B is completely depleted. Thus, as mentioned before, the total tower in the case of countercurrent operation can be separated into two sections: an upper "instantaneous reaction" section (Region I) and a lower purely "physical absorption" section (Region II).

In Region I, where the instantaneous chemical reaction occurs, the material balances for A, B, and P over a differential section dz of the tower can be expressed as

$$G\frac{dy_A^{bulk}}{dz} = R_A a A_c \tag{7.6}$$

$$L\frac{dc_B^{bulk}}{dz} = R_B a A_c = -bR_A a A_c \tag{7.7}$$

and

$$L\frac{dc_P^{bulk}}{dz} = R_P a A_c = pR_A a A_c \tag{7.8}$$

respectively, where the rate of absorption R_A of solute A at the gas–liquid interface is given by eq. 7.4, R_B is the rate of consumption of reagent B (equal to $-bR_A$), and R_P is the rate of production of P (equal to pR_A), a is the gas–liquid interfacial area of contact per unit volume of packing, A_c is the area of cross section of the tower, L and G are the volumetric flow rates of gas and liquid in the tower (assumed to be constant), respectively, and z is the location in the tower measured from the top of the packing. We note that there is no dissolved A present in the bulk liquid in Region 1.

In Region II, in which solely occurs the physical absorption of A, mass balances for A in both phases can be expressed as

$$G\frac{dy_A^{bulk}}{dz} = R_A^{phy} aA_c \tag{7.9}$$

and

$$L\frac{dc_A^{bulk}}{dz} = R_A^{phy} aA_c \tag{7.10}$$

where c_A^{bulk} is the bulk-liquid concentration of A in Region II, where its rate of physical absorption R_A^{phy} is represented by the equation

$$R_A^{phy} = k_{OL}\Delta c \tag{7.11}$$

with k_{OL} being an overall mass-transfer coefficient (see eq. 7.54) and Δc the overall concentration driving force for physical absorption.

7.2.1.1 Zero Gas-Phase Mass-Transfer Resistance

The concentration of the solute gas in the bulk gas phase is equal to its value at the gas–liquid interface (see Figure 7.2) provided the gas-phase mass-transfer resistance is zero, that is, the gas-phase mass-transfer coefficient is infinite as in the absorption of a pure gas. It is to be noted that the liquid-phase mass-transfer resistance is, nevertheless, finite because solute A, after absorption at the gas–liquid interface, has to diffuse to the reaction plane located at a distance β from the interface inside the liquid film of thickness δ, as depicted in Figure 7.2. The rate of absorption of A accompanied by an instantaneous chemical reaction in Region I can be shown to be given by (Sherwood and Pigford, 1952)

$$R_A = k_L\left(\frac{y_A^{bulk}}{H} + \frac{D_B}{bD_A}c_B^{bulk}\right) \tag{7.12}$$

whereas the rate of physical absorption of A in Region II can be expressed as

$$R_A^{phy} = k_L\Delta c = k_L\left(\frac{y_A^{bulk}}{H} - c_A^{bulk}\right) \tag{7.13}$$

where H, Henry's law constant of solute A in the liquid solvent, is taken to be the same in the two regions because of the assumed dilute nature of the gas–liquid system. Also, D_B is the binary diffusion coefficient of reagent B in the liquid solvent.

Substituting eqs. 7.12 and 7.13 into eqs. 7.6 to 7.10 yields the following system of ODEs:

$$G\frac{dy_A^{bulk}}{dz} = aA_c k_L\left(\frac{y_A^{bulk}}{H} + \frac{D_B}{bD_A}c_B^{bulk}\right) \tag{7.14}$$

$$L\frac{dc_B^{bulk}}{dz} = -abA_c k_L\left(\frac{y_A^{bulk}}{H} + \frac{D_B}{bD_A}c_B^{bulk}\right) \tag{7.15}$$

$$L\frac{dc_P^{bulk}}{dz} = apA_c k_L\left(\frac{y_A^{bulk}}{H} + \frac{D_B}{bD_A}c_B^{bulk}\right) \tag{7.16}$$

$$G\frac{dy_A^{bulk}}{dz} = aA_c k_L\left(\frac{y_A^{bulk}}{H} - c_A^{bulk}\right) \tag{7.17}$$

and

$$L\frac{dc_A^{bulk}}{dz} = aA_c k_L\left(\frac{y_A^{bulk}}{H} - c_A^{bulk}\right) \tag{7.18}$$

Equations 7.14 to 7.16 hold in Region I, whereas eqs. 7.17 and 7.18 hold in Region II.

We note that the mass balance for product P in the liquid phase, as expressed by eq. 7.16, does not involve its diffusion coefficient, D_P

The boundary conditions at the top and bottom of the tower are given by

$$c_B^{bulk}\Big|_{z=0} = c_{B,in}^{bulk} \tag{7.19}$$

$$c_P^{bulk}\Big|_{z=0} = 0 \tag{7.20}$$

and

$$y_A^{bulk}\Big|_{z=Z} = y_{A,in}^{bulk} \tag{7.21}$$

where $c_{B,in}^{bulk}$ is the molar concentration of B in the incoming liquid at the top of the tower, $y_{A,in}^{bulk}$ is the molar concentration of A in the gas fed at the tower bottom, and Z is the height of the packed section in the tower.

At a critical distance z_c measured from the top of the tower, that is, where Region I ends and Region II begins, the following continuity relations hold:

$$y_{A,I}^{bulk}\Big|_{z=z_c} = y_{A,II}^{bulk}\Big|_{z=z_c} \tag{7.22}$$

$$c_B^{bulk}\Big|_{z=z_c} = 0 \tag{7.23}$$

and

$$c_A^{bulk}\Big|_{z=z_c} = 0 \tag{7.24}$$

For ease of calculation and convenience, we introduce the following dimensionless quantities and drop the superscript "bulk" for both phases, that is,

$$z^* = \frac{z}{Z},\ y_A^* = \frac{y_A}{y_{A,in}},\ c_B^* = \frac{c_B}{c_{B,in}},\ c_A^* = \frac{Hc_A}{y_{A,in}},\ c_P^* = \frac{c_P}{pc_{B,in}}$$

where $c_{B,in}$ and $y_{A,in}$ are the concentration of B in the liquid solvent entering at the top of the tower and that of solute A in the gas entering at the tower bottom, respectively, which correspond to the boundary conditions specified by eqs. 7.19 and 7.21, respectively. Equations 7.14 to 7.18 are then transformed into the following:

$$\frac{dy_A^*}{dz^*} = N_G\left(\frac{y_A^*}{H} + D^*F^*c_B^*\right) \tag{7.25}$$

$$\frac{dc_B^*}{dz^*} = -N_L\left(\frac{y_A^*}{F^*H} + D^*c_B^*\right) \tag{7.26}$$

$$\frac{dc_P^*}{dz^*} = \frac{N_L}{bF^*}\left(\frac{y_A^*}{H} + D^*F^*c_B^*\right) \tag{7.27}$$

$$\frac{dy_A^*}{dz^*} = \frac{N_G}{H}\left(y_A^* - c_A^*\right) \tag{7.28}$$

and

$$\frac{dc_A^*}{dz^*} = N_L\left(y_A^* - c_A^*\right) \tag{7.29}$$

Here, D^* is a ratio of diffusion coefficients defined by

$$D^* = \frac{D_B}{D_A} \tag{7.30}$$

and

$$F^* = \frac{c_{B,in}}{by_{A,in}} \tag{7.31}$$

Also, N_L and N_G are, respectively, the numbers of liquid- and gas-phase mass-transfer units defined by

$$N_L = \frac{Z}{H_L},\ N_G = \frac{Z}{H_G} \tag{7.32}$$

where H_L and H_G are, respectively, the heights of the liquid- and gas-phase mass-transfer units for physical absorption, that is:

$$H_L = \frac{L}{aA_ck_L} \tag{7.33}$$

and

$$H_G = \frac{G}{L}\frac{L}{aA_c k_L} = \frac{G}{L}H_L \tag{7.34}$$

Thus, H_L and H_G are not independent quantities.

The boundary conditions given by eqs. 7.19 and 7.21 can be expressed in dimensionless form as follows:

$$c_B^*\big|_{z^*=0} = \frac{c_{B,in}^{bulk}}{c_{B,in}^{bulk}} = 1 \tag{7.35}$$

and

$$y_A^*\big|_{z^*=1} = \frac{y_{A,in}^{bulk}}{y_{A,in}^{bulk}} = 1 \tag{7.36}$$

The dimensionless boundary or continuity conditions at the critical location, which separates Region I from Region II in the tower, follow from eqs. 7.22 to 7.24, that is,

$$y_{A,I}^*\Big|_{z^*=z_c^*} = y_{A,II}^*\Big|_{z^*=z_c^*} \tag{7.37}$$

$$c_B^*\Big|_{z^*=z_c^*} = 0 \tag{7.38}$$

and

$$c_A^*\Big|_{z^*=z_c^*} = 0 \tag{7.39}$$

where $z_c^* = z_c/Z$ is the dimensionless critical location in the tower measured from the top of the packing.

The differential equations for the two regions can be concisely expressed using matrix notation as follows:

$$\text{Region I:}\quad \begin{bmatrix} \frac{dy_A^*}{dz^*} \\ \frac{dc_B^*}{dz^*} \end{bmatrix} = \begin{bmatrix} \frac{N_G}{H} & D^*F^*N_G \\ -\frac{N_L}{F^*H} & -D^*N_L \end{bmatrix} \begin{bmatrix} y_A^* \\ c_B^* \end{bmatrix} = \begin{bmatrix} a_1 & b_1 \\ c_1 & d_1 \end{bmatrix} \begin{bmatrix} y_A^* \\ c_B^* \end{bmatrix} \tag{7.40}$$

$$\text{Region II:}\quad \begin{bmatrix} \frac{dy_A^*}{dz^*} \\ \frac{dc_A^*}{dz^*} \end{bmatrix} = \begin{bmatrix} \frac{N_G}{H} & -\frac{N_G}{H} \\ N_L & N_L \end{bmatrix} \begin{bmatrix} y_A^* \\ c_A^* \end{bmatrix} = \begin{bmatrix} a_2 & b_2 \\ c_2 & d_2 \end{bmatrix} \begin{bmatrix} y_A^* \\ c_A^* \end{bmatrix} \tag{7.41}$$

$$\text{Region I:}\quad \frac{dc_P^*}{dz^*} = \frac{N_L}{bH}\left(\frac{y_A^*}{F^*} + D^*Hc_B^*\right) \tag{7.42}$$

The solution of the above system of ODEs subject to the boundary and continuity conditions given by eqs. 7.35 to 7.39 is:

$$\text{Region I:}\ y_A^* = X_1\,e^{\lambda_1 z^*} + X_2\,e^{\lambda_2 z^*} \tag{7.43}$$

$$\text{Region I:}\ c_B^* = X_1\frac{\lambda_1 - a_1}{b_1}e^{\lambda_1 z^*} + X_2\frac{\lambda_2 - a_1}{b_1}e^{\lambda_2 z^*} \tag{7.44}$$

where λ_1 and λ_2 are the eigenvalues of the matrix of coefficients $\begin{bmatrix} a_1 & b_1 \\ c_1 & d_1 \end{bmatrix}$ that are given by

$$\lambda_{1,2} = \frac{1}{2}\left(a_1 + d_1 \pm \sqrt{a_1^2 + d_1^2 - 2a_1 d_1 + 4b_1 c_1}\right) \tag{7.45}$$

and

$$\text{Region II: } y_A^* = X_3\, e^{\lambda_3 z^*} + X_4\, e^{\lambda_4 z^*} \tag{7.46}$$

$$\text{Region II: } c_A^* = X_3 \frac{\lambda_3 - a_2}{b_2} e^{\lambda_3 z^*} + X_4 \frac{\lambda_4 - a_2}{b_2} e^{\lambda_4 z^*} \tag{7.47}$$

where, in an analogous manner, λ_3 and λ_4 are the eigenvalues of the matrix of coefficients $\begin{bmatrix} a_2 & b_2 \\ c_2 & d_2 \end{bmatrix}$ given by

$$\lambda_{3,4} = \frac{1}{2}\left(a_2 + d_2 \pm \sqrt{a_2^2 + d_2^2 - 2a_2 d_2 + 4b_2 c_2}\right) \tag{7.48}$$

The four arbitrary constants (X_1 to X_4) and the dimensionless critical location (z_c^*) can be obtained by applying the five conditions given by eqs. 7.35 to 7.39. A trial-and-error method can be used to numerically determine the critical location z_c^*, as described in the following steps:

1) Divide the total dimensionless height of the packed tower into 10 equally spaced points.
2) Determine 10 groups of the four arbitrary constants (X_1 to X_4) using the boundary conditions given by eqs. 7.35, 7.36, 7.38, and 7.39 and eqs. 7.43, 7.44, 7.46, and 7.47 for the 10 points, assuming each of them to be the critical location.
3) Determine the minimum absolute value of the difference between the left- and right-hand sides of eq. 7.37 for the above 10 dimensionless critical locations to obtain a reasonable estimate of the actual critical location.
4) Conceptually divide the dimensionless height of the tower in the neighborhood of the critical location determined in step 3 into 10 new points spaced uniformly apart. For instance, if the ith point is the target, the interval of $[z_{i-1}^*, z_{i+1}^*]$ is to be separated into 10 new points spaced uniformly apart.
5) Repeat steps 2–4 until the absolute value of the difference between the left- and right-hand sides of eq. 7.37 is less than a very small, specified tolerance ε, that is, $\left| y_{A,I}^* \big|_{z^*=z_c^*} - y_{A,II}^* \big|_{z^*=z_c^*} \right| < \varepsilon$.

This procedure will allow us to eventually determine the values of the four constants (X_1 to X_4) that appear in eqs. 7.43, 7.44, 7.46, and 7.47 and also the critical dimensionless location (z_c^*) in the packed tower.

Finally, the expression for the product concentration in Region I can be obtained by integrating eq. 7.42, subject to the inlet boundary condition specified by eq. 7.20; it is given by

$$c_{\mathrm{P}}^{*}=\frac{N_{\mathrm{L}}X_{1}}{b\lambda_{1}HF^{*}}\left(1+HD^{*}F^{*}\frac{\lambda_{1}-a_{1}}{b_{1}}\right)(e^{\lambda_{1}z^{*}}-1)+\frac{N_{\mathrm{L}}X_{2}}{b\lambda_{2}HF^{*}}\left(1+HD^{*}F^{*}\frac{\lambda_{2}-a_{1}}{b_{1}}\right)(e^{\lambda_{2}z^{*}}-1)\tag{7.49}$$

It is seen from eq. 7.49 that the dimensionless product concentration c_{P}^{*} is independent of the stoichiometric coefficient p.

7.2.1.2 Finite Gas-Phase Mass-Transfer Resistance

For this case, the rate of absorption of A at the gas–liquid interface at a specific location z in the tower is given by

$$R_{\mathrm{A}}=-D_{\mathrm{A}}\left(\frac{\mathrm{d}c_{\mathrm{A}}}{\mathrm{d}x}\right)_{x=0}=k_{\mathrm{G}}\left(y_{\mathrm{A}}^{\mathrm{bulk}}-y_{\mathrm{A}}^{\mathrm{i}}\right)\tag{7.50}$$

and thus (see Figure 7.3)

$$\frac{D_{\mathrm{A}}c_{\mathrm{A}}^{\mathrm{i}}}{\beta}=k_{\mathrm{G}}\left(y_{\mathrm{A}}^{\mathrm{bulk}}-y_{\mathrm{A}}^{\mathrm{i}}\right)\tag{7.51}$$

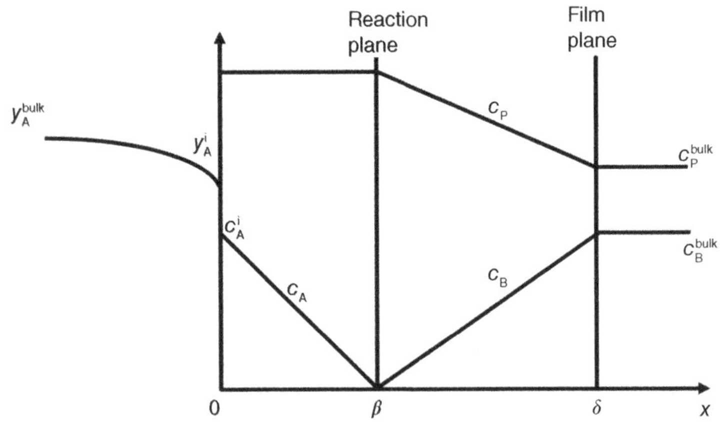

Figure 7.3 Concentration profiles in gas absorption with instantaneous chemical reaction (A = solute gas, B = liquid-phase reagent, and P = product) for the film model in the case of finite gas-phase mass-transfer resistance. β and δ are the location of the reaction front and liquid film thickness, respectively.

Further, it can be shown by combining eq. 7.51 with eqs. 7.2, 7.4, and 7.5 that the rate of absorption of A in Region I is

$$R_A = \frac{\frac{y_A^{bulk}}{H} + \frac{D_B}{bD_A} c_B^{bulk}}{\frac{1}{Hk_G} + \frac{1}{k_L}} \tag{7.52}$$

which is similar in form to the expressions given by Sherwood and Pigford (1952) and Ramachandran and Sharma (1969). Equation 7.52 can be rewritten as

$$R_A = k_{OL}\left(\frac{y_A^{bulk}}{H} + \frac{D_B}{bD_A} c_B^{bulk}\right) \tag{7.53}$$

where k_{OL} is an overall mass-transfer coefficient defined by

$$\frac{1}{k_{OL}} = \frac{1}{Hk_G} + \frac{1}{k_L} \tag{7.54}$$

Equation 7.53 can be compared with eq. 7.12 for the zero gas-phase mass-transfer resistance case.

The rate of physical absorption of A in Region II is given by eq. 7.13 with k_L replaced by k_{OL}. The rest of the development is the same as that for the zero gas-phase mass-transfer resistance case.

7.2.2 Surface Renewal Model

As seen in earlier chapters, the classical surface renewal model postulates that the gas–liquid interface is replenished with fresh liquid elements that arrive continuously from the bulk liquid. During its dwell or residence time at the interface, the solute gas is transferred in a liquid element by the mechanism of unsteady-state molecular diffusion, which is expressed by the diffusion equation, that is,

$$\frac{\partial c}{\partial \theta} = D\frac{\partial^2 c}{\partial x^2} \tag{7.55}$$

where c is the concentration of the dissolved (gaseous) species in a liquid element at location x (measured from the surface of the element), θ is the exposure time of the liquid element at the gas–liquid interface, and D is the diffusion coefficient of the dissolved species in the liquid.

The fundamental assumption in the surface renewal model is that the replacement events of surface elements, regardless of their different individual ages, have the same probability of occurrence. This leads to an exponential age distribution (i.e., the conventional Danckwerts age distribution), which contains a unique parameter, that is, the frequency or rate of surface renewal S, which depends on the extent of turbulence at the gas–liquid interface.

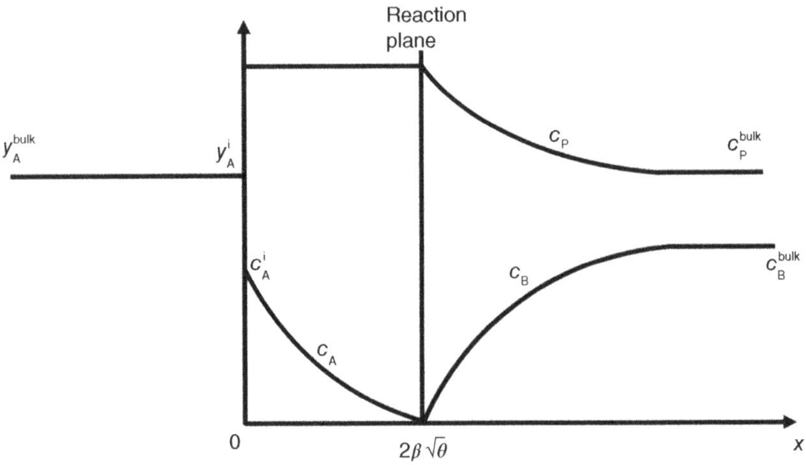

Figure 7.4 Concentration profiles in gas absorption with instantaneous chemical reaction (A = solute gas, B = liquid-phase reagent, and P = product) in a surface element for the case of zero gas-phase mass-transfer resistance at a particular location z in the tower. The location of the reaction front $\delta(\theta)$ in the liquid element is a function of the exposure time θ and is equal to $2\beta\sqrt{\theta}$.

7.2.2.1 Zero Gas-Phase Mass-Transfer Resistance

Similar to Section 7.2.1.1, the gas-phase mass-transfer resistance is assumed to be negligible in this section, as in the absorption of a pure gas. This case, shown schematically in Figure 7.4, has often been investigated in the literature since analytical solutions to the interfacial mass-transfer problem are possible (Astarita, 1967; Danckwerts, 1970; Onda *et al.*, 1971). Thus, it can be regarded as a limiting case against which the solution for the finite mass-transfer resistance case can be assessed, as will be discussed later.

Consider an element with an age of τ at a specific location z in the tower; thus, the exposure time θ of the element will vary from zero to τ. At the very first moment, the element arrives at the gas–liquid interface from the bulk liquid, the rate of absorption occurring in it will be infinite because of the absence of gas-phase mass-transfer resistance and no liquid-phase resistance due to the reaction plane being located at the surface of the element, with a subsequent rapid drop off in the rate as the exposure time θ increases (see eq. 7.67). Thus, an element with an age of $\tau = 0$ will have an infinite rate of absorption, which, however, poses no problem since we are interested in the age-averaged steady rate of absorption, which is obtained by multiplying the instantaneous rate with the age-distribution function and integrating from $\tau = 0$ to $\tau = \infty$, and this latter rate is finite (see eq. 7.70).

At a particular location z in the tower, the concentration distributions of dissolved A, liquid reagent B, and product P (which have stoichiometric coefficients of 1, b and p, respectively – see eq. 7.1) in a surface element can be obtained by solving the following diffusion equations:

$$\frac{\partial c_A}{\partial \theta} = D_A \frac{\partial^2 c_A}{\partial x^2}; \quad 0 < x < \delta(\theta) \tag{7.56}$$

$$\frac{\partial c_B}{\partial \theta} = D_B \frac{\partial^2 c_B}{\partial x^2}; \quad x \geq \delta(\theta) \tag{7.57}$$

$$\frac{\partial c_P}{\partial \theta} = D_P \frac{\partial^2 c_P}{\partial x^2}; \quad 0 < x < \delta(\theta) \text{ and } x \geq \delta(\theta) \tag{7.58}$$

where θ is the exposure time of a surface element, which varies from zero to τ, with τ being the age or residence time of the element at the gas–liquid interface before it returns to mix with the bulk liquid, whereas x is the distance into the element measured from the interface. We note that τ itself varies from zero to infinity.

The boundary conditions for eqs. 7.56 to 7.58 are given by

$$\begin{cases} c_A = c_A^i; & x = 0 \quad \text{for } \theta > 0 \\ c_B = c_B^{bulk}; & x \to \infty \quad \text{for } \theta > 0 \\ c_P = c_P^{bulk}; & x \to \infty \quad \text{for } \theta > 0 \end{cases} \tag{7.59}$$

At a specific location z in the tower, the interfacial concentration of A and bulk-liquid concentrations of B and P will be constant under steady-state operation; this constancy will be assumed in what follows.

The conservation of mass flux at the moving reaction plane can be expressed as

$$-D_A \left(\frac{\partial c_A}{\partial x} \right) = \frac{1}{b} D_B \left(\frac{\partial c_B}{\partial x} \right) = -\frac{1}{p} D_P \left(\frac{\partial c_P}{\partial x} \right) \quad \text{at } x = \delta(\theta) \tag{7.60}$$

where $\delta(\theta)$ is the location of the reaction front, which is equal to $2\beta\sqrt{\theta}$, with the constant β to be obtained from eq. 7.66, as will be seen later.

Finally, the initial conditions are

$$\begin{cases} c_A = 0; & \theta = 0 \text{ for } x > 0 \\ c_B = c_B^{bulk}; & \theta = 0 \text{ for } x > 0 \\ c_P = c_P^{bulk}; & \theta = 0 \text{ for } x > 0 \end{cases} \tag{7.61}$$

The above system (eqs. 7.56–7.61) can be solved by applying the method of similarity transformation. The solutions are given by (Astarita, 1967; Danckwerts, 1970)

$$c_A(x, \theta) = c_A^i \frac{\operatorname{erfc}\left(\frac{x}{2\sqrt{D_A \theta}} \right) - \operatorname{erfc}\left(\frac{\beta}{\sqrt{D_A}} \right)}{\operatorname{erf}\left(\frac{\beta}{\sqrt{D_A}} \right)} \tag{7.62}$$

and

$$c_{\mathrm{B}}(x,\theta) = c_{\mathrm{B}}^{\mathrm{bulk}} \frac{\operatorname{erf}\left(\frac{x}{2\sqrt{D_{\mathrm{B}}\theta}}\right) - \operatorname{erf}\left(\frac{\beta}{\sqrt{D_{\mathrm{B}}}}\right)}{\operatorname{erfc}\left(\frac{\beta}{\sqrt{D_{\mathrm{B}}}}\right)} \tag{7.63}$$

The product concentration profile consists of two parts, one from the gas–liquid interface to the reaction plane and the other from the reaction plane to the bulk liquid. Since the product P is assumed to be nonvolatile (i.e., $\partial c_{\mathrm{P}}/\partial x = 0$ at $x = 0$), it can be demonstrated by mathematical analysis that (see eq. 10 in the paper of Onda *et al.*, 1971)

$$c_{\mathrm{P}}(x,\theta) = c_{\mathrm{P}}^{\mathrm{bulk}} + \frac{p}{b}\sqrt{\frac{D_{\mathrm{B}}}{D_{\mathrm{P}}}} \frac{\exp\left(-\frac{\beta^2}{D_{\mathrm{B}}}\right)}{\exp\left(-\frac{\beta^2}{D_{\mathrm{P}}}\right)} \frac{\operatorname{erfc}\left(\frac{\beta}{\sqrt{D_{\mathrm{P}}}}\right)}{\operatorname{erfc}\left(\frac{\beta}{\sqrt{D_{\mathrm{B}}}}\right)} c_{\mathrm{B}}^{\mathrm{bulk}};\quad 0 < x < \delta(\theta) \tag{7.64}$$

and

$$c_{\mathrm{P}}(x,\theta) = c_{\mathrm{P}}^{\mathrm{bulk}} + \frac{p}{b}\sqrt{\frac{D_{\mathrm{B}}}{D_{\mathrm{P}}}} \frac{\exp\left(-\frac{\beta^2}{D_{\mathrm{B}}}\right)}{\exp\left(-\frac{\beta^2}{D_{\mathrm{P}}}\right)} \frac{\operatorname{erfc}\left(\frac{x}{2\sqrt{D_{\mathrm{P}}\theta}}\right)}{\operatorname{erfc}\left(\frac{\beta}{\sqrt{D_{\mathrm{B}}}}\right)} c_{\mathrm{B}}^{\mathrm{bulk}};\quad x \geq \delta(\theta) \tag{7.65}$$

where β is to be obtained from the nonlinear equation (Astarita, 1967; Danckwerts, 1970)

$$\exp\left(\frac{\beta^2}{D_{\mathrm{B}}}\right)\operatorname{erfc}\left(\frac{\beta}{\sqrt{D_{\mathrm{B}}}}\right) = \frac{1}{b}\frac{c_{\mathrm{B}}^{\mathrm{bulk}}}{c_{\mathrm{A}}^{\mathrm{i}}}\sqrt{\frac{D_{\mathrm{B}}}{D_{\mathrm{A}}}}\exp\left(\frac{\beta^2}{D_{\mathrm{A}}}\right)\operatorname{erf}\left(\frac{\beta}{\sqrt{D_{\mathrm{A}}}}\right) \tag{7.66}$$

As observed from eq. 7.64, the concentration of P in the surface element is uniform in front of the reaction plane, this uniformity being established immediately upon first contact of the surface element with the gas (see also Figure 1 in Onda *et al.*, 1971). This is because of (1) the absence of gas-phase mass-transfer resistance, and (2) the nonvolatility of the product, which prevents it from escaping from the interface to the gas phase. In this "static" region, which expands with the passage of time θ, there is no diffusion of P, whose concentration remains at the same level as that established at $\theta = 0$. In contrast, as shown by eq. 7.65, the product concentration is a function of θ and declines with distance behind the reaction front, that is, P diffuses from the reaction plane onward into the liquid in this "dynamic" region.

The instantaneous rate of absorption of solute A at the gas–liquid interface can then be obtained from eq. 7.62 and is given by

$$R_{\mathrm{inst}}^{\mathrm{A}}(\theta) = -D_{\mathrm{A}}\left.\frac{\partial c_{\mathrm{A}}}{\partial x}\right|_{x=0} = c_{\mathrm{A}}^{\mathrm{i}}\sqrt{\frac{D_{\mathrm{A}}}{\pi\theta}}\frac{1}{\operatorname{erf}\left(\frac{\beta}{\sqrt{D_{\mathrm{A}}}}\right)} \tag{7.67}$$

We next derive expressions for the steady-state rates of absorption of A at the gas–liquid interface, transfer of dissolved A and product P from the interface to the bulk liquid due to the surface renewal mechanism, and consumption of B in the interfacial region at a specific location z in the tower, all of which are required in tower mass balances for A, B, and P to be presented later.

Consider an element with an age of τ at a specific location z in the tower. The concentration profiles in such an element and the rate of absorption of A can be obtained by replacing θ with τ in eqs. 7.62 to 7.65 and 7.67, respectively. Also, the steady-state Danckwerts age-distribution function is given by (Danckwerts, 1970)

$$f_{\rm D}(\tau) - Se^{-S\tau} \tag{7.68}$$

where S is the frequency or rate of surface renewal, which is given by

$$S = \frac{k_{\rm L}^2}{D_{\rm A}} \tag{7.69}$$

with $k_{\rm L}$ being the mass-transfer coefficient for physical absorption of A in the liquid (Danckwerts, 1970). The steady-state age-averaged rate of absorption of solute A can be expressed as

$$R_{\rm abs}^{\rm A,ss} = \int_0^{\infty} R_{\rm inst}^{\rm A}(\tau) f_{\rm D}(\tau) {\rm d}\tau = \frac{\sqrt{D_{\rm A} S}\, c_{\rm A}^{\rm i}}{{\rm erf}\left(\frac{\beta}{\sqrt{D_{\rm A}}}\right)} \tag{7.70}$$

where "ss" denotes steady state.

To obtain the expression for the rate of transfer of dissolved A from the gas–liquid interface to the bulk liquid due to the surface renewal mechanism, we first determine the amount of A present in a surface element of age τ, that is, the area under the concentration profile of A in Figure 7.4 (with θ replaced by τ), that is,

$$\begin{aligned} Q^{\rm A}(\tau) &= \int_0^{\delta(\tau)} c_{\rm A}(x,\tau){\rm d}x \\ &= \frac{2c_{\rm A}^{\rm i}\sqrt{D_{\rm A}}\left(1 - {\rm e}^{\frac{-\beta^2}{D_{\rm A}}}\right)}{\sqrt{\pi}\,{\rm erf}\left(\frac{\beta}{\sqrt{D_{\rm A}}}\right)}\sqrt{\tau} \end{aligned} \tag{7.71}$$

The steady-state age-averaged rate of transfer of dissolved A from the interface to the bulk liquid can then be obtained from

$$R_{\rm trans}^{\rm A,ss} = S\int_0^{\infty} Q^{\rm A}(\tau)\, Se^{-S\tau}\, {\rm d}\tau = \frac{\sqrt{D_A S}\left(1 - e^{\frac{-\beta^2}{D_{\rm A}}}\right) c_{\rm A}^{\rm i}}{{\rm erf}\left(\frac{\beta}{\sqrt{D_{\rm A}}}\right)} \tag{7.72}$$

In a similar fashion, in order to determine the average rate of consumption of reagent B in the surface elements (with ages varying from $\tau = 0$ to $\tau = \infty$), we first calculate the amount of B consumed in a surface element of age τ, that is, the area between the horizontal line of height equal to c_B^{bulk} and the concentration profile of c_B (see Figure 7.4) that is given by eq. 7.63 with θ replaced by τ, that is,

$$Q^B(\tau) = \int_0^\infty \left[c_B^{bulk} - c_B(x,\tau)\right] dx$$

$$= c_B^{bulk}\delta(\tau) + \int_{\delta(\tau)}^\infty \left[c_B^{bulk} - c_B(x,\tau)\right] dx = 2\sqrt{\frac{D_B\tau}{\pi}} \frac{e^{\frac{-\beta^2}{D_B}}}{\text{erfc}\left(\frac{\beta}{\sqrt{D_B}}\right)} c_B^{bulk} \quad (7.73)$$

The instantaneous rate of consumption of B in the surface element is

$$R^B_{consump.inst}(\tau) = \frac{dQ^B(\tau)}{d\tau} = \sqrt{\frac{D_B}{\pi\tau}} \frac{e^{\frac{-\beta^2}{D_B}}}{\text{erfc}\left(\frac{\beta}{\sqrt{D_B}}\right)} c_B^{bulk} \quad (7.74)$$

Therefore, the steady-state average rate of consumption of reactant B in the surface elements that comprise the gas–liquid interface is given by

$$R^{B,ss}_{consump.ave} = \int_0^\infty R^B_{consump.inst}(\tau)\, Se^{-S\tau}\, d\tau$$

$$= \sqrt{D_B S} \frac{e^{\frac{-\beta^2}{D_B}}}{\text{erfc}\left(\frac{\beta}{\sqrt{D_B}}\right)} c_B^{bulk} \quad (7.75)$$

The net average rate of transfer of the product P (due to the mechanism of surface renewal) from the gas–liquid interface to the bulk liquid can be obtained from

$$R^{P,ss}_{trans}(t) = S\int_0^\infty \int_0^\infty \left[c_p(x,\tau) - c_P^{bulk}\right] dx\, Se^{-S\tau}\, d\tau$$

$$= \frac{p}{b} \frac{\sqrt{D_B S}\, e^{\frac{-\beta^2}{D_B}}}{\text{erfc}\left(\frac{\beta}{\sqrt{D_B}}\right)} c_B^{bulk} \quad (7.76)$$

where we have used eqs. 7.64 and 7.65 for $c_p(x, \tau)$.

Schema 7.1 offers a visual illustration of the convective exchange of liquid elements between the interface and bulk liquid caused by the surface renewal mechanism at one specific location in the tower, which leads to the four types of rates given by eqs. 7.70, 7.72, 7.75, and 7.76. As mentioned in earlier chapters, this convective exchange has no parallel in the film model, according to which the transfer of dissolved species between the interfacial region and bulk liquid occurs solely by molecular diffusion in the hypothetical liquid film, which can lead to divergence between the models, as will be seen later.

Appendix C presents the general equations for transient absorption with instantaneous reaction using the unsteady-state Danckwerts age-distribution function

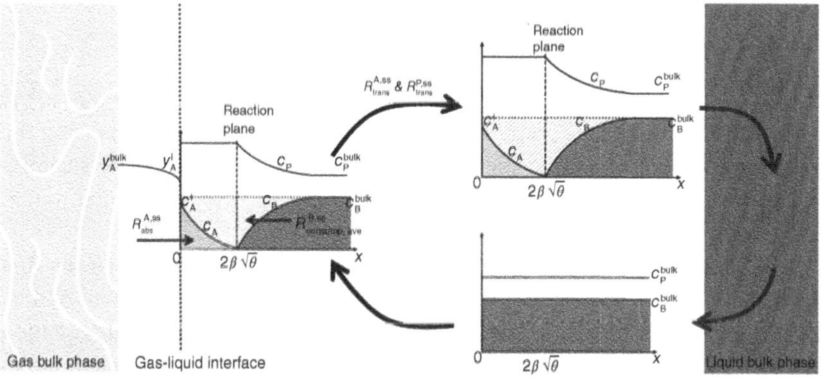

Schema 7.1 A visual picture of the convective exchange of different species between the liquid bulk and gas–liquid interface due to the surface renewal mechanism at a specific location in a packed tower.

(see eq. 2.10 in Chapter 2 or Huang and Chatterjee, 2021) in which the bulk-liquid concentrations of A, B, and P are presumed to be constant.

From eqs. 7.70 and 7.72, it follows that the fraction of dissolved A transferred from the interfacial region to the bulk liquid at steady state because of the surface renewal mechanism is equal to $R_{\text{trans}}^{\text{A,ss}}/R_{\text{abs}}^{\text{A,ss}} = 1 - e^{-\beta^2/D_{\text{A}}}$. Thus, the higher the value of β and lower the value of the diffusivity of A in the liquid, the greater is the fraction of absorbed A transferred to the bulk liquid due to surface renewal. As noted in Chapter 1, this subtle but vital aspect of the surface renewal model, that is, the distinction between the rate of absorption and rate of transfer, has been overlooked in much of the literature on gas absorption.

The gas-phase mass balance for solute A in the packed tower is given by

$$G\frac{\mathrm{d}y_{\text{A}}^{\text{bulk}}}{\mathrm{d}z} = R_{\text{abs}}^{\text{A,ss}} a A_{\text{c}} \tag{7.77}$$

The mass balance for reactant B in the liquid phase consists of two parts: (1) B consumed in the surface elements by reaction with dissolved A during their residence at the gas–liquid interface, and (2) B consumed in the bulk liquid due to its reaction with dissolved A transferred to the bulk liquid from the gas–liquid interface as a result of the surface renewal mechanism, that is,

$$L\frac{\mathrm{d}c_{\text{B}}^{\text{bulk}}}{\mathrm{d}z} = -\left(R_{\text{consump.ave}}^{\text{B,ss}} + bR_{\text{trans}}^{\text{A,ss}}\right) a A_{\text{c}} \tag{7.78}$$

The mass balance for product P in the liquid phase is also composed of two parts: (1) P transferred from the gas–liquid interface to the bulk liquid due to the surface renewal mechanism, and (2) P generated in the bulk liquid by instantaneous reaction between B and dissolved A transferred to the bulk liquid from the interface because of surface renewal, that is,

$$L\frac{dc_{\mathrm{P}}^{\mathrm{bulk}}}{dz} = \left(R_{\mathrm{trans}}^{\mathrm{P,ss}} + pR_{\mathrm{trans}}^{\mathrm{A,ss}}\right) aA_{\mathrm{c}} \tag{7.79}$$

By substituting eqs. 7.72 and 7.76 into eq. 7.79, and nondimensionalizing the resulting equation, it can be shown that the dimensionless product concentration c_{P}^{*} does not involve the stoichiometric coefficient p, which is similar to that observed for the film model (Section 7.2.1).

Equations 7.77 to 7.79 and eq. 7.66 form a system of DAEs in which the first three are ODEs, involving four unknowns, that is, the bulk concentrations of A, B, and P, and the parameter β (to be obtained from the nonlinear algebraic equation given by eq. 7.66), all of which vary along the tower. A trial-and-error method can be used to numerically solve for the value of the gas-phase concentration of A, $y_{\mathrm{A,I}}^{\mathrm{bulk}}$, at the top of the packed tower, as described in the following steps:

1) In the instantaneous reaction section of the tower (Region I), guess an initial value of $y_{\mathrm{A,I}}^{\mathrm{bulk}}$ at the top of the tower and solve the above system of DAEs, subject to the two boundary conditions at the top given by eqs. 7.19 and 7.20. The critical location z_{c} in the tower is then given by the value of z, where the bulk-liquid concentration of B $\left(c_{\mathrm{B}}^{\mathrm{bulk}}\right)$ falls to zero (see eq. 7.23). In the purely physical absorption section of the tower (Region II), use the concentration profiles given by eqs. 7.46 and 7.47 in Section 7.2.1.1, since in this region the predictions of the film and surface renewal models coincide.
2) Step 1 will yield values of $y_{\mathrm{A,I}}^{\mathrm{bulk}}\Big|_{z=z_{\mathrm{c}}}$ and $y_{\mathrm{A,II}}^{\mathrm{bulk}}\Big|_{z=z_{\mathrm{c}}}$, which in general will not be equal. Applying the method of bisection to these values will then yield a new estimate for $y_{\mathrm{A,I}}^{\mathrm{bulk}}$ at the top of the tower, after which Steps 1 and 2 are repeated until the absolute value of the difference between the left- and right-hand sides of eq. 7.22 is less than a very small specified tolerance ε, that is, $\left| y_{\mathrm{A,I}}^{\mathrm{bulk}}\Big|_{z=z_{\mathrm{c}}} - y_{\mathrm{A,II}}^{\mathrm{bulk}}\Big|_{z=z_{\mathrm{c}}} \right| < \varepsilon$.

7.2.2.2 Finite Gas-Phase Mass-Transfer Resistance

When a liquid element first arrives at the gas–liquid interface from the bulk liquid at a particular location in the tower, the reaction front is located right at the surface of the element, that is, there is no liquid-phase mass-transfer resistance, and the entire resistance is offered by the gas phase only. Thereafter, with the passage of exposure time, the front moves into the surface element, and the liquid-phase mass-transfer resistance starts becoming important. The rate of movement of the reaction plane, which will slow down with the exposure time of the liquid element, will also depend upon the level of the concentration of reagent B in it. The greater this value, the smaller will be the velocity of the reaction plane. This will be the case in the upper section of the tower, where the concentration of B in the liquid is relatively high. As the liquid travels down through the packing, B will get progressively depleted, and the depth of penetration of the reaction front in the surface elements will increase. At the critical location in the tower where B is completely exhausted, the reaction plane in the surface elements will disappear, after which there will only be physical absorption of solute A in the liquid.

As seen previously in Section 7.2.1.2, the expression for the rate of absorption of the solute gas A depends upon the mass-transfer driving force in the gas phase. Equation 7.50 is used instead of the boundary condition given by the first equation of eq. 7.59 for the case of zero gas-phase mass-transfer, that is,

$$\begin{cases} -D_{\mathrm{A}} \dfrac{\partial c_{\mathrm{A}}}{\partial x} = k_{\mathrm{G}} \left(y_{\mathrm{A}}^{\mathrm{bulk}} - H c_{\mathrm{A}}^{\mathrm{i}} \right); \quad \theta > 0 \text{ and } x = 0 \\ c_{\mathrm{B}} = c_{\mathrm{B}}^{\mathrm{bulk}}; \quad \theta = 0 \text{ and } x > 0 \\ c_{\mathrm{P}} = c_{\mathrm{P}}^{\mathrm{bulk}}; \quad \theta = 0 \text{ and } x > 0 \end{cases} \tag{7.80}$$

An analytical solution to the problem for the finite mass-transfer case is not possible in general. However, such a solution is feasible under two restrictive assumptions: (1) The location of the reaction front $\delta(\theta)$ in the liquid element is equal to $2\beta\sqrt{\theta}$ (i.e., the same as that in the zero mass-transfer resistance case) with the value of β to be determined as will be explained later, and (2) all species have identical binary diffusion coefficients in the liquid solvent, that is, $D_{\mathrm{A}} = D_{\mathrm{B}} = D_{\mathrm{P}} = D$, where D is a generic diffusion coefficient.

Invoking the concept of a fictitious "negative concentration" (Sada and Ameno, 1973; Garg *et al.*, 2000) for the reaction between dissolved A and reactant B in the liquid, the concentration of B is regarded as a fictitious "negative concentration" of A (i.e., $c_{\mathrm{B}} = -c_{\mathrm{A}}, x > \delta$), and assuming identical diffusion coefficients then makes the diffusion equations given by eqs. 7.56 to 7.58 equivalent. In addition, to smoothen the moving boundary, we assume unitary chemical stoichiometric numbers (i.e., $b = p = 1$). The governing equations given by eqs. 7.56 to 7.58 can then be represented by the following single diffusion equation in the generic concentration c and diffusion coefficient D:

$$\frac{\partial c}{\partial \theta} = D \frac{\partial^2 c}{\partial x^2} \tag{7.81}$$

The boundary and initial conditions for c can be written as (see eq. 7.80)

$$\begin{cases} -D \dfrac{\partial c}{\partial x} = k_{\mathrm{G}} \left(y_{\mathrm{A}}^{\mathrm{bulk}} - H c^{\mathrm{i}} \right); \quad \theta > 0 \text{ and } x = 0 \\ c = -c_{\mathrm{B}}^{\mathrm{bulk}}; \quad \theta = 0 \text{ and } x > 0 \\ c = -c_{\mathrm{B}}^{\mathrm{bulk}}; \quad \theta > 0 \text{ and } x \to \infty \end{cases} \tag{7.82}$$

Applying the method of Laplace transformation to the above initial-boundary-value problem yields (see also Carslaw and Jaeger, 1959; Crank, 1964)

$$c(x,\theta) = \left(\frac{y_{\mathrm{A}}^{\mathrm{bulk}}}{H} + c_{\mathrm{B}}^{\mathrm{bulk}}\right)\left[\mathrm{erfc}\left(\frac{x}{2\sqrt{D\theta}}\right) - \mathrm{e}^{hx+h^2D\theta}\mathrm{erfc}\left(h\sqrt{D\theta} + \frac{x}{2\sqrt{D\theta}}\right)\right] - c_{\mathrm{B}}^{\mathrm{bulk}} \tag{7.83}$$

where the constant h is defined by

$$h = \frac{k_{\mathrm{G}}H}{D} \tag{7.84}$$

Therefore, the liquid-phase concentration distributions of A and B in a surface element are

$$c_{\mathrm{A}}(x,\theta) = c(x,\theta);\quad 0 < x \le \delta(\theta) \tag{7.85}$$

and

$$c_{\mathrm{B}}(x,\theta) = -c(x,\theta);\quad x > \delta(\theta) \tag{7.86}$$

According to eq. 7.86, the concentration profiles of A and B in a surface element are mirror images of one another after the moving reaction front, as can be seen by mentally extending the c_{A} curve in Figure 7.5 beyond the point $x = \delta$, that is, they are symmetrical about the x-axis in this figure. This follows from the assumption of equal diffusion coefficients and unitary stoichiometric coefficients for all species. The concentration profile of the product P in the surface element is therefore given by

$$c_{\mathrm{P}}(x,\theta) = \begin{cases} c(x,\theta) + c_{\mathrm{B}}^{\mathrm{bulk}} + c_{\mathrm{P}}^{\mathrm{bulk}};\quad x > \delta(\theta) \\ c(\delta,\theta) + c_{\mathrm{B}}^{\mathrm{bulk}} + c_{\mathrm{P}}^{\mathrm{bulk}};\quad 0 < x \le \delta(\theta) \end{cases} \tag{7.87}$$

The instantaneous rate of absorption of the solute gas A at the gas–liquid interface can be obtained from eqs. 7.83 and 7.85, that is,

$$R_{\mathrm{inst}}^{\mathrm{A}}(\theta) = -D_{\mathrm{A}}\left.\frac{\partial c_{\mathrm{A}}}{\partial x}\right|_{x=0} = Dh\left(\frac{y_{\mathrm{A}}^{\mathrm{bulk}}}{H} + c_{\mathrm{B}}^{\mathrm{bulk}}\right)\mathrm{e}^{h^2D\theta}\mathrm{erfc}(h\sqrt{D\theta}) \tag{7.88}$$

The steady-state age-averaged rate of absorption of A at a specific location z in the tower may then be evaluated from

$$R_{\mathrm{abs,ss}}^{\mathrm{A}} = \int_0^\infty R_{\mathrm{inst}}^{\mathrm{A}}(\tau)f_{\mathrm{D}}(\tau)\mathrm{d}\tau = \frac{DSh\left(\frac{y_{\mathrm{A}}^{\mathrm{bulk}}}{H} + c_{\mathrm{B}}^{\mathrm{bulk}}\right)}{S - Dh^2}\left(1 - h\sqrt{\frac{D}{S}}\right) \tag{7.89}$$

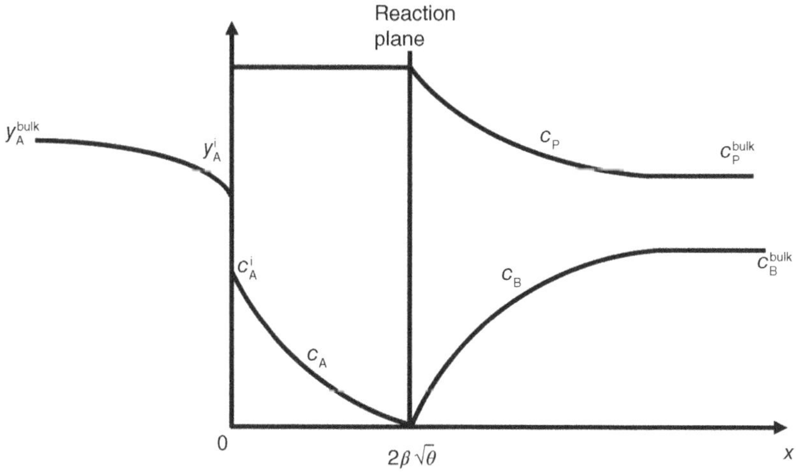

Figure 7.5 Concentration profiles in gas absorption with instantaneous chemical reaction (A = solute gas, B = liquid-phase reagent, and P = product) in a surface element for the case of finite gas-phase mass-transfer resistance at a particular location z in the tower. The location of the reaction front $\delta(\theta)$ in the liquid element is a function of the exposure time θ and is assumed to be equal to $2\beta\sqrt{\theta}$.

where the exposure time θ has been replaced with the age τ of the surface element as was done earlier in Section 7.2.2.1 for the zero gas-phase mass-transfer resistance case.

To obtain the expression for the rate of transfer of dissolved A from the gas–liquid interface to the bulk liquid due to the mechanism of surface renewal, we first determine the amount of A in a surface element of age τ, that is, the area under the concentration profile of A (see Figure 7.5), which is given by

$$
\begin{aligned}
Q^{A}(\tau) = \int_0^{\delta(\tau)} c_A(\tau)\mathrm{d}x = \left(\frac{y_A^{bulk}}{H} + c_B^{bulk}\right) \\
\times \left[\delta\,\mathrm{erfc}\left(\frac{\delta}{2\sqrt{D\tau}}\right) - 2\sqrt{\frac{D\theta}{\pi}}\left(e^{\frac{-\delta^2}{4D\tau}} - 1\right)\right. \\
- \frac{1}{h}e^{h\delta+h^2D\tau}\,\mathrm{erfc}\left(h\sqrt{D\tau} + \frac{\delta}{2\sqrt{D\tau}}\right) - \frac{1}{h}\mathrm{erf}\left(\frac{\delta}{2\sqrt{D\tau}}\right) \\
\left. + \frac{1}{h}e^{h^2D\tau}\,\mathrm{erfc}(h\sqrt{D\tau})\right] - \delta c_B^{bulk}
\end{aligned}
\tag{7.90}
$$

If we adopt the hypothesis, similar to that made for the zero gas-phase mass-transfer resistance case (Section 7.2.2.1), that the location of the reaction front in a surface element is proportional to the square root of its age τ, that is, $\delta = 2\beta\sqrt{\tau}$, the above expression can be rewritten as

$$Q^{\mathrm{A}}(\tau) = \left(\frac{y_{\mathrm{A}}^{\mathrm{bulk}}}{H} + c_{\mathrm{B}}^{\mathrm{bulk}}\right)\left[2\beta\sqrt{\tau}\,\mathrm{erfc}\left(\frac{\beta}{\sqrt{D}}\right) - 2\sqrt{\frac{D\tau}{\pi}}\left(e^{\frac{-\beta^2}{D}} - 1\right)\right.$$
$$- \frac{1}{h}e^{2\beta h\sqrt{\tau}+h^2D\tau}\mathrm{erfc}\left(h\sqrt{D\tau} + \frac{\beta}{\sqrt{D}}\right) - \frac{1}{h}\mathrm{erf}\left(\frac{\beta}{\sqrt{D}}\right)$$
$$\left. + \frac{1}{h}e^{h^2D\tau}\mathrm{erfc}(h\sqrt{D\tau})\right] - 2\beta\sqrt{\tau}\,c_{\mathrm{B}}^{\mathrm{bulk}} \tag{7.91}$$

The steady-state rate of the transfer of A to the bulk liquid can be obtained from

$$R_{\mathrm{trans}}^{\mathrm{A,ss}} = S\int_0^{\infty} Q^{\mathrm{A}}(\tau)\,Se^{-S\tau}\,\mathrm{d}\tau$$
$$= \left(\frac{y_{\mathrm{A}}^{\mathrm{bulk}}}{H} + c_{\mathrm{B}}^{\mathrm{bulk}}\right)\left[\sqrt{\pi S}\,\beta\,\mathrm{erfc}\left(\frac{\beta}{\sqrt{D}}\right) - \sqrt{DS}\left(e^{\frac{-\beta^2}{D}} - 1\right) - \frac{S^2}{h}I_1\right.$$
$$\left. - \frac{S}{h}\mathrm{erf}\left(\frac{\beta}{\sqrt{D}}\right) + \frac{S^2\left(1 - h\sqrt{\frac{D}{S}}\right)}{h(S - Dh^2)}\right] - \sqrt{\pi S}\,\beta c_{\mathrm{B}}^{\mathrm{bulk}} \tag{7.92}$$

where the transcendental integral I_1 is defined by

$$I_1 = \int_0^{\infty} \mathrm{e}^{2\beta h\sqrt{\tau}+(h^2D-S)\tau}\mathrm{erfc}\left(h\sqrt{D\tau} + \frac{\beta}{\sqrt{D}}\right)\mathrm{d}\tau \tag{7.93}$$

In a similar fashion to what was done in Section 7.2.2.1, in order to determine the average rate of consumption of reagent B in the surface elements (with ages varying from $\tau = 0$ to $\tau = \infty$), we first determine the amount of B consumed in a surface element of age τ, that is, the area between the horizontal line of height equal to $c_{\mathrm{B}}^{\mathrm{bulk}}$ and the concentration profile of c_{B} (see Figure 7.5) given by eq. (7.86) with θ replaced by τ, that is,

$$Q^{\mathrm{B}}(\tau) = \int_0^{\infty}\left[c_{\mathrm{B}}^{\mathrm{bulk}} - c_{\mathrm{B}}(x,\tau)\right]\mathrm{d}x = c_{\mathrm{B}}^{\mathrm{bulk}}\delta(\tau) + \int_{\delta(\tau)}^{\infty}\left[c_{\mathrm{B}}^{\mathrm{bulk}} - c_{\mathrm{B}}(x,\tau)\right]\mathrm{d}x$$
$$= \left(\frac{y_{\mathrm{A}}^{\mathrm{bulk}}}{H} + c_{\mathrm{B}}^{\mathrm{bulk}}\right)\left[-2\beta\sqrt{\tau}\,\mathrm{erfc}\left(\frac{\beta}{\sqrt{D}}\right) + 2\sqrt{\frac{D\tau}{\pi}}e^{\frac{-\beta^2}{D}} + \frac{1}{h}e^{2\beta h\sqrt{\tau}+h^2D\tau}\right.$$
$$\left.\times\,\mathrm{erfc}\left(h\sqrt{D\tau} + \frac{\beta}{\sqrt{D}}\right) - \frac{1}{h}\mathrm{erfc}\left(\frac{\beta}{\sqrt{D}}\right)\right] + 2\beta\sqrt{\tau}\,c_{\mathrm{B}}^{\mathrm{bulk}} \tag{7.94}$$

The instantaneous rate of consumption of B can be obtained from

$$R^{\mathrm{B}}_{\text{consump.inst}}(\tau) = \frac{\mathrm{d}Q^{\mathrm{B}}(\tau)}{\mathrm{d}\tau} = \frac{\beta}{\sqrt{\tau}} c^{\text{bulk}}_{\mathrm{B}} + \left(\frac{y^{\text{bulk}}_{\mathrm{A}}}{H} + c^{\text{bulk}}_{\mathrm{B}}\right)\left[-\frac{\beta}{\sqrt{\tau}}\operatorname{erfc}\left(\frac{\beta}{\sqrt{D}}\right)\right.$$
$$\left. + \frac{1}{h} e^{2\beta h\sqrt{\tau}+h^2 D\tau} \operatorname{erfc}\left(h\sqrt{D\tau} + \frac{\beta}{\sqrt{D}}\right)\left(h^2 D + \frac{\beta h}{\sqrt{\tau}}\right)\right] \tag{7.95}$$

Therefore, the average steady-state consumption rate of reactant B is given by

$$\begin{aligned} R^{\mathrm{B,ss}}_{\text{consump.ave}} &= \int_0^{\infty} R^{\mathrm{B}}_{\text{consump.inst}}(\tau) S e^{-S\tau}\, \mathrm{d}\tau \\ &= \left(\frac{y^{\text{bulk}}_{\mathrm{A}}}{H} + c^{\text{bulk}}_{\mathrm{B}}\right)\left[-\sqrt{\pi S}\,\beta \operatorname{erfc}\left(\frac{\beta}{\sqrt{D}}\right) + hDSI_1 + \beta S I_2\right] \\ &\quad + \sqrt{\pi S}\,\beta c^{\text{bulk}}_{\mathrm{B}} \end{aligned} \tag{7.96}$$

where the transcendental integral I_2 is defined by

$$I_2 = \int_0^{\infty} e^{2\beta h\sqrt{\tau}+(h^2 D-S)\tau} \operatorname{erfc}\left(h\sqrt{D\tau} + \frac{\beta}{\sqrt{D}}\right)\frac{1}{\sqrt{\tau}}\, \mathrm{d}\tau \tag{7.97}$$

The steady-state net rate of transfer of product P from the interface to the bulk liquid (due to the mechanism of surface renewal) can be obtained from

$$\begin{aligned} R^{\mathrm{P,ss}}_{\text{trans}} &= S\int_0^{\infty}\int_0^{\infty} \left[c_{\mathrm{p}}(x,\tau) - c^{\text{bulk}}_{\mathrm{P}}\right] \mathrm{d}x\, S e^{-S\tau}\, \mathrm{d}\tau \\ &= \left(\frac{y^{\text{bulk}}_{\mathrm{A}}}{H} + c^{\text{bulk}}_{\mathrm{B}}\right)\left[-2\beta S^2 I_3 + \sqrt{DS}\, e^{\frac{-\beta^2}{D}} + \frac{S^2}{h} I_1 - \frac{S}{h}\operatorname{erfc}\left(\frac{\beta}{\sqrt{D}}\right)\right] \end{aligned} \tag{7.98}$$

where the transcendental integral I_3 is defined by

$$I_3 = \int_0^{\infty} e^{2\beta h\sqrt{\tau}+(h^2 D-S)\tau} \operatorname{erfc}\left(h\sqrt{D\tau} + \frac{\beta}{\sqrt{D}}\right)\sqrt{\tau}\, \mathrm{d}\tau \tag{7.99}$$

Substituting eqs. 7.89, 7.92, 7.96, and 7.98 into eqs. 7.77 to 7.79 yields an ODE system (containing the parameter β), which must be solved numerically. The value of β at a specific location z in the tower can be determined as follows: Replacing θ with τ in eq. 7.83 and setting $c(\delta, \tau) = 0$ will result in an implicit relation between δ and τ, which can be solved numerically to obtain the location of the reaction front δ as a function of τ. Since we had previously assumed that $\delta = 2\beta\sqrt{\tau}$, which has experimental confirmation at least for the zero gas-phase mass-transfer resistance case, as will be seen later, the value of β can be estimated by fitting this equation to the above calculated values of δ (i.e., the location of the reaction front). It should be noted that the value of β depends upon the levels of $y^{\text{bulk}}_{\mathrm{A}}$ and $c^{\text{bulk}}_{\mathrm{B}}$ as shown by eq. 7.83 and hence will vary along the length of the column.

The equations presented thus far for the surface renewal model hold in the upper "instantaneous reaction" region of the tower, whereas the lower "physical absorption" region is governed by eq. 7.13 for the zero gas-phase mass-transfer resistance case, which also holds for the case of finite gas-phase mass-transfer resistance provided k_L is replaced with k_{OL} (defined by eq. 7.54). Finally, the boundary conditions at the top and bottom of the tower are specified by eqs. 7.19 to 7.21, while the conditions at the interface of the two regions are given by eqs. 7.22 to 7.24. The same trial-and-error numerical method presented earlier for the zero gas-phase mass-transfer resistance case in Section 7.2.2.1 can be used to solve the DAE system represented by the ODEs given by eqs. 7.77 to 7.79 and the algebraic equation $c(\delta, \tau) = 0$ (see eq. 7.83).

7.3 Illustrative Calculations

SO_2 is generated by the combustion of fossil fuels in industrial production and transportation and is of primary concern as regards the global environment. Besides causing acid rain, it has been reported that SO_2 promotes ozone-depleting reactions in the stratosphere (Brogren and Karlsson, 1997).

This section compares the theoretical predictions of the film and surface renewal models for the absorption of gaseous sulfur dioxide (SO_2) into an aqueous solution of sodium sulfite (Na_2SO_3) in a packed tower with and without consideration of gas-phase mass-transfer resistance. Onda *et al.* (1971) have shown that the reaction between sulfur dioxide and the sulfite ion is a one-step instantaneous chemical reaction and is thus an appropriate example for a comparison to be made between the two models. A 0.5-m diameter packed column, filled with ceramic Raschig rings with a nominal size of 0.0254 m and a 3-m packed height, is assumed to be operating at steady state. Further, the absorption is assumed to occur at a column pressure of one standard atmosphere and 25 °C. Values of the physical properties and parameters, some of which were used in the calculations, are reported in Table 7.1. Many of these were derived from Perry *et al.* (1984) or the literature; for example, the correlations of Onda *et al.* (1968) were used to estimate the liquid-phase mass-transfer coefficient k_L and gas–liquid interfacial contact area a.

7.3.1 Zero Gas-Phase Mass-Transfer Resistance

It has been demonstrated that the reaction between sulfur dioxide and sulfite ions (SO_3^{2-}) is a one-step instantaneous irreversible chemical reaction (Hikita *et al.*, 1968; Onda *et al.*, 1971) and can be represented as

$$SO_2(aq.) + SO_3^{2-} + H_2O \rightarrow 2HSO_3^-$$

Table 7.1 Values of physical properties and parameters.

Physical property or parameter	SO_2–Na_2SO_3 system
type of packing	1-inch ceramic Raschig rings
critical surface tension of packing (mN m^{-1})	61
nominal packing size (m)	0.0254
packing surface area (m^2 m^{-3})	190
packing factor (m^{-1})	510
column temperature (°C)	25
column pressure (atm)	1
packed height (Z, m)	3
column diameter (m)	0.5
liquid mass velocity (kg m^{-2} s^{-1}; Perry *et al.*, 1984)	2.14
vapor mass velocity (kg m^{-2} s^{-1})	0.09
liquid density (kg m^{-3})	997.045
vapor density (kg m^{-3})	1.184
liquid viscosity (kg m^{-1} s^{-1})	8.937×10^{-4}
vapor viscosity (kg m^{-1} s^{-1})	1.837×10^{-5}
inlet-liquid concentration of reagent B ($c_{B,in}^{bulk}$, kgmol m^{-3})	1 to 10
inlet concentration of solute gas ($y_{A,in}^{bulk}$, kgmol m^{-3})	1
diffusivity of dissolved A in the liquid solvent, i.e., water (D_A, m^2 s^{-1}; Ebrahimi *et al.*, 2003)	1.83×10^{-9}
diffusivity of reagent B in the liquid solvent, i.e., water (D_B, m^2 s^{-1}; Ebrahimi *et al.*, 2003)	0.96×10^{-9}
diffusivity of product P in the liquid solvent, i.e., water (D_P, m^2 s^{-1})	[a] 1.55×10^{-9} [b] 1.83×10^{-9}
Henry's law constant (H)	[c] 2.431×10^{-2}
surface tension of liquid (mN m^{-1})	72.75
gas-phase mass-transfer coefficient (k_G, m s^{-1})[d]	5.033×10^{-4}, 1.007×10^{-3}, and 1.007×10^{-2}

(Continued)

Table 7.1 (Continued)

Physical property or parameter	SO_2–Na_2SO_3 system
Calculated parameters	
liquid-phase mass-transfer coefficient (k_L, m s^{-1})—estimated from the correlation of Onda *et al.* (1968)	8.77×10^{-5}
surface renewal rate (S, s^{-1}); calculated by eq. 7.69	4.205
gas–liquid interfacial area of contact (a, m^{-1}) – estimated from the correlation of Onda *et al.* (1968)	77.70
liquid volumetric flow rate (L, m^3 s^{-1})	4.21×10^{-4}
vapor volumetric flow rate (G, m^3 s^{-1})	1.49×10^{-2}

a) From Ebrahimi *et al.* (2003); used in calculations with the surface renewal model for the zero gas-phase mass-transfer resistance case.
b) Assumed equal to the value of D_A; used in calculations with the surface renewal model for the finite gas-phase mass-transfer resistance case (see Section 7.2.2.2).
c) Value for SO_2 absorption in water at 20 °C; no temperature correction is made. This value is 7.4% lower than the value of H for the same gas–liquid system at an average temperature of 25 °C as derived from the data of Terraglio and Manganelli (1967).
d) The values of k_G were chosen to span two orders of magnitude and had their basis in the correlation of Onda *et al.* (1968).

Hence, values of the stoichiometric coefficients b and p are equal to 1 and 2, respectively (except in Sections 7.3.2.2 and 7.3.2.3, where both are assumed to be equal to 1).

7.3.1.1 Tower Concentration Profiles

In the following, we compare dimensionless concentration profiles of the gas- and liquid-phases along the length of the packed tower as predicted by the film and surface renewal models, assuming negligible gas-phase mass-transfer resistance (Figure 7.6). The liquid-phase diffusion coefficients of dissolved A [i.e., SO_2(aq.)] and reagent B [i.e., SO_3^{2-}] were obtained from the investigation of Ebrahimi *et al.* (2003), whereas the values of the remaining parameters were either assumed or taken from the literature, as indicated earlier (see Table 7.1).

It is seen in Figure 7.6 that the bulk concentration profiles along the length of the tower for all components (i.e., A and B), except product P, and the critical location as predicted by film and surface renewal models exhibit excellent agreement with each other for the zero gas-phase mass-transfer resistance case. Besides, both models reveal that Region I (region of absorption with instantaneous chemical reaction) occupies only about 7% of the volume of the packed tower, that is, Region II (region of purely physical absorption) occupies 93% of the tower, which indicates inefficient tower operation, that is, most of the tower is utilized for

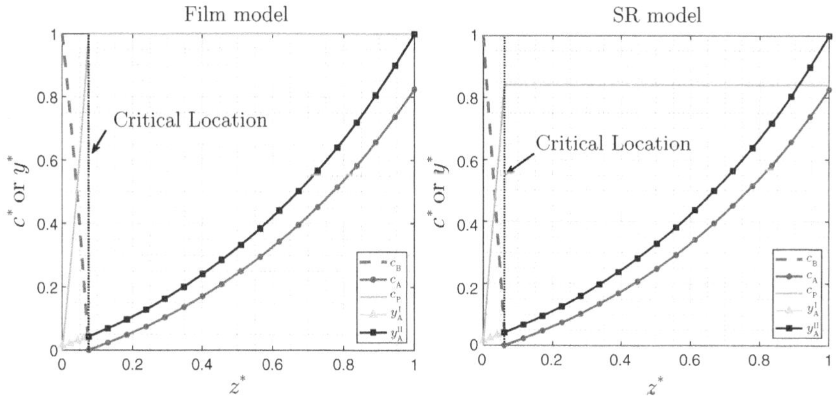

Figure 7.6 Dimensionless bulk concentration profiles of the gas and liquid phases along the dimensionless length of the packed tower (measured from its top) as calculated by the film (left) and surface renewal (SR; right) models assuming zero gas-phase mass-transfer resistance. In each plot, the region of instantaneous chemical reaction (Region I) is to the left of the critical location, whereas that on its right is that of pure physical absorption (Region II). $c_{B,in}^{bulk} = 1\ kgmol\,m^{-3}$ and other parameter values are given in Table 7.1. Note: The superscript "*" has been dropped in the legends.

physical absorption. The dimensionless concentration of solute A in the gas phase varies from 100% at the tower bottom to 4% at the critical location, that is, about 96% of A in the gas phase is removed by physical absorption in Region II. In comparison, the dimensionless concentration of dissolved gas A in the bulk liquid increases from zero at the critical location to about 83% at the bottom of the tower (i.e., in the same region).

For product P, the film model predicts a 100% increase in the dimensionless bulk-liquid concentration from the top of the tower to the critical location (i.e., in Region I), whereas this is only 84% for the surface renewal model. A plausible explanation for the discrepancy between the film and surface renewal models as regards the prediction of the concentration of P in the liquid at the critical location is the different mass-transfer mechanisms assumed in them. According to eq. 7.8 of the film model, the rate of change in the bulk-liquid concentration of P along the tower is directly proportional to the rate of absorption of A (R_A) at the gas–liquid interface (at a given location in the tower), the proportionality factor being the stoichiometric coefficient p. From this, it follows that for every mole of dissolved A consumed by chemical reaction with b moles of B at a particular location in Region I of the tower, p moles of P will be produced, and therefore the nondimensional concentration of P at the critical location, $c_P^{crit*} = \frac{c_P^{crit}}{pc_{B,in}^{bulk}}$, will always be equal to 1, where c_P^{crit} is the dimensional liquid-phase product concentration at the critical location. For example, for the film model results shown in Figure 7.6,

$b = 1$, $p = 2$, and $c_{\mathrm{B,in}}^{\mathrm{bulk}} = 1$ kgmol m^{-3}. Hence, $c_{\mathrm{P}}^{\mathrm{crit}} = 2$ kgmol m^{-3} and therefore $c_{\mathrm{P}}^{\mathrm{crit}*} = 1$. In sharp contrast, according to eq. 7.79 of the surface renewal model, this rate of change depends on the (convective) transfer rates of both P (created at the reaction front in the interfacial region) and dissolved gas A from the interfacial region to the bulk liquid due to surface renewal. This can be seen more clearly when one compares the tower mass balances for P in the two models. Expanding eq. 7.16 of the film model and inserting eqs. 7.72 and 7.76 into eq. 7.79 of the surface renewal model, we obtain, respectively, the following expressions:

$$L\frac{dc_{\mathrm{P}}^{\mathrm{bulk}}}{dz} = apA_{\mathrm{c}}k_{\mathrm{L}}\frac{y_{\mathrm{A}}^{\mathrm{bulk}}}{H} + a\frac{p}{b}A_{\mathrm{c}}k_{\mathrm{L}}\frac{D_{\mathrm{B}}}{D_{\mathrm{A}}}c_{\mathrm{B}}^{\mathrm{bulk}} \tag{7.100}$$

and

$$L\frac{dc_{\mathrm{P}}^{\mathrm{bulk}}}{dz} = apA_{\mathrm{c}}k_{\mathrm{L}}\frac{\left(1 - e^{-\frac{\beta^2}{D_{\mathrm{A}}}}\right)}{\mathrm{erf}\left(\frac{\beta}{\sqrt{D_{\mathrm{A}}}}\right)}\frac{y_{\mathrm{A}}^{\mathrm{bulk}}}{H} + a\frac{p}{b}A_{\mathrm{c}}k_{\mathrm{L}}\sqrt{\frac{D_{\mathrm{B}}}{D_{\mathrm{A}}}}\frac{e^{-\frac{\beta^2}{D_{\mathrm{B}}}}}{\mathrm{erfc}\left(\frac{\beta}{\sqrt{D_{\mathrm{B}}}}\right)}c_{\mathrm{B}}^{\mathrm{bulk}} \tag{7.101}$$

Two observations immediately follow upon comparison of eq. 7.100 of the film model with eq. 7.101 of the surface renewal model. On the right-hand sides of these expressions, (1) the first term of eq. 7.101 contains an extra factor $\frac{\left(1-e^{-\frac{\beta^2}{D_{\mathrm{A}}}}\right)}{\mathrm{erf}\left(\frac{\beta}{\sqrt{D_{\mathrm{A}}}}\right)}$ compared to eq. 7.100, and (2) the diffusivity ratio $\frac{D_{\mathrm{B}}}{D_{\mathrm{A}}}$ in the second term of eq. 7.100 is replaced by $\sqrt{\frac{D_{\mathrm{B}}}{D_{\mathrm{A}}}}\frac{e^{-\frac{\beta^2}{D_{\mathrm{B}}}}}{\mathrm{erfc}\left(\frac{\beta}{\sqrt{D_{\mathrm{B}}}}\right)}$ in the corresponding term of eq. 7.101. This difference between the two models follows directly from the convective exchange of liquid elements between the gas–liquid interface and the bulk liquid as postulated in the surface renewal model, which is governed by the surface renewal frequency S, and which, as mentioned in earlier chapters, has no parallel in the film model according to which the transport of dissolved species (A, B, and P) occurs solely by molecular diffusion in the hypothetical liquid film. We also note that the diffusion coefficient of the product, D_{P}, does not appear in either eq. 7.100 or eq. 7.101.

7.3.1.2 Movement of Reaction Front

Onda *et al.* (1971) experimentally measured the motion of the reaction front for SO_2 absorption in Na_2SO_3 solution, for which the value of the interfacial concentration $c_{\mathrm{A}}^{\mathrm{i}} = 1.35$ kgmol m^{-3}, neglecting the ionic strength of the solution (see Table 1 of Onda *et al.*, 1971). They utilized methyl red as the indicator to detect the movement of the reaction plane in a 0.5–0.7 weight percent agar-agar jelly at four different concentrations of Na_2SO_3; their data are shown in Figure 7.7. This figure shows excellent agreement for the location of the reaction front determined

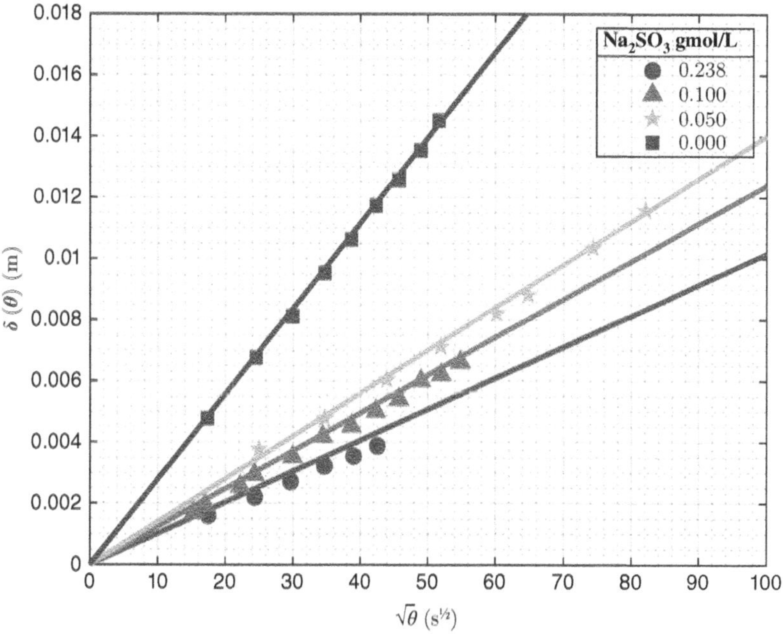

Figure 7.7 Variation of the location of the reaction front with the square root of exposure time for four different concentrations of Na_2SO_3 solution (0, 0.05, 0.10, and 0.238 gmol L^{-1}). Solid lines represent theoretical values calculated from the equation $\delta(\theta) = 2\beta\sqrt{\theta}$, with β being determined from eq. 7.66 whereas symbols represent the experimental measurements of Onda *et al.* (1971).

experimentally with its theoretically predicted value, that is, $\delta(\theta) = 2\beta\sqrt{\theta}$, where the constant β is obtained by numerically solving eq. 7.66. Also, Figure 7.7 shows that increasing the concentration of the Na_2SO_3 solution retards the movement of the reaction front as expected from physical considerations.

According to the comparison between experiment and theory shown in Table 7.2, the relative error between the slope of the experimental data (k) shown in Figure 7.7 and that calculated theoretically (2β) from eq. 7.66 varies from 0.9 to 10%, as the Na_2SO_3 concentration increases from 0 to 0.238 gmol L^{-1}, that is, the agreement is quite good.

7.3.2 Finite Gas-Phase Mass-Transfer Resistance

7.3.2.1 Validity of the Solution of the Surface Renewal Model

To examine the validity of the approach taken in Section 7.2.2.2 for the case of finite gas-phase mass-transfer resistance under the special condition of equal diffusion coefficients (i.e., $D_A = D_B = D_P = D = 1.83 \times 10^{-9}$ m^2 s^{-1}) and chemical

Table 7.2 Comparison of the theoretical values of 2β calculated from eq. 7.66, with values of the slope k obtained by fitting the expression $\delta(\theta) = k\sqrt{\theta}$ to the experimental data of Onda *et al.* (1971) shown in Figure 7.7 using the method of least squares.

Concentration c_{B,Na_2SO_3} (gmol L^{-1})	2β (theory, m s$^{-0.5}$) eq. 7.66 $\delta = 2\beta\sqrt{\theta}$	k (expt., m s$^{-0.5}$) Least-squares method $\delta = k\sqrt{\theta}$	Relative error (%) $\left\lvert \frac{x_{exp} - x_{theory}}{x_{theory}} \right\rvert \times 100$
0.238	1.016×10^{-4}	0.915×10^{-4}	9.999
0.100	1.239×10^{-4}	1.194×10^{-4}	3.631
0.050	1.401×10^{-4}	1.386×10^{-4}	1.046
0.000	2.789×10^{-4}	2.764×10^{-4}	0.905

stoichiometric numbers of unity (i.e., $b = p = 1$), the transient concentration profiles of the solute A and reagent B in a surface element for the zero gas-phase mass-transfer resistance case are compared against those for the finite gas-phase mass-transfer resistance case in Figure 7.8, assuming a very high value for the gas-phase mass-transfer coefficient k_G. In the calculations, both the bulk gas-phase concentration of A $\left(y_A^{bulk}\right)$ and the bulk-liquid concentration of B $\left(c_B^{bulk}\right)$ were assumed to be 1 kgmol m^{-3}. The interfacial concentration of dissolved A $\left(c_A^i\right)$ when there is no mass-transfer resistance in the gas phase is equal to $y_A^{bulk}/H = 41.14$ kgmol m^{-3} (see Table 7.1 for the value of H). Equations 7.62 and 7.63, along with eq. 7.66 for the determination of β, were used in the calculations for the zero gas-phase mass-transfer resistance case. For the case of finite gas-phase mass-transfer resistance, eqs. 7.83 to 7.86 were used to determine the concentration profiles of A and B in the liquid element, with the location of the reaction front δ being calculated using the procedure outlined in the paragraph below eq. 7.99.

As Figure 7.8 shows, the concentration profiles of A for the two cases cannot be visually distinguished from one another. However, there is a slight discrepancy between the concentration profiles of B on the right-hand side of the advancing reaction front. As regards the movement of this front, the value of β obtained from eq. 7.66 is 7.361×10^{-5} m s$^{-0.5}$ that can be compared against the value of 6.840×10^{-5} m s$^{-0.5}$, which is half of the slope obtained by fitting the roots of eq. 7.83 (by setting $c(x, \theta) = 0$) against the square root of the exposure time θ. This leads to a minor incongruity in the predicted location of the moving reaction plane at a large exposure time of 50,000 s (about 13.9 h). The difference in the location of the moving reaction plane predicted by the two solutions is only 2.33 mm, which is 7.8% of the total width of the reaction region of 30 mm for this large exposure time.

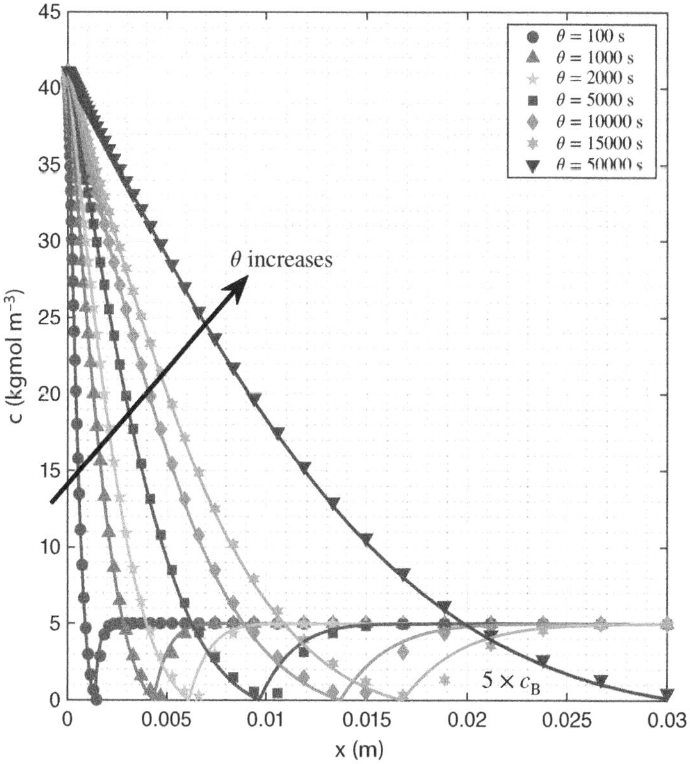

Figure 7.8 Transient concentration distributions of *A* and *B* in a surface element along the distance measured from the gas–liquid interface at a specific location in a packed tower as predicted by the surface renewal model. The bulk gas-phase concentration of *A* (y_A^{bulk}) and bulk-liquid concentration of *B* (c_B^{bulk}) are assumed to be equal to 1 kgmol m^{-3}. The solid lines represent the solution for the finite gas-phase mass-transfer resistance case using an extremely high value of the gas-phase mass transfer coefficient (k_G), whereas symbols represent the solution for the zero gas-phase mass-transfer resistance case. The liquid-phase diffusion coefficients of *A* and *B* are assumed to be identical ($= 1.83 \times 10^{-9}$ m^2 s^{-1}), and the stoichiometric numbers are taken to be unity.

Thus, it can be concluded that, at least under the special situation of identical liquid-phase diffusion coefficients of the components *A* and *B* and stoichiometric numbers of unity, the approach taken for deriving the solution of the surface renewal model for the case of finite gas-phase mass-transfer resistance is valid.

7.3.2.2 Tower Concentration Profiles

Figure 7.9 compares the dimensionless bulk gas- and liquid-phase concentration profiles of *A*, *B*, and *P* along the tower height as predicted by the film and surface renewal models for three different values of k_G (5.033×10^{-4}, 1.007×10^{-3},

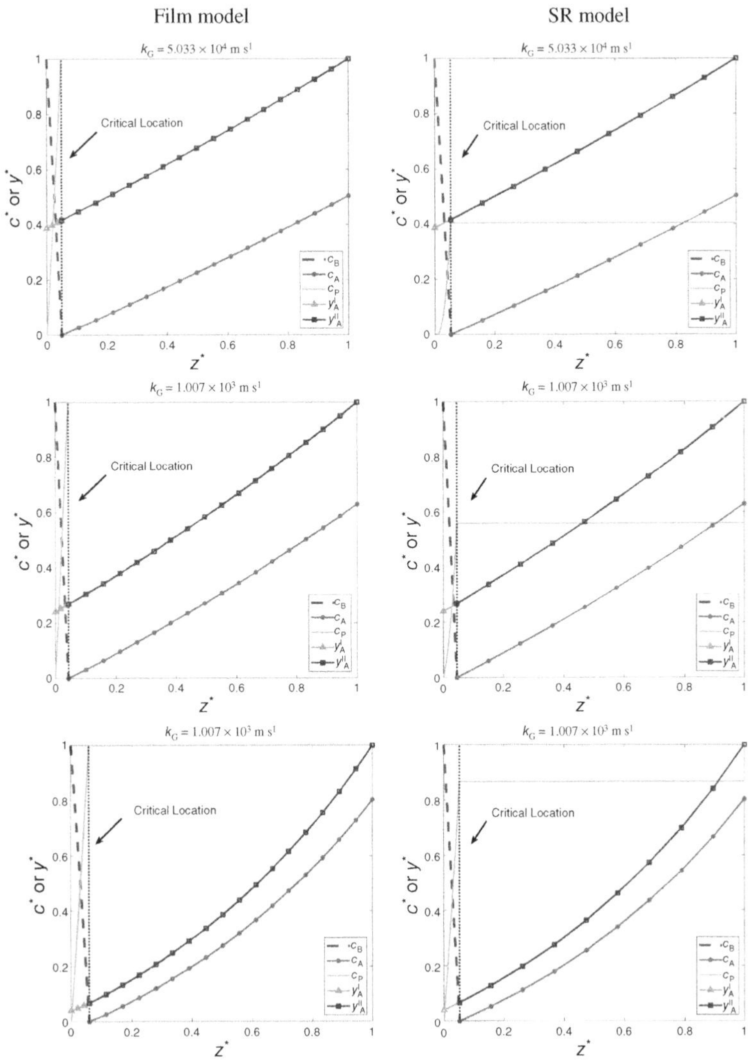

Figure 7.9 Dimensionless bulk concentration profiles in the gas and liquid phases along the dimensionless length of the packed tower as predicted by the film (left) and surface renewal (SR, right) models, assuming finite gas-phase mass-transfer resistance. In each plot, the region of instantaneous chemical reaction (Region I) is to the left of the critical location, whereas that on its right is that of pure physical absorption (Region II). $c_{\mathrm{B,in}}^{\mathrm{bulk}} = 1$ kgmol m^{-3} and other parameter values are given in Table 7.1. Equal diffusion coefficients ($= 1.83 \times 10^{-9}$ m^2 s^{-1}; only for the surface renewal model) and stoichiometric numbers of unity for all species are assumed. The superscript "*" has been dropped in the legends.

and 1.007×10^{-2} m s^{-1}). As in Section 7.3.2.1, equal diffusion coefficients ($= 1.83 \times 10^{-9}$ m^2 s^{-1}) are assumed in the surface renewal model, whereas individual diffusion coefficients are used in the film model. Also, stoichiometric numbers of unity for all species are assumed to be equal to 1. Similar to the case of zero gas-phase mass-transfer resistance, for both models, all concentration profiles (except for product P) compare very well at the same value of k_G (see Figure 7.9). Further, both models predict that Region I (instantaneous reaction region) occupies about 5–6% of the total packed height and is independent of k_G, whereas Region II (physical absorption region) occupies the remaining 94%. For a (large) value of $k_G = 1.007 \times 10^{-2}$ m s^{-1}, the dimensionless concentration of A in the gas phase decreases from 100% at the tower bottom to 6% at the critical location (i.e., a reduction of 94% occurring in Region II). Correspondingly, in the same region, the dimensionless concentration of solute A in the liquid increases from zero at the critical location to 80% at the tower bottom (i.e., a gain of 80%).

The efficiency of removal of A from the bulk gas decreases from 96.4 to 59.1% as k_G decreases from 1.007×10^{-2} to 5.033×10^{-4} m s^{-1} (i.e., approximately over two orders of magnitude) in the tower, and correspondingly, the dimensionless concentration of solute A in the liquid leaving the bottom of the tower decreases from 80% to 50% over this same range of variation in k_G, indicating the significant influence of the gas-phase mass-transfer resistance.

The dimensionless concentration of the product P in the liquid leaving at the bottom of the tower as predicted by the film model is 100% for all three values of k_G. In striking contrast, the value of this concentration as predicted by the surface renewal model depends upon the magnitude of k_G, that is, it increases from 40 to 86% as k_G increases from 5.033×10^{-4} to 1.007×10^{-2} m s^{-1}, the latter being the highest value of k_G used in Figure 7.9. We recall from Section 7.3.1.1 that the maximum concentration of P in the liquid leaving the tower at its bottom was 84% for the zero gas-phase mass-transfer resistance case (albeit for unequal diffusion and stoichiometric coefficients), which is one more confirmation of the validity of the approach taken in Section 7.2.2.2 for the case of finite gas-phase mass-transfer resistance. As discussed previously in Section 7.3.1.1, a likely reason for this discrepancy between the two models as regards prediction of the concentration of P is the different mass-transfer mechanisms assumed in them.

7.3.2.3 Effect of Concentration of *B* in the Incoming Liquid on Tower Performance

This section offers a comparison between the film and surface renewal models for the case of finite gas-phase mass-transfer resistance when the concentration of B in the incoming liquid at the top of the tower $c_{B,in}^{bulk}$ increases from 2 to 10 kgmol m^{-3} for a value of $k_G = 1.007 \times 10^{-3}$ m s^{-1}, with values of the other parameters being given in Table 7.1. Also, equal diffusion coefficients for all species

($=1.83\times10^{-9}$ m^2 s^{-1}) are assumed in the calculations with the surface renewal model besides using stoichiometric coefficients of unity for all species in both models, that is, $b=p=1$.

Figure 7.10 shows particularly good agreement between the predictions of the two models as regards the gas- and liquid-phase concentration profiles of all components, except for product P, at the three values of $c_{\mathrm{B,in}}^{\mathrm{bulk}}$ (2, 5, and 10 kgmol m^{-3}) used in the calculations. It can also be seen from Figures 7.9, 7.10 and Table 7.3 that both models predict Region I (instantaneous reaction section) to extend approximately to 4, 9, 21, and 40% of the tower packed height as $c_{\mathrm{B,in}}^{\mathrm{bulk}}$ takes on the values of 1, 2, 5, and 10 kgmol m^{-3}, respectively. Correspondingly, the dimensionless concentration of A in the gas phase decreases from 100% at the tower bottom to 27, 28, 32, and 38% at the critical location with increasing $c_{\mathrm{B,in}}^{\mathrm{bulk}}$, that is, the absorption of A decreases by 73, 72, 68, and 62%, respectively, in Region II (physical absorption section) as Region I (instantaneous reaction section) starts playing a more important role in the removal of A from the gas stream. In Region II, the dimensionless concentration of solute A in the liquid phase increases from zero at the critical location to 63, 62, 59, and 53% at the bottom of tower, respectively, that is, the concentration of dissolved A in the liquid leaving the tower bottom decreases with an increase in $c_{\mathrm{B,in}}^{\mathrm{bulk}}$ due to the greater extension of Region I into the tower. In addition, the dimensionless concentration of A in the gas stream leaving at the top of tower decreases in the order 24, 22, 17, and 7% (the last two values being averages of the two models – see Table 7.3), when $c_{\mathrm{B,in}}^{\mathrm{bulk}}$ increases in the order 1, 2, 5, and 10 kgmol m^{-3}, respectively. However, a further increase in the value of $c_{\mathrm{B,in}}^{\mathrm{bulk}}$, say to 20 kgmol m^{-3}, gives physically unrealistic results, which indicates that an extra region, that is, a surface reaction zone, may need to be considered in the upper section of the tower as done previously by Secor and Southworth (1961) in their film model. This will make the analysis more complex, especially in the case of the surface renewal model.

In summary, the removal of A from the gas stream and creation of P is enhanced with an increase in the inlet concentration of reagent B in the incoming liquid, thereby improving tower operational efficiency.

The last column of Table 7.3 shows the behavior of the dimensionless product concentration at the critical location $c_{\mathrm{P}}^{\mathrm{crit}*}$ as a function of $c_{\mathrm{B,in}}^{\mathrm{bulk}}$. For the film model, the value of this concentration is equal to 1 (an explanation for which was offered in Section 7.3.1.1) and is independent of $c_{\mathrm{B,in}}^{\mathrm{bulk}}$. In contrast, in the case of the surface renewal model, this value never reaches 1 and depends upon the level of $c_{\mathrm{B,in}}^{\mathrm{bulk}}$; that is, it first decreases and then increases as the latter concentration goes up (see also Figure 7.10). However, the dimensional concentration of P at the critical location, $c_{\mathrm{P}}^{\mathrm{crit}}$, increases for both models with an increase in $c_{\mathrm{B,in}}^{\mathrm{bulk}}$. For the film model, $c_{\mathrm{P}}^{\mathrm{crit}} = pc_{\mathrm{B,in}}^{\mathrm{bulk}}c_{\mathrm{P}}^{\mathrm{crit}*} = c_{\mathrm{B,in}}^{\mathrm{bulk}}$ since p and $c_{\mathrm{P}}^{\mathrm{crit}*}$ are both equal to 1. Thus, for every mole of B consumed by chemical reaction (with one mole of dissolved A)

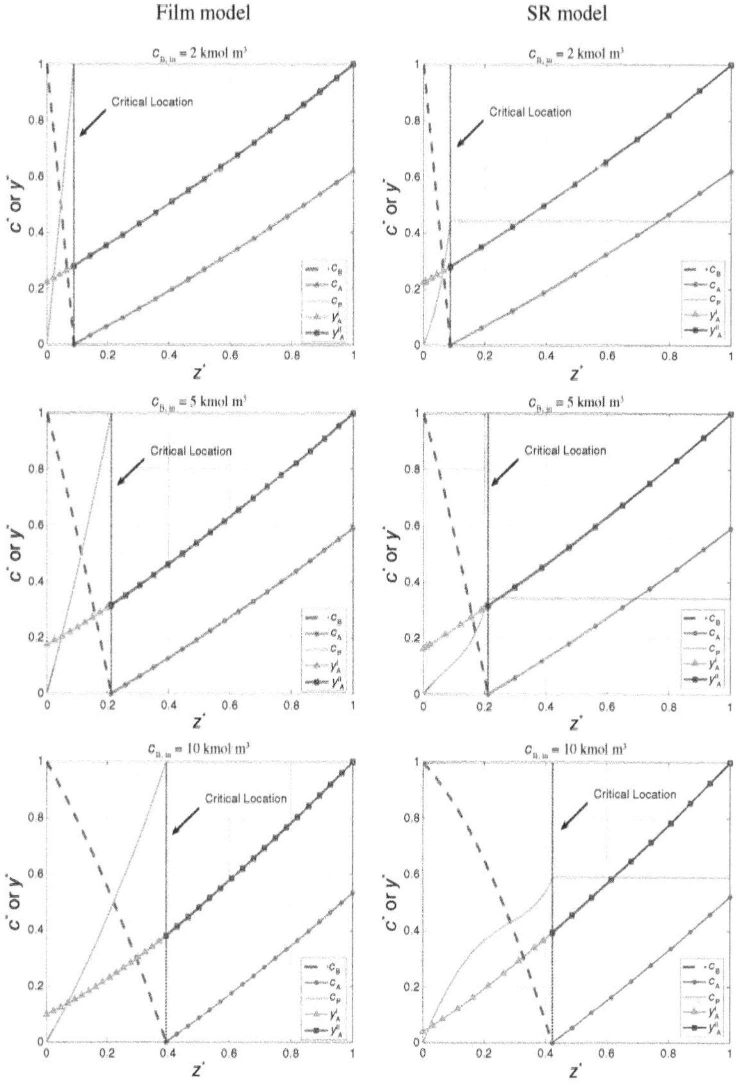

Figure 7.10 Dimensionless bulk concentration profiles in the gas and liquid phases along the dimensionless length of the packed tower as predicted by the film (left) and surface renewal (SR; right) models for $k_G = 1.007 \times 10^{-3}$ m s^{-1} and $c^{bulk}_{B,in} = 2$, 5, and 10 kgmol m^{-3}. In each plot, the region of instantaneous chemical reaction (Region I) is to the left of the critical location, whereas that on its right is that of pure physical absorption (Region II). Parameter values used in the calculations are shown in Table 7.1. Equal diffusion coefficients (= 1.83×10^{-9} m^2 s^{-1}; only for the surface renewal model) and stoichiometric numbers of unity for all species are assumed. The superscript "*" has been dropped in the legends.

Table 7.3 Comparison of the dimensionless critical location (z_c^*), bulk gas-phase dimensionless concentrations of A at the top of the tower (y_A^{top*}) and at the critical location (y_A^{crit*}), and dimensionless and dimensional concentration of P at the critical location (c_P^{crit*} and c_P^{crit}) as predicted by the film and surface renewal (SR) models as a function of the concentration of B in the incoming liquid $\left(c_{B,in}^{bulk}\right)$.

$c_{B,in}^{bulk}$	z_c^*		y_A^{top*}		y_A^{crit*}		c_P^{crit*} $\left(c_P^{crit}\right)$[c]	
(kgmol m^{-3})	Film	SR	Film	SR	Film	SR	Film	SR
1[a]	0.045	0.044	0.241	0.241	0.269	0.269	1 (1)	0.554 (0.554)
2[b]	0.089	0.087	0.223	0.222	0.280	0.279	1 (2)	0.442 (0.884)
3	0.131	0.129	0.207	0.203	0.291	0.290	1 (3)	0.379 (1.137)
4	0.173	0.170	0.190	0.183	0.303	0.302	1 (4)	0.347 (1.388)
5[b]	0.213	0.211	0.174	0.161	0.315	0.315	1 (5)	0.341 (1.705)
6	0.252	0.253	0.158	0.139	0.328	0.328	1 (6)	0.356 (2.136)
7	0.290	0.295	0.143	0.116	0.341	0.343	1 (7)	0.390 (2.730)
8	0.326	0.338	0.128	0.091	0.354	0.358	1 (8)	0.443 (3.544)
9	0.360	0.381	0.113	0.065	0.367	0.376	1 (9)	0.510 (4.590)
10[b]	0.394	0.422	0.099	0.039	0.381	0.394	1 (10)	0.590 (5.900)

Values of $k_G = 1.007 \times 10^{-3}$ m s^{-1} equal diffusion coefficients (= 1.83×10^{-9} m^2 s^{-1}; only for the surface renewal model), and stoichiometric coefficients of unity for all species were used in the calculations. All other parameter values are given in Table 7.1.

a) Gas- and liquid-phase concentration profiles in the tower are shown in Figure 7.9 (middle plot).
b) Gas- and liquid-phase concentration profiles in the tower are shown in Figure 7.10.
c) The units of the numbers within parentheses in the last column of the table are kgmol m^{-3}.

in the instantaneous reaction section of the tower (Region I), exactly 1 mol of P is produced, that is, the intrinsic stoichiometry of the reaction is preserved. In striking contrast, the extrinsic stoichiometry of the reaction is different from the intrinsic one for the surface renewal model, that is, only 0.435 mol of P are produced on average for every mole of B consumed in Region I of the tower, as can be deduced from Table 7.3. This counterintuitive result is due to the radically different mass-transfer mechanisms postulated in the surface renewal model, that is, diffusion and convection, compared to the transport by pure diffusion assumed in the film model. An analogy is the deleterious effect of diffusional resistances on the intrinsic rate of a gas–solid reaction occurring in a porous catalyst, which is captured in the concept of the effectiveness factor.

Finally, Table 7.3 reveals that the film and surface renewal models are in good agreement as regards the prediction of the critical location and gas-phase concentration of A at that location. However, the agreement between them as regards the concentration of A in the gas leaving the top of the tower deteriorates as the inlet concentration of B increases.

Notation

a	gas–liquid interfacial area of contact ($m^2\ m^{-3}$)
a_1	N_G/H
a_2	N_G/H
A	solute gas
A_c	area of cross-section of the packed tower (m^2)
b	stoichiometric coefficient of B
B	liquid-phase reagent
b_1	$D^*F^*N_G$
b_2	$-N_G/H$
c	generic species concentration in the liquid phase (kgmol m^{-3})
c^i	generic species concentration at the gas–liquid interface in the liquid phase (kgmol m^{-3})
c_1	$-\frac{N_L}{F^*H}$
c_2	N_L
c_A^*	dimensionless concentration of A in the bulk liquid at a specific location z^* in the tower, $\frac{Hc_A}{y_{A,in}}$
c_A^{bulk} or c_A	concentration of A in the bulk liquid at a specific location z in the tower (kgmol m^{-3})
c_A^i	concentration of dissolved A at the gas–liquid interface at a specific location z in the tower (kgmol m^{-3})
c_B^*	dimensionless concentration of B in the bulk liquid at a specific location z^* in the tower, $\frac{c_B}{c_{B,in}}$
c_B^{bulk} or c_B	concentration of B in the bulk liquid at a specific location z in the tower (kgmol m^{-3})
$c_{B,in}^{bulk}$ or $c_{B,in}$	concentration of B in the incoming liquid at the top of the tower (kgmol m^{-3})
c_P^*	dimensionless concentration of P in the bulk liquid at a specific location z^* in the tower, $\frac{c_P}{pc_{B,in}}$
c_P^{bulk} or c_P	concentration of P in the bulk liquid at a specific location z in the tower (kgmol m^{-3})
c_P^{crit}	concentration of P in the bulk liquid at the critical location z_c in the tower (kgmol m^{-3})

$c_{\mathrm{P}}^{\mathrm{crit*}}$	dimensionless concentration of P in the bulk liquid at the critical location z_c^* in the tower, $\frac{c_{\mathrm{P}}^{\mathrm{crit}}}{pc_{\mathrm{B,in}}^{\mathrm{bulk}}}$
d_1	$-D^*N_{\mathrm{L}}$
d_2	N_{L}
D	binary diffusion coefficient of generic species in the liquid solvent ($=D_{\mathrm{A}}=D_{\mathrm{B}}=D_{\mathrm{P}}$; $\mathrm{m^2\,s^{-1}}$)
D^*	ratio of diffusion coefficients defined by eq. 7.30
D_{A}	binary diffusion coefficient of dissolved A in the liquid solvent ($\mathrm{m^2\,s^{-1}}$)
D_{B}	binary diffusion coefficient of reagent B in the liquid solvent ($\mathrm{m^2\,s^{-1}}$)
D_{P}	binary diffusion coefficient of product P in the liquid solvent ($\mathrm{m^2\,s^{-1}}$)
f_{D}	steady-state Danckwerts age-distribution function given by eq. 7.68
F^*	dimensionless quantity defined by eq. 7.31
G	volumetric flow rate of gas in the tower ($\mathrm{m^3\,s^{-1}}$)
h	constant defined by eq. 7.84 ($\mathrm{m^{-1}}$)
H	Henry's constant (dimensionless)
H_{G}	height of gas-phase mass-transfer unit defined by eq. 7.34 (m)
H_{L}	height of liquid-phase mass-transfer unit defined by eq. 7.33 (m)
i	index
I_1, I_2 and I_3	transcendental integrals defined by eqs. 7.93, 7.97 and 7.99, respectively (s, $\mathrm{s^{0.5}}$ and $\mathrm{s^{1.5}}$, respectively)
k	slope of the plot of $\delta(\theta)$ versus $\sqrt{\theta}$ ($\mathrm{m\,s^{-0.5}}$)
k_{OL}	overall mass-transfer coefficient ($\mathrm{m\,s^{-1}}$)
k_{G}	gas-phase mass-transfer coefficient ($\mathrm{m\,s^{-1}}$)
k_{L}	liquid-phase mass-transfer coefficient ($\mathrm{m\,s^{-1}}$)
L	volumetric flow rate of liquid in the tower ($\mathrm{m^3\,s^{-1}}$)
N_{G}	number of gas-phase mass-transfer units, $\frac{Z}{H_{\mathrm{G}}}$
N_{L}	number of liquid-phase mass-transfer units, $\frac{Z}{H_{\mathrm{L}}}$
p	stoichiometric coefficient of product P
P	product
Q^{A}	amount of A present in a surface element of age τ ($\mathrm{kgmol\,m^{-2}}$)
Q^{B}	amount of B consumed in a surface element of age τ ($\mathrm{kgmol\,m^{-2}}$)
R_{A}	rate of absorption of solute A at the gas–liquid interface in the film model at a specific location z in the tower ($\mathrm{kgmol\,m^{-2}\,s^{-1}}$)
$R_{\mathrm{abs}}^{\mathrm{A,ss}}$	steady-state age-averaged absorption rate of A at a specific location z in the tower ($\mathrm{kgmol\,m^{-2}\,s^{-1}}$)

$R_{\text{inst}}^{\text{A}}$	instantaneous rate of absorption of A in a surface element at a specific location z in the tower (kgmol m^{-2} s^{-1})
$R_{\text{trans}}^{\text{A,ss}}$	steady-state age-averaged rate of transfer of dissolved A from the gas–liquid interface to the bulk liquid due to surface renewal at a specific location z in the tower (kgmol m^{-2} s^{-1})
$R_{\text{consump.inst}}^{\text{B}}$	instantaneous consumption rate of B in a surface element at a specific location z in the tower (kgmol m^{-2} s^{-1})
$R_{\text{consump.ave}}^{\text{B,ss}}$	steady-state age-averaged consumption rate of B at a specific location z in the tower (kgmol m^{-2} s^{-1})
$R_{\text{trans}}^{\text{P,ss}}$	steady-state age-averaged rate of transfer of product P due to surface renewal from the gas–liquid interface to the bulk liquid at a specific location z in the tower (kgmol m^{-2} s^{-1})
$R_{\text{A}}^{\text{phy}}$	rate of physical absorption of A in the film model in Region II (kgmol m^{-2} s^{-1})
R_{B}	rate of consumption of reagent B (equal to $-bR_A$) in the film model at a specific location z in the tower (kgmol m^{-2} s^{-1})
R_{P}	rate of production of P (equal to pR_A) in the film model at a specific location z in the tower (kgmol m^{-2} s^{-1})
S	frequency or rate of surface renewal (Hz or s^{-1})
t	process (i.e., "clock") time elapsed since the beginning of absorption (s)
x	distance into a surface element measured from the gas–liquid interface (m)
X_1, X_2, X_3, X_4	arbitrary constants
$y_{\text{A}}^{\text{bulk}}$ or y_{A}	concentration of solute A in the bulk gas at a specific location z in the tower (kgmol m^{-3})
$y_{\text{A,in}}$	concentration of solute A in the bulk gas entering at the bottom of the tower (kgmol m^{-3})
$y_{\text{A}}^{\text{crit.}}, y_{\text{A}}^{\text{top}}$	bulk gas-phase concentrations of A at the critical location and at the top of the tower, respectively (kgmol m^{-3})
$y_{\text{A,I}}^{\text{bulk}}, y_{\text{A,II}}^{\text{bulk}}$	bulk gas-phase concentrations of A in Region I and Region II of the tower, respectively (kgmol m^{-3})
$y_{\text{A,in}}^{\text{bulk}}$ or $y_{\text{A,in}}$	concentration of A in the gas fed at the tower bottom (kgmol m^{-3})
y_{A}^{i}	concentration of A at the gas–liquid interface on the gas side at a specific location z in the tower (kgmol m^{-3})
y_{A}^{*}	dimensionless concentration of solute A in the bulk gas at a specific location z in the tower, $\frac{y_{\text{A}}}{y_{\text{A,in}}}$
$y_{\text{A}}^{\text{crit*}}$	dimensionless concentration of solute A in the bulk gas at the critical location, $\frac{y_{\text{A}}^{\text{crit}}}{y_{\text{A,in}}}$

y_A^{top*}	dimensionless concentration of solute A in the bulk gas at the top of the tower, $\frac{y_A^{top}}{y_{A,in}}$
$y^*_{A,I}, y^*_{A,II}$	dimensionless bulk gas-phase concentrations of A in Region I and Region II of the tower, respectively
z	location in the packed tower measured from its top (m)
Z	height of packed section (m)
z^*	dimensionless location in the tower measured from its top, $\frac{z}{Z}$
z_c	critical location in the tower measured from its top (m)
z_c^*	dimensionless critical location in the tower measured from its top, $\frac{z_c}{Z}$

Greek Letters

β	thickness of reaction zone in the film model (m) or a constant in the expression for the location of the reaction plane in the surface renewal model (m s$^{-0.5}$)
δ	thickness of liquid film in the film model or location of the reaction plane in the surface renewal model equal to $2\beta\sqrt{\theta}$ (m)
Δc	concentration driving force for physical absorption (kgmol m^{-3})
ε	very small tolerance
$\lambda_1, \lambda_2, \lambda_3, \lambda_4$	eigenvalues of the matrix of coefficients defined by eqs. 7.45 and 7.48
τ	age or residence time of a liquid element at the gas–liquid interface (s)
θ	exposure time of a liquid element at the gas–liquid interface (s)

Acronyms

DAEs	differential algebraic equations
ODEs	ordinary differential equations

Homework Exercises

i) Using appropriate boundary conditions for the system specified by eqs. 7.40–7.42 show that its solution is given by eqs. 7.43, 7.44, 7.46, 7.47, and 7.49.

ii) Develop an alternative method for determining the critical location and tower concentration profiles for the film model when there is no gas-phase mass-transfer resistance (see Section 7.2.1.1). In your method, start with an initial guess value for the critical location and refine it progressively. Can you reproduce the results for the film model shown in Figure 7.6 using your algorithm?

iii) Using appropriate initial and boundary conditions for the system specified by eqs. 7.56–7.58 show that its solution is given by eqs. 7.62–7.65.

iv) Check the validity of eqs. 7.70, 7.72, 7.75, and 7.76.

v) Develop equivalent versions of eqs. 7.25, 7.26, 7.27 for the surface renewal model with no gas-phase mass-transfer resistance, which is presented in Section 7.2.2.1. Comment on the similarities and differences between the film and surface renewal models.

vi) Describe how you would implement the algorithm for solving the DAE system discussed in Section 7.2.2.1. Draw a sketch to illustrate the algorithm.

vii) For $D_A = D_B = 1.83 \times 10^{-9}$ m^2 s^{-1}, $b = 1$, and the concentration ratio c_B^{bulk}/c_A^i varying from 0 to 0.5, calculate values of β from eq. 7.66 and plot them against each other. Comment on your result.

References

Astarita, G. (1967). *Mass transfer with chemical reaction*. Amsterdam: Elsevier.

Astarita, G. and Gioia, F. (1965). "Simultaneous absorption of hydrogen sulfide and carbon dioxide in aqueous hydroxide solutions," *Industrial & Engineering Chemistry Fundamentals*, 4(3), pp. 317–320. Available at: http://dx.doi.org/10.1021/i160015a014.

Brian, P.L.T., Hurley, J.F. and Hasseltine, E.H. (1961). "Penetration theory for gas absorption accompanied by a second order chemical reaction," *AIChE Journal*, 7(2), pp. 226–231. Available at: http://dx.doi.org/10.1002/aic.690070212.

Brogren, C. and Karlsson, H.T. (1997). "Modeling the absorption of SO_2 in a spray scrubber using the penetration theory," *Chemical Engineering Science*, 52(18), pp. 3085–3099. Available at: http://dx.doi.org/10.1016/S0009-2509(97)00126-7.

Carslaw, H.S. and Jaeger, J.C. (1959). *Conduction of heat in solids*. London: Oxford University Press.

Crank, J. (1964). *The mathematics of diffusion*. London: Oxford University Press.

Danckwerts, P.V. (1950). "Unsteady-state diffusion or heat-conduction with moving boundary," *Transactions of the Faraday Society*, 46, p. 701 Available at: http://dx.doi.org/10.1039/tf9504600701.

Danckwerts, P.V. (1970). *Gas-liquid reactions*. New York: McGraw-Hill.

DeCoursey, W.J. (1974). "Absorption with chemical reaction: development of a new relation for the Danckwerts model," *Chemical Engineering Science*, 29(9), pp. 1867–1872. Available at: http://dx.doi.org/10.1016/0009-2509(74)85003-7.

Ebrahimi, S., Picioreanu, C., Kleerebezem, R., Heijnen, J.J. and van Loosdrecht, M.C.M. (2003). "Rate-based modelling of SO_2 absorption into aqueous $NaHCO_3/Na_2CO_3$ solutions accompanied by the desorption of CO_2," *Chemical Engineering Science*, 58(16), pp. 3589–3600. Available at: http://dx.doi.org/10.1016/S0009-2509(03)00231-8.

Garg, R., Nair, S. and Bhaskarwar, A.N. (2000). "Mass transfer with instantaneous chemical reaction in finite gas–liquid systems," *Chemical Engineering Journal*, 76(2), pp. 89–98. Available at: http://dx.doi.org/10.1016/S1385-8947(99)00090-X.

Hikita, H., Asai, S., Himukashi, Y. and Ishikawa, T. (1968). "Absorption of hydrogen sulfide into aqueous ammonia and sodium hydroxide solutions," *Journal of Chemical Engineering (Japan)*, 32, pp. 278–285.

Hikita, H., Ishimi, K. and Koroyasu, S. (1985). "Gas absorption accompanied by an instantaneous irreversible reaction in laminar falling films of pseudoplastic liquids," *The*," *Canadian Journal of Chemical Engineering*, 63(3), pp. 522–524. Available at: http://dx.doi.org/10.1002/cjce.5450630323.

Huang, H. and Chatterjee, S.G. (2021). "Transient physical gas absorption in a stirred liquid using the surface renewal model of mass transfer," *Chemical Engineering Science*, 234, 116449 Available at: http://dx.doi.org/10.1016/j.ces.2021.116449.

van Krevelen, D.W. and Hoftijzer, P.J. (1948). "Kinetics of gas-liquid reactions part I. General theory," *Recueil des Travaux Chimiques des Pays-Bas*, 67(7), pp. 563–586.

Onda, K., Kobayashi, T., Fujine, M. and Takahashi, M. (1971). "Behavior of the reaction plane movement in gas absorption accompanied by instantaneous chemical reactions," *Chemical Engineering Science*, 26(12), pp. 2009–2026. Available at: http://dx.doi.org/10.1016/0009-2509(71)80039-8.

Onda, K., Takeuchi, H., Kobayashi, T. and Yokota, K. (1972). "Simultaneous absorption of hydrogen sulphide and carbon dioxide in aqueous sodium hydroxide solutions," *Journal of Chemical Engineering of Japan*, 5(1), pp. 27–33. Available at: http://dx.doi.org/10.1252/jcej.5.27.

Onda, K., Takeuchi, H. and Okumoto, Y. (1968). "Mass transfer coefficients between gas and liquid phases in packed columns," *Journal of Chemical Engineering of Japan*, 1(1), pp. 56–62. Available at: http://dx.doi.org/10.1252/jcej.1.56.

Peng, J., Yao, W. and Yang, Z. (2020). "Prediction of SO2 removal efficiency for ammonia-based wet flue gas desulfurization in a packed tower," *SN Applied Sciences*, 2(3), p. 360. Available at: http://dx.doi.org/10.1007/s42452-020-2147-z.

Perry, R.H., Green, D.W. and Maloney, J.O. (eds.) (1984). *Perry's chemical engineers' handbook*. Sixth edition. New York: McGraw-Hill.

Ramachandran, P.A. and Sharma, M.M. (1969). "Absorption with fast reaction in a slurry containing sparingly soluble fine particles," *Chemical Engineering Science*, 24(11), pp. 1681–1686. Available at: http://dx.doi.org/10.1016/0009-2509(69)87033-8.

Sada, E. and Ameno, T. (1973). "Gas absorption accompanied by an instantaneous irreversible chemical reaction in liquid of a finite depth," *Journal of Chemical Engineering of Japan*, 6(3), pp. 247–251. Available at: http://dx.doi.org/10.1252/jcej.6.247.

Secor, R.M. and Southworth, R.W. (1961). "Absorption with an infinitely rapid chemical reaction in packed towers," *AIChE Journal*, 7(4), pp. 705–707. Available at: http://dx.doi.org/10.1002/aic.690070438.

Sherwood, T.K. and Pigford, R.L. (1952). *Absorption and extraction*. Second edition. New York: McGraw-Hill.

Terraglio, F.P. and Manganelli, R.M. (1967). "The absorption of atmospheric sulfur dioxide by water solutions," *Journal of the Air Pollution Control Association*, 17(6), pp. 403–406. Available at: http://dx.doi.org/10.1080/00022470.1967.10468999.

8

Constant Pressure Crossflow Ultrafiltration (UF)*

8.1 Introduction

In crossflow membrane filtration like microfiltration (MF), ultrafiltration (UF), nanofiltration (NF), and reverse osmosis (RO), an incoming liquid feed containing suspended/dissolved material flows across the surface of a porous membrane and the permeate flow is that portion of the liquid, which passes through the membrane in a direction perpendicular to that of the main flow. The purpose of the membrane (polymeric or ceramic) is to remove suspended colloidal particles (MF), macromolecules (UF), sugars, divalent salts, and dissociated acids (NF), and monovalent salts and undissociated acids (RO) from the liquid (Cheryan, 1998). The chief distinction between these types of filtration is the pore size of the membrane used for the separation. UF retains particles and macromolecules in the size range of 10–200 Å, and the molecular weight cutoff (MWCO) of UF membranes varies from 1000 to 500,000 Da (Cheryan, 1998). The magnitude of the permeate flux is affected by the liquid velocity over the membrane surface, permeate viscosity, dissolved solids concentration, applied transmembrane pressure drop (TMP), temperature, membrane characteristics, and membrane fouling.

Crossflow membrane filtration can be operated in two different modes: The first, where the TMP is held constant and which is more common, and second, in which the permeate flux is maintained at a constant level by changing the TMP over time. Both types of operation suffer from fouling of the membrane due to pore blocking, concentration polarization (UF), and cake buildup on the membrane surface (MF), which causes the permeation flux to decline with the advance of process time in constant TMP operation.

For comprehensive reviews of some of the theoretical and experimental work performed on the phenomenon of concentration polarization in UF and RO, the

* Some of the material in this chapter has been adapted from Chatterjee (2010).

Applications of the Surface Renewal Model of Mass Transfer, First Edition.
Hengshuo Huang and Siddharth G. Chatterjee.

Companion website: www.wiley.com/go/huang/surfacerenewalmodel

reader is referred to the papers of Sablani *et al.* (2001) and Quezada *et al.* (2021). According to Trettin and Doshi (1980), the film model underestimates the value of the permeate flux compared to the penetration model. The surface renewal concept to describe crossflow UF under constant TMP conditions was first applied by Koltuniewicz and coworkers in the 1990s (Koltuniewicz, 1992; Koltuniewicz and Noworyta, 1994, 1995). Examples of other studies, which have used this concept to model crossflow constant TMP and constant flux UF and MF, are those of Constenla and Lozano (1996), Arnot *et al.* (2000), Chatterjee (2010), Sarkar *et al.* (2011), Hasan *et al.* (2013), Jiang and Chatterjee (2015), Zhang and Chatterjee (2015), Idan and Chatterjee (2015), and Dąbkowska-Susfał *et al.* (2024). As noted in Chapter 1, in contrast to the film and boundary layer models of membrane filtration, the surface renewal model has the capacity to describe the transfer of dissolved/suspended solids due to hydrodynamic impulses originating at the membrane surface owing to membrane roughness, the use of spacers and turbulence promoters, or the deliberate introduction of controlled centrifugal instabilities (called Dean vortices) into the main flow, which induce back migration of accumulated solute molecules or particulates away from the membrane surface, thereby significantly enhancing permeation rates (see Chapter 9).

The chief objective of this chapter is to present surface renewal and film models for constant-pressure crossflow UF. Using UF of cheese whey as an example, values of the steady-state or limiting permeate flux as a function of TMP calculated by the two models are compared against each other. The models are also compared against literature data on the UF of dextran and skim milk solutions.

8.2 Surface Renewal Model

A schematic of crossflow UF is shown in Figure 8.1. The surface renewal model, sketched in Figure 8.2, visualizes fresh liquid elements to be continuously brought from the bulk liquid to the membrane–solution interface as a result of flow instability or turbulence. A liquid element is assumed to remain at the interface for a certain amount of time τ, after which it returns and remixes with the bulk liquid. At any time t after the start of filtration, the surface of the membrane is visualized as being populated by a mosaic of liquid elements, which have ages that range from $\tau = 0$ to $\tau = t$. Let $J(\tau)$ and $f_D(\tau, t)$ denote the permeate flux in a surface element of age τ and age-distribution function of the surface elements, respectively. The age-averaged permeate flux (i.e., process flux) $J_a(t)$ at process time t is then given by

$$J_a(t) = \int_0^t J(\tau) f_D(\tau, t) d\tau \tag{8.1}$$

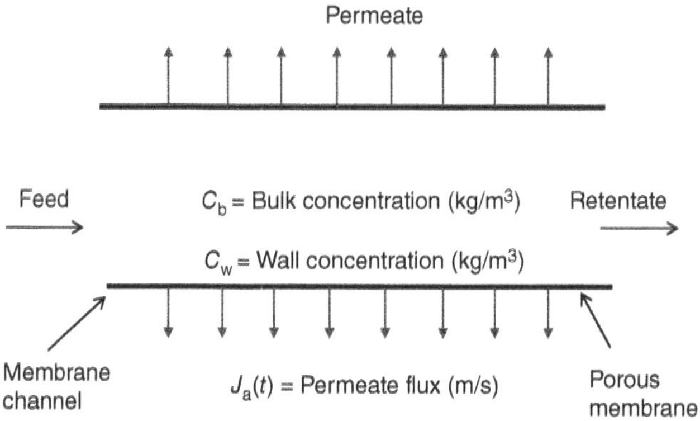

Figure 8.1 Schematic of crossflow ultrafiltration. Source: Chatterjee (2010) / Taylor & Francis Group.

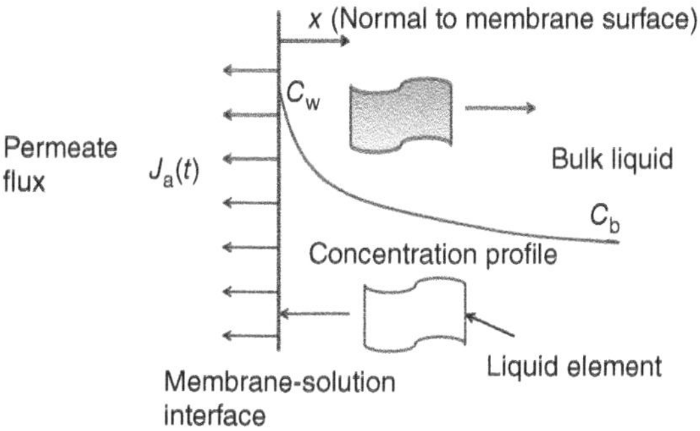

Figure 8.2 Surface renewal model of crossflow ultrafiltration. Source: Chatterjee (2010) / Taylor & Francis Group.

where $f_D(\tau, t)$ is assumed to be the transient Danckwerts age-distribution function, which is given by eq. 2.10 of Chapter 2, that is,

$$f_D(\tau, t) = \frac{Se^{-S\tau}}{1 - e^{-St}} \tag{8.2}$$

and S is the frequency of renewal of liquid elements at the membrane wall.

The surface renewal model developed by Koltuniewicz and coworkers (Koltuniewicz, 1992; Koltuniewicz and Noworyta, 1994, 1995) for crossflow UF is based on an empirical equation for $J(\tau)$ that predicts the decline of permeate flux with time in unstirred batch or dead-end UF. In contrast, in the surface renewal model proposed by Chatterjee (2010), $J(\tau)$ is determined from the unsteady-state diffusion equation. Both approaches are presented in the next two sections.

8.2.1 Empirical Equation

The surface renewal model of Koltuniewicz and coworkers is based upon the assumption that the unsteady-state permeate flux $J(\tau)$ in a liquid element with an age of τ at the membrane surface is the same as that occurring in UF in an unstirred batch cell, which is represented by the empirical equation

$$J(\tau) = (J_0 - J^*)e^{-A\tau} + J^* \tag{8.3}$$

where J_0 is assumed to be either the initial value of the flux during batch ultrafiltration (Koltuniewicz, 1992) or pure solvent (e.g., water) flux (Koltuniewicz and Noworyta, 1994), J^* is the permeate flux after infinite time in batch ultrafiltration, and A is a parameter that expresses the rate of flux decline. From eqs. 8.1 to 8.3, it follows that the permeate flux in crossflow UF under unsteady-state conditions is given by

$$J_a(t) = (J_0 - J^*)\frac{S}{S+A}\frac{1-e^{-(S+A)t}}{1-e^{-St}} + J^* \tag{8.4}$$

where $J_a(t)$ is the age-averaged permeate flux when the process time is t. The limiting or steady-state flux $J_{\lim}$ in crossflow UF is obtained by letting $t \to \infty$ in eq. 8.4, that is,

$$J_{\lim} = J_a(t \to \infty) = (J_0 - J^*)\frac{S}{S+A} + J^* \tag{8.5}$$

Figure 8.3 shows the qualitative behavior of eq. 8.4. Starting from an initial value of J_0, the permeate flux declines in an exponential fashion with the passage of time, approaching an asymptotic value $J_{\lim}$ at large times. Koltuniewicz (1992) and Koltuniewicz and Noworyta (1995) applied eq. 8.5 to crossflow filtration of skim milk, bovine serum albumin solutions, and oil–water emulsions in hollow-fiber and flat-plate membrane modules. Koltuniewicz and Noworyta (1994) used eq. 8.4 to successfully correlate data on permeate flux decline in the ultrafiltration of skim milk in a hollow-fiber module under dynamic conditions;

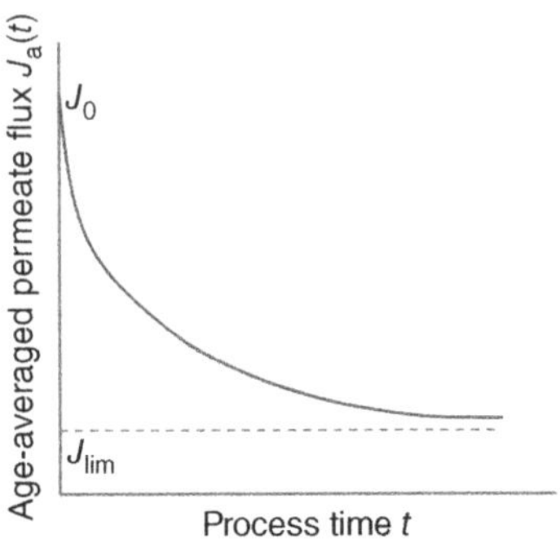

Figure 8.3 Qualitative behavior of the age-averaged permeate flux as a function of process time in crossflow UF according to eq. 8.4. Source: Chatterjee (2010) / Taylor & Francis Group.

Table 8.1 Parameters of eq. 8.4 reported by Koltuniewicz and Noworyta (1994) for skim milk UF at 303 K in a hollow-fiber module.

Expt. No.	Δp kPa	c_b g L^{-1}	u m s^{-1}	$J_0 \times 10^6$ m s^{-1}	$J^* \times 10^6$ m s^{-1}	$A \times 10^3$ s^{-1}	$S \times 10^3$ s^{-1}	$J_{lim} \times 10^6$ m s^{-1}
1	161	13.4	0.622	66.41	0.168	20.89	2.542	7.35
2	161	13.4	0.252	66.41	0.168	20.89	1.399	4.32
3	161	85	0.622	66.41	0.018	59.56	2.542	2.74
4	161	85	0.252	66.41	0.018	59.56	1.399	1.54
5	40	13.4	0.622	16.5	0.168	5.192	2.542	5.536
6	40	13.4	0.252	16.5	0.168	5.192	1.399	3.635
7	40	85	0.622	16.5	0.018	14.79	2.542	2.430
8	40	85	0.252	16.5	0.018	14.79	1.399	1.447

c_b and u denote the feed concentration and velocity, respectively, and Δp is the TMP.
Source: Chatterjee (2010) / Taylor & Francis Group.

values of the parameters obtained from their experimental runs are reported in Table 8.1.

The following are the main features of the surface renewal model of Koltuniewicz and Noworyta (1994, 1995): (1) There are four parameters in the flux equation (eq. 8.4) – J_0, J^*, A, and S; (2) the flux equation is of a semiempirical nature, with A and J^* being obtained from the flux decline observed in a batch or dead-end UF experiment under similar conditions as those in the crossflow run; and (3) $J_0 = J_0(\Delta p)$, $J^* = J^*(c_b)$, $A = A(\Delta p, c_b)$, and $S = S(u)$. Constenla and Lozano (1996) used the surface renewal model of Koltuniewicz (1992) to correlate the behavior of J_{lim} with Δp in UF of apple juice in a hollow-fiber module.

8.2.2 Diffusion Equation

The surface renewal model developed by Chatterjee (2010) assumes that above the liquid elements on the membrane surface, the liquid is well mixed, and the concentration of the dissolved solute is held constant at its bulk value c_b due to the high rate of turbulent mass transfer from this point to the bulk liquid. During the residence time τ of a liquid element at the membrane–solution interface, mass transport of solute in the direction x (measured normal to the membrane surface; see Figure 8.2) is assumed to occur within it via the mechanism of unsteady-state molecular diffusion in an infinitely deep medium, that is, the element possesses no characteristic size. The unsteady-state diffusion of solute in a specific liquid element whose exposure time at the membrane–solution interface is θ is described by the diffusion equation (Trettin and Doshi, 1980):

$$\frac{\partial c}{\partial \theta} + [-J(\theta)]\frac{\partial c}{\partial x} = D\frac{\partial^2 c}{\partial x^2} \tag{8.6}$$

whose initial and boundary conditions are given by

$$c = c_b \quad \text{at} \quad \theta = 0 \quad \text{for} \quad x \geq 0 \tag{8.7}$$

$$c = c_b \quad \text{for} \quad \theta > 0 \quad \text{at} \quad x = \infty \tag{8.8}$$

and

$$c = c_w \quad \text{for} \quad \theta > 0 \quad \text{at} \quad x = 0 \tag{8.9}$$

In the preceding equations, c is the solute concentration in the liquid element at location x and exposure time θ, which varies from 0 to τ, whereas c_b and c_w are the concentrations of the solute in the bulk liquid and at the membrane wall, respectively, both of which are taken to be constant. Thus, it is assumed that the concentration of the solute at the membrane wall, c_w,[1] is attained instantaneously and the diffusion coefficient D of the solute in the liquid is constant at some average value although the solute concentration varies from c_w to c_b, the difference between whose magnitudes can be very great as will be seen later. These assumptions, which are commonly invoked in the film and boundary layer models of membrane filtration (Rautenbach and Albrecht, 1989), were also made by Trettin and Doshi (1980), who provided exact and approximate expressions for the transient permeate flux and solute concentration profile for the penetration model represented by eqs. 8.6 to 8.9, which describes UF of macromolecular solutions in an unstirred batch cell. These equations are similar in form to those that describe nonisothermal gas absorption with volume change of the liquid phase (Chiang and Toor, 1964) and batch reverse osmosis or hyperfiltration (Nakano *et al.*, 1967).

Following Trettin and Doshi (1980), let us postulate that

$$J(\theta) = \frac{K}{2\sqrt{D\theta}} \tag{8.10}$$

where K is a constant. We also introduce a similarity variable η defined by

$$\eta = \frac{x}{2\sqrt{D\theta}} \tag{8.11}$$

It can be shown that the solution of eq. 8.6, subject to eqs. 8.7 to 8.9, is given by

$$c(x, \theta) = c_w - (c_w - c_b)\left[\frac{\text{erf}\,(\eta + \phi) - \text{erf}\,\phi}{1 - \text{erf}\,\phi}\right] \tag{8.12}$$

1 The maximum value that c_w can attain is c_{gel}, the gel concentration. Table 4.2 in Cheryan (1998) lists values of c_{gel} for different types of feed.

where

$$\phi = \frac{K}{2D} \tag{8.13}$$

The value of the constant ϕ (or K) can be determined by balancing the convective flux of solute to the membrane wall caused by the imposed TMP, which creates a concentration gradient of the solute at the wall, against the back diffusion of solute from the wall toward the bulk solution. This mass balance can be expressed as

$$J(\theta)c_{\mathrm{w}} = -D\left(\frac{\partial c}{\partial x}\right)_{x=0} \tag{8.14}$$

Equation 8.14 assumes that the membrane rejects the solute perfectly, since any practically useful membrane should approach this limiting behavior as closely as possible. The equation also contains a pseudo steady-state assumption that has been imposed on the permeate flux, that is, there is no accumulation of solute at the membrane wall (Trettin and Doshi, 1980).

Substituting eqs. 8.12 and 8.13 into eq. 8.14 yields the equation

$$\sqrt{\pi}\phi(1 - \mathrm{erf}\ \phi)\ \exp(\phi^2) = 1 - \frac{1}{F_{\mathrm{w}}} \tag{8.15}$$

where F_{w} is the concentration polarization factor defined by

$$F_{\mathrm{w}} = \frac{c_{\mathrm{w}}}{c_{\mathrm{b}}} \tag{8.16}$$

Equation 8.15, which is an implicit equation for ϕ, shows that it depends only on the factor, F_{w}. An explicit expression for ϕ can be obtained from the approximate solution of Trettin and Doshi (1980); this solution is given by

$$\phi = (F_{\mathrm{w}} - 1)\left[\frac{F_{\mathrm{w}} + \sqrt{F_{\mathrm{w}}{}^2 + 8F_{\mathrm{w}}}}{2F_{\mathrm{w}}\left(F_{\mathrm{w}} + \sqrt{F_{\mathrm{w}}{}^2 + 8F_{\mathrm{w}}} + 2\right)}\right]^{1/2} \tag{8.17}$$

Table 8.2 shows values of ϕ calculated by eqs. 8.15 and 8.17 for different values of $c_{\mathrm{b}}/c_{\mathrm{w}}$. It is observed that (1) there is excellent agreement between the exact and approximate solutions (Trettin and Doshi, 1980), and (2) when the bulk concentration c_{b} is less than 10% of the wall concentration c_{w}, ϕ increases rapidly, whereas it tends toward zero when these two concentrations approach one another.

Replacing θ with τ in eq. 8.10, substituting it into eq. 8.1 and using eq. 8.2 yield the following expression for the age-averaged permeate flux at process time t:

$$J_{\mathrm{a}}(t) = \frac{J_{\mathrm{lim}}}{1 - e^{-St}}\mathrm{erf}(\sqrt{St}) \tag{8.18}$$

where J_{lim} is the limiting or steady-state permeate flux given by

$$J_{\mathrm{lim}} = J_{\mathrm{a}}(t \to \infty) = 0.5K\sqrt{\frac{\pi S}{D}} = k_{\mathrm{f}}\phi\sqrt{\pi} \tag{8.19}$$

Table 8.2 Variation of ϕ with c_b/c_w.

c_b/c_w	ϕ (exact) (eq. 8.15)	ϕ (approximate) (eq. 8.17)
0.05	2.9372	2.9376
0.10	1.9305	1.9317
0.15	1.4639	1.4659
0.20	1.1752	1.1779
0.25	0.9719	0.9752
0.30	0.8174	0.8212
0.35	0.6944	0.6984
0.40	0.5928	0.5971
0.45	0.5069	0.5112
0.50	0.4328	0.4370
0.55	0.3678	0.3719
0.60	0.3101	0.3140
0.65	0.2584	0.2620
0.70	0.2116	0.2149
0.75	0.1690	0.1718
0.80	0.1299	0.1322
0.85	0.0938	0.0956
0.90	0.0603	0.0616
0.95	0.0291	0.0298
1.00	0.0000	0.0000

Source: Chatterjee (2010) / Taylor & Francis Group.

with k_f being the mass-transfer coefficient of the surface renewal model, which is defined by

$$k_f = \sqrt{DS} \tag{8.20}$$

Equation 8.18, which is theoretical and contains only one parameter, that is, S (assuming all physical properties are known or can be estimated), can be compared to the four-parameter semiempirical flux equation of Koltuniewicz and Noworyta (1994, 1995) given by eq. 8.4.

The age-averaged concentration profile of the solute can be obtained from

$$c_a(x,t) = \int_0^t c(x,\tau) f_D(\tau,t) d\tau \tag{8.21}$$

which upon substitution in eqs. 8.2 and 8.12 yields

$$c_a(x,t) = c_w - \frac{(c_w - c_b)}{1 - \operatorname{erf}\phi}\left[\int_0^t \operatorname{erf}\left(\frac{x}{2\sqrt{D\tau}} + \phi\right)\frac{Se^{-S\tau}}{1 - e^{-St}}d\tau - \operatorname{erf}\phi\right] \tag{8.22}$$

Letting $t \to \infty$ in eq. 8.22 gives the steady-state concentration profile, that is,

$$c_a(x, t \to \infty) = c_w - \frac{(c_w - c_b)}{1 - \operatorname{erf}\phi}\left[\int_0^\infty \operatorname{erf}\left(\frac{x}{2\sqrt{D\tau}} + \phi\right)Se^{-S\tau}d\tau - \operatorname{erf}\phi\right] \tag{8.23}$$

which can be transformed into a more convenient form by the following mathematical procedure. Let

$$g(x,\tau) = \operatorname{erf}\left(\frac{x}{2\sqrt{D\tau}} + \phi\right) = \frac{2}{\sqrt{\pi}}\int_0^{\frac{x}{2\sqrt{D\tau}}+\phi} e^{-\xi^2}d\xi \tag{8.24}$$

which upon differentiating with respect to τ gives

$$\frac{\partial g(x,\tau)}{\partial \tau} = -\frac{x}{2\tau\sqrt{\pi D\tau}}\exp\left[-\left(\frac{x}{2\sqrt{D\tau}} + \phi\right)^2\right] \tag{8.25}$$

By taking the Laplace transform of eq. 8.25 and using eqs. 8.23 and 8.24, it can be shown that

$$c_a(x, t \to \infty) = c_w - \frac{(c_w - c_b)}{1 - \operatorname{erf}\phi}\left[1 - \frac{x}{2\sqrt{\pi D}}\int_0^\infty \exp\left[-\left(\frac{x}{2\sqrt{D\tau}} + \phi\right)^2\right]\frac{e^{-S\tau}}{\tau\sqrt{\tau}}d\tau - \operatorname{erf}\phi\right] \tag{8.26}$$

From eqs. 8.15 or 8.17 (see also Table 8.2), it is seen that as the ratio $c_b/c_w \to 1$, $\phi \to 0$, eq. 8.26 simplifies to

$$c_a(x, t \to \infty, c_b \to c_w) = c_w - (c_w - c_b)(1 - e^{-x\sqrt{S/D}}) \tag{8.27}$$

In the derivation of eq. 8.27, the following relation has been used (Abramowitz and Stegun, 1965):

$$\frac{x}{2\sqrt{\pi D}}\int_0^\infty \exp\left[-\left(\frac{x}{2\sqrt{D\tau}}\right)^2\right]\frac{e^{-S\tau}}{\tau\sqrt{\tau}}d\tau = e^{-x\sqrt{S/D}} \tag{8.28}$$

8.3 Film Model

The limiting permeate flux predicted by the film model, assuming complete solute rejection by the membrane, is given by (McCabe *et al.*, 1993)

$$J_{\text{lim}} = k_{\text{f}} \ln F_{\text{w}} \tag{8.29}$$

where

$$k_{\text{f}} = \frac{D}{\delta} \tag{8.30}$$

with δ being the film thickness.

If the value of the mass-transfer coefficient k_{f} is known by comparing eqs. 8.19 and 8.29, it is seen that the permeate flux of the surface renewal model varies from 1.74 to 1 times that of the film model as F_{w} decreases from 20 to 1. Trettin and Doshi (1980) also found that the film model consistently underpredicted the penetration model, with the agreement between them getting progressively better as F_{w} approached 1.

The solute concentration profile $c(x)$ for a perfectly rejecting membrane is

$$c(x) = c_{\text{w}} e^{-(J_{\text{lim}}/D)x} \tag{8.31}$$

which can be compared against eq. 8.27 of the surface renewal model. In eq. 8.31, x varies from zero to δ, where the solute concentration is c_{b}.

Zydney (1997) has provided a theoretical justification for the film model.

8.4 Mass-Transfer Coefficient

The mass-transfer coefficient k_{f} in the surface renewal and film models can be estimated from a suitable correlation that typically has the form (Cheryan, 1998)

$$\text{Sh} = f(\text{Re}, \text{Sc}) \tag{8.32}$$

where Sh, Re, and Sc are the Sherwood, Reynolds, and Schmidt numbers, respectively. These are defined by

$$\text{Sh} = \frac{k_{\text{f}} d_{\text{h}}}{D} \tag{8.33}$$

$$\text{Re} = \frac{d_{\text{h}} u \rho_{\text{f}}}{\mu_{\text{f}}} \tag{8.34}$$

and

$$\text{Sc} = \frac{\mu_{\text{f}}}{\rho_{\text{f}} D} \tag{8.35}$$

Here, d_{h} is the hydraulic diameter of the membrane channel, u is the liquid velocity in the channel, μ_{f} is the viscosity of the feed, and ρ_{f} is its density.

8.5 Relation Between Limiting Flux and Transmembrane Pressure (TMP)

The relation between the limiting flux and TMP can be obtained from the osmotic pressure model of membrane filtration, that is,

$$J_{\text{lim}} = \frac{\Delta p - \Pi(c_{\text{w}})}{\mu(R_{\text{m}} + R_{\text{gel}})} \tag{8.36}$$

where Π is the osmotic pressure, R_{m} and R_{gel} are the hydraulic resistances of the membrane and gel layer that may form on the surface of the membrane when the permeate flux is high, respectively, and μ is the viscosity of the permeate. Equation 8.36 assumes that there is perfect rejection of the solute by the membrane.

For a clean membrane (i.e., $R_{\text{gel}} = 0$), eq. 8.36 reduces to

$$J_{\text{lim}} = Q_{\text{m}}[\Delta p - \Pi(c_{\text{w}})] \tag{8.37}$$

where Q_{m} is the solvent permeability of the clean membrane given by

$$Q_{\text{m}} = \frac{1}{\mu R_{\text{m}}} \tag{8.38}$$

8.6 Application of the Surface Renewal and Film Models

The surface renewal and film models will be illustrated by applying them to UF of cheese whey (McCabe *et al.*, 1993) and experimental data on UF of Dextran solutions (Wijmans *et al.*, 1985) and skim milk (Koltuniewicz and Noworyta, 1994). It will be assumed in the following that all bulk solutions have the same density and viscosity as water.

8.6.1 UF of Cheese Whey

This example on crossflow UF of cheese whey, whose specifications are shown in Table 8.3, is adapted from McCabe *et al.* (1993).

The following calculation procedure is adopted for determining the relationship between the steady-state permeate flux (J_{lim}) and transmembrane pressure drop (Δp) for the two models:

1) Calculate the Reynolds number Re and Schmidt number Sc using eqs. 8.34 and 8.35, respectively, and then evaluate the mass-transfer coefficient k_{f} from eqs. 8.32 and 8.33 – see Table 8.3 for an example of a Sherwood number correlation. For the surface renewal model, the frequency of surface renewal S can be established from eq. 8.20.

Table 8.3 UF of cheese whey (McCabe *et al.*, 1993).

Parameter	Description or value
membrane type	tubular with 100% rejection
diameter of membrane channel (d_h)	2 cm
water permeability of clean membrane (Q_m)	250 LMH atm^{-1}
average diffusivity of whey proteins (D)	4×10^{-7} cm^2 s^{-1}
osmotic pressure (Π)	$4.4\times10^{-3}\,c - 1.7\times10^{-6}\,c^2 + 7.9\times10^{-8}\,c^3$, where c is the protein concentration in g L^{-1} and Π is in atmospheres
gel concentration of whey proteins (c_{gel})	400 g L^{-1}
concentration of whey proteins in feed (c_b)	10, 20, and 40 g L^{-1}
feed velocity (u)	1.5 m s^{-1}
operating temperature (assumed)	20 °C
feed density (ρ_f)	1 g cm^{-3}
feed viscosity (μ_f)	1 cp
permeate viscosity (μ)	1 cp
Sherwood number (Sh)	$0.0096\ \mathrm{Re}^{0.913}\ \mathrm{Sc}^{0.346}$

LMH = L m^{-2} h^{-1}.

2) For a specified value of the feed concentration c_b, assume a value for the limiting permeate flux J_{lim} and calculate the solute concentration at the membrane wall c_w from eqs. 8.19, 8.15, and 8.16 (surface renewal model) or eqs. 8.29 and 8.16 (film model).
3) Calculate the osmotic pressure at the membrane wall from an appropriate correlation for $\Pi(c_w)$; Table 8.3 provides an example of such a correlation for skim milk, which is also assumed to hold for cheese whey.
4) Using the known value of the pure solvent permeability of the membrane Q_m (or evaluating it from eq. 8.38), calculate the required transmembrane pressure drop Δp by eq. 8.37 (assuming a clean membrane).
5) Calculate the maximum value of the concentration polarization factor, denoted by F_w^{max}, by setting $c_w = c_{gel}$ (gel concentration value) in eq. 8.16.
6) Use the value of F_w^{max} from step 5 along with eqs. 8.17 and 8.19 (surface renewal model) or eq. 8.29 (film model) to determine the maximum possible value of the permeate flux J_{lim}^{max} when gel polarization sets in.
7) Calculate the transmembrane pressure drop Δp_{max} (TMP at the onset of gel polarization) from eq. 8.37 using the values of J_{lim}^{max} (from step 6), Q_m, and $\Pi(c_{gel})$.
8) Assume another value of $J_{lim} < J_{lim}^{max}$ and repeat steps 2–4 to determine J_{lim} as a function of Δp. The permeate flux will remain constant at J_{lim}^{max} for all values of Δp greater than Δp_{max}.

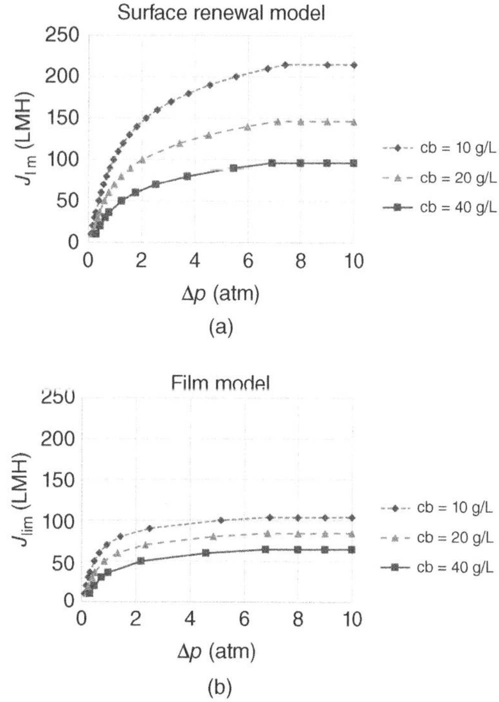

Figure 8.4 Variation of the steady-state permeate flux (J_{lim}) with transmembrane pressure drop (Δp) as predicted by the (a) surface renewal and (b) film models for cheese whey UF. Parameters used in the calculations are provided in Table 8.3. LMH = L m^{-2} h^{-1}.

Figure 8.4 shows the behavior of the limiting or steady-state permeate flux (J_{lim}) as a function of TMP (Δp) predicted by the surface renewal and film models for UF of cheese whey at three different levels of feed concentration (10, 20, and 40 g L^{-1}) using the input parameters of Table 8.3. The high value of the Reynolds number indicates turbulent flow in the membrane channel. Both models exhibit the same qualitative behavior of a rising permeate flux with an increase in TMP, which stabilizes as gel polarization is approached. Also, at the same TMP, both models predict a flux decline with an increase in the bulk concentration of the solute, which is confirmed by experimental data in the literature to be presented later. Theoretically, as regards the surface renewal model, this reduction in flux is due to a decrease in the magnitude of ϕ as the bulk and wall concentrations of the solute approach one another (see Table 8.2 and eq. 8.19). In the case of the film model, this flux decrease is clearly evident from eq. 8.29 as $c_w \rightarrow c_b$. However, quantitatively, at a given level of TMP, the theoretical permeate flux of the surface renewal model is much greater than the corresponding one predicted by the film model. This can be observed more clearly in Table 8.4, which shows that the ratio of the maximum permeate flux J_{lim}^{max} of the surface renewal model to that of the film model varies from about 2 to 1.5 as the bulk concentration of the solute increases fourfold. However, the effect of this great concentration change on the ratio of the

Table 8.4 Calculated parameters of the film and surface renewal models for cheese whey UF. Re = 30,000, Sc = 25,000, Sh = 3904.54, k_f = 28.11 LMH, and S = 1.52 Hz.

	Concentration of whey proteins in feed (c_b)					
	10 g L^{-1}		20 g L^{-1}		40 g L^{-1}	
Parameter	**Film**	**SR**	**Film**	**SR**	**Film**	**SR**
F_w^{max}	40	40	20	20	10	10
J_{lim}^{max} (LMH)	103.70	214.72	84.22	146.38	64.73	96.25
Δp_{max} (atm)	6.96	7.40	6.88	7.13	6.80	6.93

SR stands for surface renewal. LMH = L m^{-2} h^{-1}.

values of Δp_{max} predicted by the two models is much smaller., that is, it decreases slightly from 1.06 to 1.02.

8.6.2 UF of Dextran Solutions

Wijmans *et al.* (1985) performed UF experiments with solutions of Dextran T70 in ultrafiltered and demineralized water at 25 °C using a thin-channel cell at different feed velocities, feed concentrations, and TMP levels. A summary of their experimental parameters is provided in Table 8.5.

The predictions of the film and surface renewal models and their comparison with the experimental data of Wijmans *et al.* (1985) for UF of Dextran T70 solutions are shown in Table 8.6 and Figure 8.5 from which the following observations can be derived:

1) The models and data both show that permeate flux is enhanced by a higher liquid velocity over the membrane surface and a lower feed concentration.
2) The liquid flows in the filtration channel are in the turbulent flow regime.
3) At a specified feed concentration, the agreement between the film and surface renewal models improves as the feed velocity rises, which reduces concentration polarization.
4) At a specified TMP, in general, both models overestimate the experimental flux significantly, but the film model predictions are closer to the data.
5) At a specified flux, both models significantly underpredict the experimental TMP (except for Runs 1b and 1c in the case of the film model), the discrepancy being less severe for the film model. Taking all the experimental runs together, the pressure correction factor by which the theoretical TMP should be multiplied to bring it in line with the experimental TMP is on average 1.44 for the film model and 2.09 for the surface renewal model. Considering both models

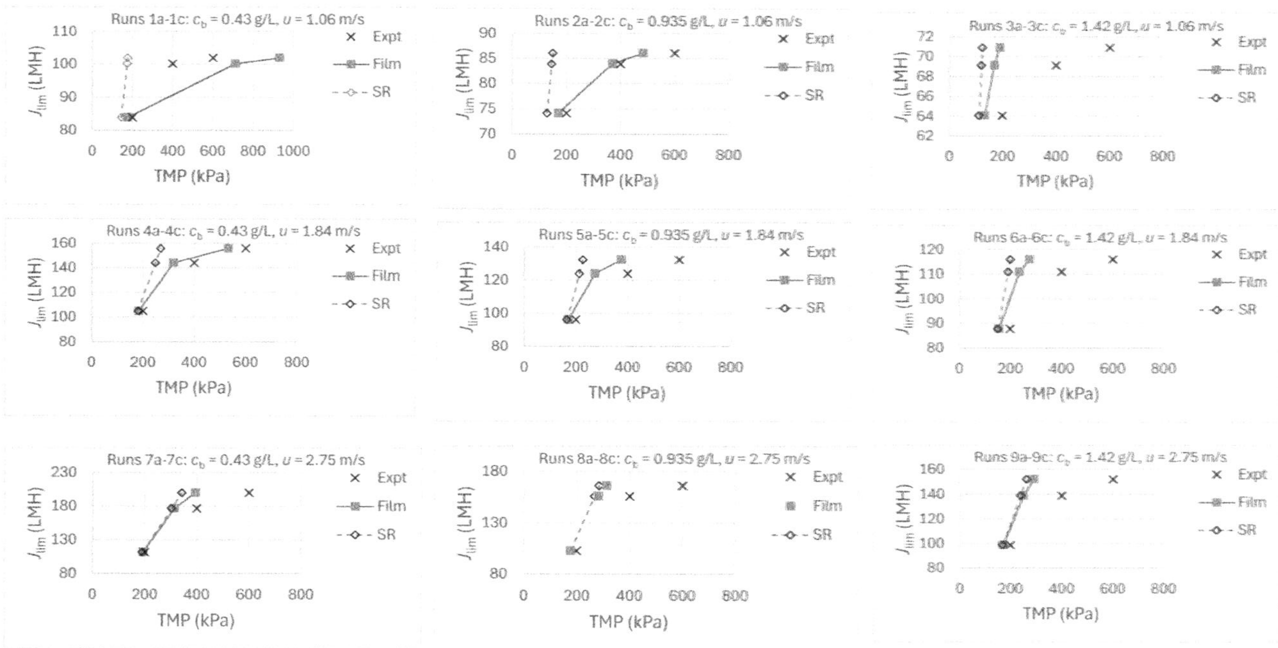

Figure 8.5 Comparison of the surface renewal and film models with experimental data on the variation of the limiting permeate flux (J_{lim}) with TMP (Δp) in UF of Dextran solutions. Parameters used in the calculations are provided in Table 8.5. LMH = L m^{-2} h^{-1}. Source: Adapted from Wijmans *et al.* (1985).

Table 8.5 Experimental parameters in UF of Dextran T70 solutions (Wijmans *et al.*, 1985).

Parameter	Description or value
membrane type	polysulfone with almost 100% rejection
dimensions of membrane channel	60 mm (width) × 5.9 mm (height) × 100 mm (length)
length of entrance region	250 mm
hydraulic diameter of membrane channel (d_h)	10.74 mm
membrane resistance (R_m)	6.94×10^{12} m^{-1}
diffusivity of Dextran in bulk solution (D)	4×10^{-7} $cm^2\ s^{-1}$
osmotic pressure (Π)	$0.375\ c + 7.52\ c^2 + 76.4\ c^3$, where c is the Dextran concentration in $g\ L^{-1}$ and Π is in bar
gel concentration of Dextran (c_{gel})	700 $g\ L^{-1}$
concentration of Dextran in feed (c_b)	0.43, 0.935, and 1.42 $g\ L^{-1}$
feed velocity (u)	1.06, 1.84, and 2.75 $m\ s^{-1}$
TMP (Δp)	200, 400, and 600 kPa
operating temperature	25 °C
feed density (ρ_f)	0.997 $g\ cm^{-3}$
feed viscosity (μ_f)	0.889 cp
permeate viscosity (μ)	0.889 cp
Sherwood number (Sh)	$0.023\ Re^{0.875}\ Sc^{0.25}$

The values of the feed density (ρ_f), feed viscosity (μ_f), and permeate viscosity (μ) were taken to be the same as those of water at the operating temperature.

individually, the correction factor is almost independent of feed velocity in the case of the film model but decreases by 45% for the surface renewal model, as the feed velocity increases by 159%.

6) The polarization factor F_w of the surface renewal model is several times smaller than that of the film model, the latter being much closer to that of the boundary layer (BL) resistance model of Wijmans *et al.* (1985).[2] All three models predict the wall concentration to be lower than the gel concentration (about 700 $g\ L^{-1}$), indicating the absence of gel polarization (Wijmans *et al.*, 1985).
7) The last column of Table 8.6 shows that the flux predicted by the BL resistance model is in excellent agreement with the experimentally measured one.

2 According to Cheryan (1998), the maximum value of the polarization factor F_w^{max} often lies in the range of 5 to 20.

Table 8.6 Comparison of theoretical predictions and experimental data of UF of Dextran T70 solutions.

Run	c_b ($g\,L^{-1}$)	u ($m\,s^{-1}$)	J_{lim}(LMH) (expt)	F_w (film)	F_w (BL)	F_w (SR)	Δp (kPa) (film)	Δp (kPa) (SR)	Δp (kPa) (expt)	J_{lim}^{calc} (LMH) (BL)
1a	0.43	1.06	83.88	294	370	23	176	144	200	83.52
1b	0.43	1.06	100.08	883	647	32	712	172	400	97.56
1c	0.43	1.06	101.88	997	809	33	931	175	600	102.24
2a	0.935	1.06	74.16	152	189	19	170	128	200	74.16
2b	0.935	1.06	83.88	294	308	23	370	145	400	84.6
2c	0.935	1.06	86.04	341	381	24	483	149	600	86.76
3a	1.42	1.06	64.08	77	137	15	133	111	200	63.72
3b	1.42	1.06	69.12	108	211	17	170	120	400	69.48
3c	1.42	1.06	70.92	122	256	17	191	123	600	71.64
k_f (LMH) = 14.75, Re = 12 771, Δp (expt)/Δp (film) = 1.43 (avg), Δp (expt)/Δp (SR) = 2.83 (avg)										
4a	0.43	1.84	105.12	81	228	15	183	180	200	104.4
4b	0.43	1.84	144	413	547	26	320	247	400	144
4c	0.43	1.84	155.88	679	737	30	532	268	600	156.96
5a	0.935	1.84	96.12	56	137	13	170	165	200	95.4
5b	0.935	1.84	123.84	178	273	20	274	213	400	124.56
5c	0.935	1.84	132.12	251	354	22	376	228	600	132.84
6a	1.42	1.84	87.84	39	104	11	156	151	200	87.84
6b	1.42	1.84	110.88	103	188	16	236	191	400	111.6
6c	1.42	1.84	115.92	128	239	18	276	200	600	117
k_f (LMH) = 23.91, Re = 22 168, Δp (expt)/Δp (film) = 1.43 (avg), Δp (expt)/Δp (SR) = 1.9 (avg)										
7a	0.43	2.75	111.96	27	119	9	192	192	200	111.96
7b	0.43	2.75	177.12	184	458	20	315	304	400	177.12
7c	0.43	2.75	200.16	362	670	25	396	344	600	200.52
8a	0.935	2.75	102.96	21	117	8	178	177	200	101.88
8b	0.935	2.75	155.88	98	239	16	283	268	400	156.6
8c	0.935	2.75	165.96	132	333	18	315	285	600	167.04
9a	1.42	2.75	99	18	87	8	171	170	200	97.92
9b	1.42	2.75	138.96	60	170	13	251	239	400	139.68
9c	1.42	2.75	151.92	87	225	15	291	262	600	153.36
k_f (LMH) = 33.98, Re = 33 132, Δp (expt)/Δp (film) = 1.45 (avg), Δp (expt)/Δp (SR) = 1.55 (avg)										

BL refers to the boundary layer resistance model. Parameters used in the calculations are provided in Table 8.5. LMH = $L\,m^{-2}\,h^{-1}$.

Source: Experimental data and values in the last column are adapted from Wijmans *et al.* (1985).

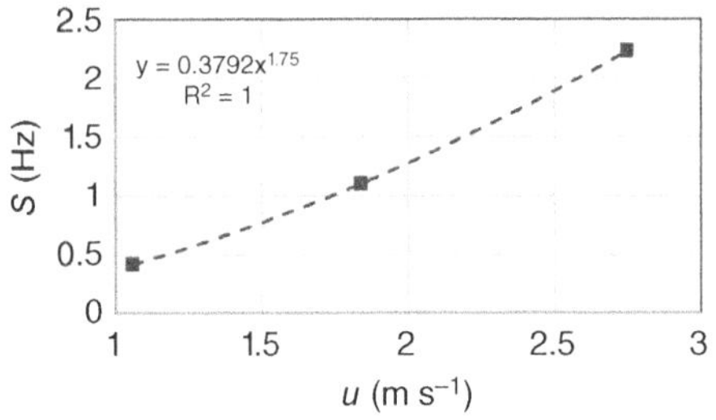

Figure 8.6 Variation of the surface renewal frequency S with liquid velocity u for the data shown in Table 8.6. Source: Adapted from Wijmans *et al.* (1985).

The increase in the surface renewal frequency with liquid velocity, as calculated from eq. 8.20, is depicted in Figure 8.6.

8.6.3 UF of Skim Milk

Koltuniewicz and Noworyta (1994) studied skim milk UF in a hollow-fiber membrane module at 293 and 303 K using varying levels of feed velocity, feed concentration, and TMP. An overview of their experimental parameters and conditions is given in Table 8.7, and predictions of the film and surface renewal models are compared against their experimental data in Table 8.8. Asterisks denote gel polarized values, that is, $J_{\text{lim}} = J_{\text{lim}}^{\text{max}}$ and $\Delta p = \Delta p_{\text{max}}$. According to the film model, eight out of the twelve experimental runs were gel polarized. In contrast, from the perspective of the surface renewal model, there were only two such runs. On average, the film model underpredicts the experimental flux and TMP by 12%, whereas the surface renewal model does so by 2% and 86%, respectively.

Some possible reasons for the discrepancy between the film and surface renewal models, which offer radically different and simplistic pictures of the hydrodynamics near the membrane surface, and experimental data are the following:

1) The assumptions of constant solute diffusivity and liquid viscosity in the models are inadequate because of large gradients in solute concentration in the vicinity of the membrane wall. Cheryan (1998) has provided an extensive discussion on the effects of physical properties and operating conditions on permeate flux in UF.
2) The film model predicts the mass-transfer coefficient k_{f} to be proportional to the solute diffusivity D (eq. 8.30), whereas the surface renewal model gives a square root dependency (eq. 8.20). Experimental correlations for k_{f} for laminar and turbulent flow in channels show that $k_{\text{f}} \propto D^{0.67}$ (Cheryan, 1998).
3) The prediction of the film model is very sensitive to the magnitude of k_{f} because of the exponential dependence of F_{w} (concentration polarization factor) on J_{lim} (see eq. 8.29). For instance, multiplying the value of k_{f} obtained from the Sherwood number correlation in Table 8.5 by a factor ranging from 0.867 to

Table 8.7 Experimental parameters and conditions in skim milk UF (Koltuniewicz and Noworyta, 1994).

Parameter	Description or value
composition of powdered skim milk	51.2% lactose, 35.7% proteins, 8.3% mineral salts, 0.8% fat, and 4.0% water
membrane module	Romicon HF-15-43-PM50: membrane surface area = 1.4 m^2, number of fibers = 682, and fiber length (l_{fib}) = 0.60 m
diameter of fiber (d_h)	1.1 mm
membrane resistance (R_m)	2.41×10^{12} m^{-1} (estimated from eq. 8.38 using a value of Q_m = 150.73 LMH atm^{-1})
diffusivity of milk proteins in bulk solution (D)	4×10^{-7} cm^2 s^{-1} (assumed)
osmotic pressure (Π)	$4.4 \times 10^{-3}\, c - 1.7 \times 10^{-6}\, c^2 + 7.9 \times 10^{-8}\, c^3$, where c is the protein concentration in g L^{-1} and Π is in atmospheres
gel concentration of skim milk (c_{gel})	220 g L^{-1}
concentration of skim milk powder in feed (c_b)	10, 13.4, 20, 34.6, 50, and 85 g L^{-1}
feed velocity (u)	0.179, 0.252, 0.443, and 0.622 m s^{-1}
TMP (Δp)	20, 40, 161, and 180 kPa (crossflow runs)
operating temperature	293 and 303 K
feed density (ρ_f)	0.998 g cm^{-3} (293 K)
feed viscosity (μ_f)	1 cp (293 K)
permeate viscosity (μ)	1 cp (293 K)
Sherwood number (Sh)	$1.86\, \mathrm{Re}^{0.33} Sc^{0.33} (d_h/l_{fib})^{0.33}$

The values of the feed density (ρ_f), feed viscosity (μ_f), and permeate viscosity (μ) were assumed to be the same as those of water at the operating temperature. LMH = L m^{-2} h^{-1}.

1.05 results in very good agreement between the theoretical and experimental TMPs for Runs 1a to 3c in Table 8.6. It should also be noted that the value of k_f obtained from two different Sherwood number (Sh) correlations can differ from one another, especially for laminar flow. This can be observed in Figure 4.30 of Cheryan (1998), which shows that Sh of the Chiang correlation, developed specifically for skim milk UF (Table 4.4 in Cheryan, 1998), can be roughly two to four times the value obtained from the Leveque correlation (given in Table 8.7) for Re lying within 100 and 2000 (i.e., laminar flow), using the film model for determining the experimental value of k_f.

Table 8.8 Comparison of the film and surface renewal models with the experimental data in Table 4 of Koltuniewicz and Noworyta (1994) for skim milk UF at 293 K and $\Delta p = 180$ kPa.

Expt	c_b (g L^{-1})	u (m s^{-1})	Re	J_{lim} (LMH) (expt)	J_{lim} (LMH) (film)	J_{lim} (LMH) (SR)	Δp (kPa) (film)	Δp (kPa) (SR)
1	13.4	0.179	196	12.74	12.74	12.744	129	53
2		0.252	276	15.66	15.41*	15.66	185*	63
3		0.443	484	21.92	18.56*	21.92	187*	85
4		0.622	680	26.75	20.76*	26.75	189*	104
5	34.6	0.179	196	7.44	7.44	7.44	102	70
6		0.252	276	9.22	9.22	9.22	133	82
7		0.443	484	13.10	12.27*	13.10	183*	111
8		0.622	680	16.15	13.72*	16.15	184*	136
9	85	0.179	196	4.47	4.47	4.47	164	137
10		0.252	276	5.56	5.24*	5.56	178*	158
11		0.443	484	7.97	6.31*	7.32*	179*	180*
12		0.622	680	9.89	7.05*	8.19*	180*	180*

Asterisks denote gel-polarized values. LMH = L m^{-2} h^{-1}.

4) Unlike microfiltration (examined in Chapters 9 and 10), in the case of UF, there is no built-in theoretical relation between the TMP and permeate flux in the basic (i.e., diffusion) equation from which the film and surface renewal models are derived. A connection between the two is assumed to hold through the osmotic pressure equation (eqs. 8.36 or 8.37) for steady-state operation.

The film and surface renewal models of UF presented in this chapter can be used for performing exploratory calculations for a particular membrane separation problem. For a specified value of the permeate flux, the theoretical TMP calculated by these models may underpredict the required TMP by a significant amount. Although the film model is closer to empirical reality, it cannot represent the transient decline in permeate flux typically observed after the start of filtration. The surface renewal model, on the other hand, can be used to correlate such dynamic experimental flux data as a function of process time through eq. 8.18, in which the only unknown theoretical parameter is the surface renewal frequency

S. This was demonstrated by Chatterjee (2010) for the unsteady-state permeate flux data on skim milk UF reported in Figure 3 of Koltuniewicz and Noworyta (1994), which were obtained in the laminar flow regime (Re = 345 and 852) at a temperature of 303 K. However, preliminary calculations using values of S obtained by regression of these transient flux data (see Table 3 in Chatterjee, 2010), eq. 8.20, and the value of the mass-transfer coefficient k_f determined from the Chiang correlation reveal that k_f (needed to calculate the limiting flux J_{lim} by eq. 8.19) estimated from this correlation is on average 22 times greater than the value obtained from eq. 8.20 (using the regressed value of S) for the eight experimental runs of Koltuniewicz and Noworyta (1994) reported in Table 8.1. If one assumes that k_f obtained from the correlation is closer to reality, then the right-hand side of eq. 8.20 will have to be multiplied by a correction factor for resolving the large discrepancy between the two mass-transfer coefficients, thereby bringing another parameter, which can be called the flux correction factor, into the analysis. Together with the pressure correction factor seen earlier in Section 8.6.2, the surface renewal model will then become a three-parameter semiempirical model. This aspect needs further investigation.

Notation

A	rate of decline of permeate flux (s^{-1})
$c(x)$	solute concentration at location x in the film model ($kg\,m^{-3}$)
$c(x, \theta)$	solute concentration at location x in a liquid element at exposure time θ ($kg\,m^{-3}$)
$c(x, \tau)$	solute concentration at location x in a liquid element with age τ ($kg\,m^{-3}$)
$c_a(x, t)$	age-averaged solute concentration at location x at process time t ($kg\,m^{-3}$)
c_b	solute concentration in bulk liquid ($g\,L^{-1}$ or $kg\,m^{-3}$)
c_{gel}	gel concentration of solute in solvent ($g\,L^{-1}$ or $kg\,m^{-3}$)
c_w	solute concentration at membrane wall ($g\,L^{-1}$ or $kg\,m^{-3}$)
d_h	hydraulic diameter of membrane channel (mm or m)
D	diffusion coefficient of solute in solvent ($cm^2\,s^{-1}$ or $m^2\,s^{-1}$)
f	function of Re and Sc in eq. 8.32
$f_D(\tau, t)$	transient Danckwerts age distribution function given by eq. 8.2 (Hz or s^{-1})
F_w	concentration polarization factor defined by eq. 8.16
$g(x, \tau)$	function defined by eq. 8.24

$J(\tau)$	permeate flux in a surface element having an age of τ (m s^{-1})
$J(\theta)$	permeate flux in a surface element at exposure time θ (m s^{-1})
J^*	permeate flux after infinite time in batch ultrafiltration (m s^{-1})
$J_a(t)$	age-averaged permeate flux at process time t (m s^{-1})
J_{lim}	limiting or steady-state permeate flux (m s^{-1} or L m^{-2} h^{-1})
J_{lim}^{calc}	limiting or steady-state permeate flux calculated from the boundary layer resistance model (L m^{-2} h^{-1})
J_{lim}^{max}	permeate flux at onset of gel polarization (m s^{-1} or L m^{-2} h^{-1})
J_0	initial permeate flux in batch ultrafiltration or pure solvent flux (m s^{-1})
K	constant (m^2 s^{-1})
k_f	mass-transfer coefficient (m s^{-1} or L m^{-2} h^{-1})
l_{fib}	fiber length (m)
Q_m	solvent permeability of clean membrane defined by eq. 8.38 (m s^{-1} Pa^{-1} or L m^{-2} h^{-1} atm^{-1})
R_{gel}	hydraulic resistance of gel layer (m^{-1})
R_m	hydraulic resistance of membrane (m^{-1})
Re	Reynolds number defined by eq. 8.34
S	surface renewal frequency (Hz or s^{-1})
Sc	Schmidt number defined by eq. 8.35
Sh	Sherwood number defined by eq. 8.33
t	process time (s)
u	liquid velocity in main flow direction in membrane channel (m s^{-1})
x	distance into liquid measured normal to membrane surface (m)

Greek Letters

δ	film thickness (m) s
Δp	transmembrane pressure drop (Pa or kPa or atm)
Δp_{max}	transmembrane pressure drop at onset of gel polarization (Pa or kPa or atm)
θ	exposure time (s)
η	defined by eq. 8.11
μ	viscosity of permeate (kg m^{-1} s^{-1} or cp)
μ_f	viscosity of feed solution (kg m^{-1} s^{-1} or cp)
ξ	variable of integration in eq. 8.24
Π	osmotic pressure (Pa or atm or bar)
ρ_f	density of feed solution (kg m^{-3} or g cm^{-3})

τ	age or residence time of a liquid element at the membrane–liquid interface (s)
ϕ	defined by eq. 8.13

Acronyms

BL	boundary layer
LMH	liter per square meter per hour
MF	microfiltration
MWCO	molecular weight cutoff
NF	nanofiltration
RO	reverse osmosis
SR	surface renewal
TMP	transmembrane pressure drop
UF	ultrafiltration

Homework Exercises

i) Derive eq. 8.4.
ii) Derive eqs. 8.18 and 8.19.
iii) Check the validity of eqs. 8.26 and 8.27.
iv) Using the data in Table 8.3 and the algorithm given in Section 8.6.1, reproduce the results shown in Figure 8.4.
v) A crossflow UF membrane system has been proposed for generating a permeate flow rate of 100 L h^{-1} continuously from a skim milk solution. A transmembrane pressure drop Δp of 40 kPa will be employed, and the system will operate at a temperature of 20 °C. Using the film and surface renewal models, develop plots showing the required membrane filtration area (m^2), permeate flux J_{lim} (L m^{-2} h^{-1}), and concentration polarization factor F_{w} as a function of the liquid velocity u in a membrane channel for u varying from 0.1 to 0.5 m s^{-1}. Use the information in Table 8.7 for your design. Perform calculations for feed concentrations of 10 and 20 g L^{-1}. Provide a brief discussion of your results.

Acknowledgment

We thank Haijiang Yu for performing exploratory calculations with the film and surface renewal models using the information in Table 8.5.

References

Abramowitz, M. and Stegun, I.A. (eds.) (1965). *Handbook of mathematical functions—with formulas, graphs, and mathematical tables*. New York: Dover Publications.

Arnot, T.C., Field, R.W. and Koltuniewicz, A.B. (2000). "Cross-flow and dead-end microfiltration of oily-water emulsions," *Journal of Membrane Science*, 169(1), pp. 1–15. Available at: https://doi.org/10.1016/S0376-7388(99)00321-X.

Chatterjee, S.G. (2010). "On the use of the surface-renewal concept to describe cross-flow ultrafiltration," *Indian Chemical Engineer*, 52(3), pp. 179–193. Available at: https://doi.org/10.1080/00194506.2010.547927.

Cheryan, M. (1998). *Ultrafiltration and microfiltration handbook*. Second edition. Boca Raton, Florida: CRC Press.

Chiang, S.H. and Toor, H.L. (1964). "Gas absorption accompanied by a large heat effect and volume change of the liquid phase," *AIChE Journal*, 10(3), pp. 398–402. Available at: https://doi.org/10.1002/aic.690100323.

Constenla, D.T. and Lozano, J.E. (1996). "Predicting stationary permeate flux in the ultrafiltration of apple juice," *Lebensm. Wiss. u. Technol.*, 29, pp. 587–592.

Dąbkowska-Susfał, K., Półgrabska, A., Sobieszuk, P. and Kołtuniewicz, A.B. (2024). "Mathematical modeling of the enzymatic hydrolysis of lignocellulosic waste in membrane bioreactor considering transport phenomena," *Chemical Engineering Research and Design*, 208, pp. 656–665. Available at: https://doi.org/10.1016/j.cherd.2024.07.034.

Hasan, A., Peluso, C.R., Hull, T.S., Fieschko, J. and Chatterjee, S.G. (2013). "A surface-renewal model of cross-flow microfiltration," *Brazilian Journal of Chemical Engineering*, 30(1), pp. 167–186. Available at: https://doi.org/10.1590/S0104-66322013000100019.

Idan, G. and Chatterjee, S.G. (2015). "Application of a surface-renewal model to permeate-flux data of constant-pressure cross-flow microfiltration with Dean vortices," *Brazilian Journal of Chemical Engineering*, 32(2), pp. 609–627. Available at: https://doi.org/10.1590/0104-6632.20150322s00003417.

Jiang, S. and Chatterjee, S.G. (2015). "A surface-renewal model for constant flux cross-flow microfiltration," *Journal of Applied Polymer Science*, 132(21). Available at: https://doi.org/10.1002/app.41778.

Kołtuniewicz, A. (1992). "Predicting permeate flux in ultrafiltration on the basis of surface renewal concept," *Journal of Membrane Science*, 68(1–2), pp. 107–118. Available at: https://doi.org/10.1016/0376-7388(92)80153-B.

Koltuniewicz, A. and Noworyta, A. (1994). "Dynamic properties of ultrafiltration systems in light of the surface renewal theory," *Industrial & Engineering Chemistry Research*, 33(7), pp. 1771–1779. Available at: https://doi.org/10.1021/ie00031a016.

Koltuniewicz, A. and Noworyta, A. (1995). "Method of yield evaluation for pressure-driven membrane processes," *Chemical Engineering Journal*, 58, pp. 175–182.

McCabe, W.L., Smith, J.C. and Harriott, P. (1993). *Unit operations of chemical engineering. Fifth*. New York: McGraw-Hill.

Nakano, Y., Tien, C. and Gill, W.N. (1967). "Nonlinear convective diffusion: A hyperfiltration application," *AIChE Journal*, 13(6), pp. 1092–1098. Available at: https://doi.org/10.1002/aic.690130611.

Quezada, C., Estay, H., Cassano, A., Troncoso, E. and Ruby-Figueroa, R. (2021). "Prediction of permeate flux in ultrafiltration processes: A review of modeling approaches," *Membranes*, 11(5), p. 368. Available at: https://doi.org/10.3390/membranes11050368.

Rautenbach, R. and Albrecht, R. (1989). *Membrane processes*. New York: John Wiley & Sons.

Sablani, S., Goosen, M., Al-Belushi, R. and Wilf, M. (2001). "Concentration polarization in ultrafiltration and reverse osmosis: A critical review," *Desalination*, 141(3), pp. 269–289. Available at: https://doi.org/10.1016/S0011-9164(01)85005-0.

Sarkar, D., Datta, D., Sen, D. and Bhattacharjee, C. (2011). "Simulation of continuous stirred rotating disk-membrane module: An approach based on surface renewal theory," *Chemical Engineering Science*, 66(12), pp. 2554–2567. Available at: https://doi.org/10.1016/j.ces.2011.02.056.

Trettin, D.R. and Doshi, M.R. (1980). "Ultrafiltration in an unstirred batch cell," *Industrial & Engineering Chemistry Fundamentals*, 19(2), pp. 189–194. Available at: https://doi.org/10.1021/i160074a010.

Wijmans, J.G., Nakao, S., Van Den Berg, J.W.A., Troelstra, F.R. and Smolders, C.A. (1985). "Hydrodynamic resistance of concentration polarization boundary layers in ultrafiltration," *Journal of Membrane Science*, 22(1), pp. 117–135. Available at: https://doi.org/10.1016/S0376-7388(00)80534-7.

Zhang, W. and Chatterjee, S.G. (2015). "Influence of residence time distribution on a surface-renewal model of constant-pressure cross-flow microfiltration," *Brazilian Journal of Chemical Engineering*, 32(1), pp. 139–154. Available at: https://doi.org/10.1590/0104-6632.20150321s00003129.

Zydney, A.L. (1997). "Stagnant film model for concentration polarization in membrane systems," *Journal of Membrane Science*, 130(1–2), pp. 275–281. Available at: https://doi.org/10.1016/S0376-7388(97)00006-9.

9

Constant Pressure Crossflow Microfiltration*

9.1 Introduction

Microfiltration is the physical separation of solid particles from a liquid by a membrane for concentrating suspensions. While being similar to ultrafiltration and nanofiltration, the difference is the pore size of the membrane used for the separation (Cheryan, 1998). This technique is frequently used to separate cells, proteins, and bacteria from a liquid suspension (e.g., fermentation broths) and also has a variety of biotech uses like enzyme recovery (Kroner *et al.*, 1984; Le *et al.*, 1984; Marston *et al.*, 1984), hormone production (Schoner *et al.*, 1985), protein recovery (Titchener-Hooker *et al.*, 1991), and cell lysate recovery (Bailey and Meagher, 1997a; 1997b). According to Gurian (2010), large-scale microfiltration for the treatment of drinking water, which was not competitive cost-wise in the past, has been adopted in plants with design flows as large as 20 MGD (million gallons per day). Besides these applications, membrane bioreactor (MBR) and reverse osmosis (RO) technologies are also finding increasing use in wastewater treatment. An MBR is a combination of an activated sludge process and a membrane separation unit – the latter replacing the secondary clarifier in the biological treatment section of a wastewater treatment plant, thereby reducing the plant footprint. For example, Dhagumudi and Yan (2012) reported the results of a pilot study undertaken over two months using an MBR and an RO unit for directly treating the wastewater of a paper mill located in the southwestern United States. The wastewater discharge was sent to a 4 GPM (gallons per minute) MBR pilot plant that consisted of a biological aeration tank and a subsequent tank in which the MBR was housed. The effluent from the MBR was sent to a 4-inch single module RO unit for further treatment to demineralize the water. During the study, the MBR was operated at various flux rates ranging from 19 GFD (gallons

*The material in this chapter has been adapted from Hasan *et al.* (2013) and Idan and Chatterjee (2015).

Applications of the Surface Renewal Model of Mass Transfer, First Edition.
Hengshuo Huang and Siddharth G. Chatterjee.

Companion website: www.wiley.com/go/huang/surfacerenewalmodel

per square feet per day) to 30 GFD. The product water from the MBR averaged a 98% BOD (biological oxygen demand) reduction, 90% COD (chemical oxygen demand) reduction, and 99% TSS (total suspended solids) removal at all flux rates. There were no fouling issues with either the MBR or RO during the pilot study, and testing showed that the product water from the MBR/RO pilot met all papermaking water quality standards and thus could be used as an alternative to fresh water or reduce its consumption in the papermaking process.

In crossflow membrane filtration (microfiltration, ultrafiltration, and RO), the incoming feed suspension flows across the surface of a porous membrane, and the permeate flow is that portion of the liquid that passes through the membrane in a direction perpendicular to the main flow. Depending on the type of application, the filtration membrane can be either polymeric or ceramic, which is contained in a module that comes in different configurations, e.g., tubular, spiral wound, and hollow fiber.

Two process schematics of crossflow microfiltration (CFMF) are shown in Figure 9.1. In Figure 9.1a, both permeate and concentrate streams are recycled back to the feed tank, which is a common practice in laboratory studies; thus, the feed concentration remains invariant with filtration or process time. The system shown in Figure 9.1b, in which only the concentrate stream is returned to the feed vessel, is used in industrial applications for concentrating a feed suspension (Rudolph and MacDonald, 1994); here, the feed concentration changes with process time. The analysis presented in this chapter is concerned with the system depicted in Figure 9.1a.

A common feature observed in any type of filtration process, whether it is microfiltration, ultrafiltration, vacuum filtration, or gravity-driven filtration, is the decline in permeate flux with process time. This decline, which looks similar to an exponential-type decay, approaches a zero value in dead-end filtration and an ultimate nonzero value in crossflow filtration. The modeling of CFMF is essential for improving the effectiveness and efficiency of membrane separation technologies, including the scale-up of such technologies from the laboratory- or pilot-scale to industrial scale.

As mentioned by Idan and Chatterjee (2015), it is generally believed that the primary reasons for permeate-flux decline are two fouling mechanisms called pore blocking and cake formation. As soon as the CFMF operation is started, some of the particles in the feed suspension become lodged within the pores of the membrane, thereby blocking them either partially or completely. According to Song (1998), this phenomenon is a rapid process, with just one layer of particles being sufficient to complete maximum blocking of the membrane, which is the cause of permeate flux decline observed during the initial moments of filtration. With the progress of time, solid or colloidal particles begin to collide, forming, at first, a few layers on the membrane wall – a phenomenon known as "gel polarization."

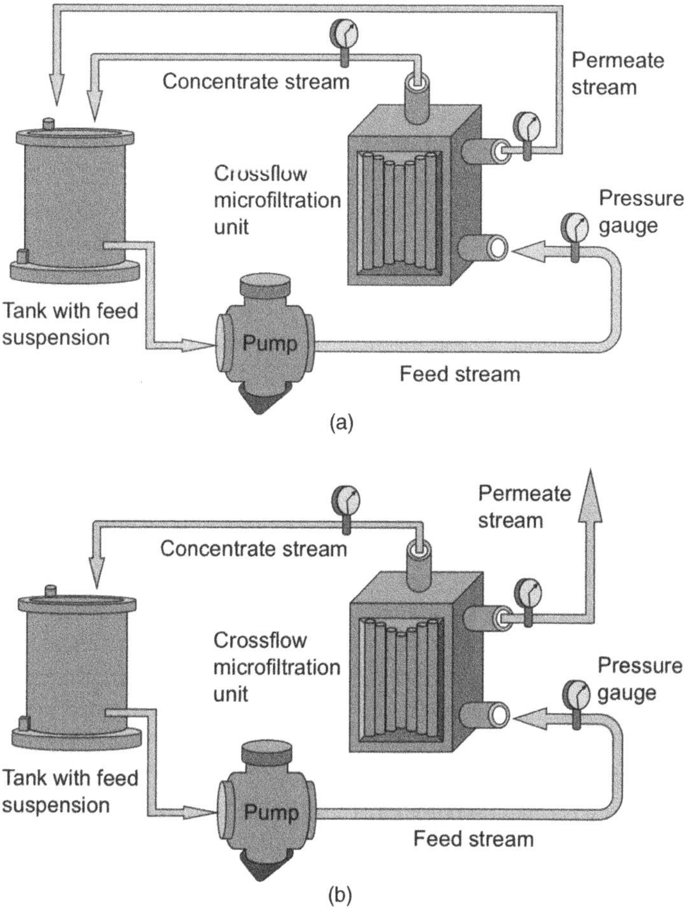

Figure 9.1 Schematics of (a) laboratory-scale microfiltration unit with both permeate and concentrate streams recycled back to the feed tank and (b) industrial-scale microfiltration unit with only the concentrate stream recycled back to the feed tank.

With the further passage of time, these layers build upon one another and form a "cake" on the membrane surface. In this phase, the cake resistance increases as the cake gets thicker, and flux decline continues until the cake layer attains an equilibrium thickness when a steady-state (nonzero) plateau in the permeate flux is established.

There is a vast literature on crossflow membrane filtration; the interested reader is referred to the reviews of Sablani *et al.* (2001) and Quezada *et al.* (2021). The chief focus, however, in this chapter will be on the application of the surface renewal concept to describe crossflow membrane filtration, which was pioneered by Koltuniewicz and coworkers in the 1990s (Kołtuniewicz, 1992; Koltuniewicz and Noworyta, 1994; 1995). Other examples where this concept has been used to model crossflow constant pressure/flux ultrafiltration and microfiltration are the studies of Constenla and Lozano (1996), Arnot *et al.* (2000), Chatterjee (2010), Sarkar *et al.* (2011), Hasan *et al.* (2013), Jiang and Chatterjee (2015), Zhang and Chatterjee (2015), Idan and Chatterjee (2015), and Dąbkowska-Susfał *et al.* (2024). As noted in Chapter 1, in contrast to the film (i.e., gel polarization) and boundary layer models of membrane filtration, the surface renewal model has the potential to describe the transfer of dissolved/suspended solids due to hydrodynamic impulses generated at the membrane surface due to membrane roughness or by the use of spacers and turbulence promoters, or by the deliberate introduction of controlled centrifugal instabilities, called Dean vortices, into the main flow, which induce the back migration of accumulated solute molecules or particulates away from the membrane surface, thereby significantly enhancing permeation rates (Chung *et al.*, 1996; Mallubhotla and Belfort, 1997; Chung *et al.*, 1998; Gehlert *et al.*, 1998; Mallubhotla *et al.*, 1998; Luque *et al.*, 1999; Mallubhotla *et al.*, 1999; Mallubhotla *et al.*, 2001).

This chapter presents a mathematical model for predicting permeate flux decline and cake accumulation on the membrane surface in CFMF, which was originally proposed by Hasan *et al.* (2013). The model is a synthesis of classical cake filtration theory and the surface renewal concept and contains the following three basic parameters: membrane resistance R_m, specific cake resistance α, and frequency of surface renewal S at the membrane wall. The surface renewal model, which has two forms, complete and approximate, is applied for correlating the experimental permeate flux data reported by Hasan *et al.* (2013) and Mallubhotla and Belfort (1997).

9.2 Surface Renewal Model of Crossflow Microfiltration

9.2.1 Complete Model

Figure 9.2a and 9.2b illustrates crossflow membrane filtration and the surface renewal concept, respectively. The surface renewal model postulates that the chief fouling mechanism that causes permeate flux decline is cake formation. The phenomenon of pore blocking is assumed to occur very rapidly, i.e., during the first moments of filtration, whose effects are included in the membrane resistance R_m and which is considered to be an empirical parameter. Flow instabilities continuously bring fresh liquid elements from the bulk liquid to the membrane–liquid interface on the feed side. A liquid element remains at the

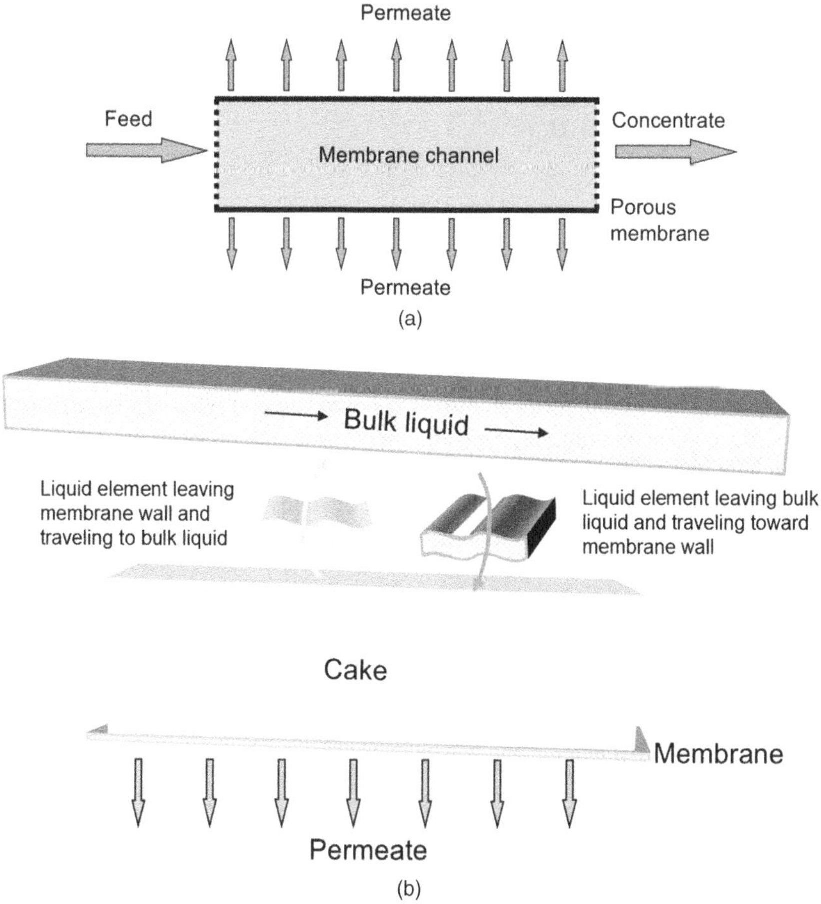

Figure 9.2 Schematics of (a) crossflow membrane filtration and (b) surface renewal mechanism.

membrane surface for a definite time τ after which it returns and remixes with the bulk liquid. Below or above the surface elements, the liquid is assumed to be well mixed where the concentration of solids is held constant at its bulk value c_b, because of a high rate of transport, due to flow instabilities, from this point to the bulk liquid and vice versa. With the progress of time, a cake layer builds up on the membrane surface, which leads to a gradual decline in the permeate flux until it attains a steady value. To model this process, it is assumed that during the residence time τ of a liquid element at the membrane surface, permeate flux and cake accumulation in the surface element can be described by classical cake filtration theory (McCabe *et al.*, 1993). It can be shown from this theory that the permeate flux in a surface element having a residence time of τ at the membrane wall is given by

$$J(\tau) = \frac{1}{\sqrt{\left(1/J_0^2\right) + 2k_c\tau}} \tag{9.1}$$

where

$$J_0 = \frac{\Delta p}{\mu R_m} \tag{9.2}$$

and

$$k_c = \frac{\mu c_b \alpha}{\Delta p} \tag{9.3}$$

Here, J_0 is the initial permeate flux in a liquid element when it first arrives at the surface of the membrane, Δp is the transmembrane pressure drop (assumed to be held constant), μ is the viscosity of the filtrate, R_m is the resistance of the membrane or filter medium, c_b is the mass of solids deposited in the filter per unit volume of filtrate[1] that is assumed to be constant, and α is the specific cake resistance.

The mass $m_c(\tau)$ of solids accumulated in the element during its residence time of τ at the membrane wall per unit area of the membrane surface is given by

$$m_c(\tau) = \int_0^{\tau} J(\lambda) c_b \, d\lambda \tag{9.4}$$

which, upon insertion of eq. 9.1, results in

$$m_c(\tau) = \frac{c_b}{k_c}\sqrt{\left(1/J_0^2\right) + 2k_c\tau} - \frac{c_b}{k_c J_0} \tag{9.5}$$

At any time t after the start of filtration, the surface of the membrane is visualized as being composed of a mosaic of liquid elements, which have different ages ranging from zero to t. Denoting the age-distribution function of the surface elements by $f(\tau, t)$, the age-averaged permeate flux (i.e., process flux) $J_a(t)$ at process time t is given by

$$J_a(t) = \int_0^t J(\tau) f(\tau, t) d\tau \tag{9.6}$$

where $f(\tau, t)$ is assumed to be the transient Danckwerts age-distribution function given by eq. 2.10 of Chapter 2, i.e.,

$$f(\tau, t) = \frac{Se^{-S\tau}}{1 - e^{-St}} \tag{9.7}$$

1 This is the same as the bulk or feed concentration of the suspension if the amount of liquid trapped in the pores of the wet cake is neglected.

where S, as stated earlier, is the frequency of renewal of liquid elements at the membrane wall. As indicated in Chapter 1, eq. (9.7) has been experimentally vindicated by Lesage *et al.* (2002) for the case of a single liquid phase flowing through a packed-bed reactor under steady-state conditions.

Substituting eq. (9.7) into eq. (9.6) yields

$$J_{\mathrm{a}}(t) = \int_0^t J(\tau)\frac{Se^{-S\tau}}{1 - e^{-St}}\mathrm{d}\tau \tag{9.8}$$

In order to describe the nature of the cake formed on the membrane surface in a theoretical analysis, either of the two following extreme assumptions may be made: (1) Flow instabilities extend right up to the membrane wall, with the cake being visualized as a dynamic suspension of particles. Hence, in a new surface element, cake buildup starts from a "clean" membrane state, and J_0 is the same for all elements and independent of the age of the element, τ. (2) Flow instabilities extend only up to the outside edge of the cake, but not into the cake layer. In a surface element of age τ, the initial flux is equal to the value of the process flux at process time $t - \tau$, which will, therefore, be different for different surface elements. An actual situation will lie somewhere between these two extremes. Using eqs. 9.1 and 9.8, and invoking these assumptions yields two alternative formulations for $J_{\mathrm{a}}(t)$, i.e.,

$$J_{\mathrm{a}}(t) = \int_0^t \frac{1}{\sqrt{\left(1/J_0^2\right) + 2k_{\mathrm{c}}\tau}}\frac{Se^{-S\tau}}{1 - e^{-St}}\mathrm{d}\tau \tag{9.9}$$

or

$$J_{\mathrm{a}}(t) = \int_0^t \frac{1}{\sqrt{\left[1/(J_{\mathrm{a}}(t-\tau))^2\right] + 2k_{\mathrm{c}}\tau}}\frac{Se^{-S\tau}}{1 - e^{-St}}\mathrm{d}\tau \tag{9.10}$$

According to eq. (9.9), which can be readily integrated to obtain $J_{\mathrm{a}}(t)$ as an explicit function of process time t (see eq. (9.14)), the process flux is a weighted average of the individual elemental fluxes, i.e., the direction of communication is from the microscopic (elemental) to the macroscopic (process) domain. In striking contrast, eq. (9.10) in which the permeate flux $J_{\mathrm{a}}(t)$ is expressed in terms of a convolution integral that has to be integrated subject to the initial condition $J_{\mathrm{a}}(t = 0) = J_0$ shows that the value of the process flux $J_{\mathrm{a}}(t)$ is governed by the entire history of this flux starting from $t = 0$ onwards, i.e., the process and elemental fluxes co-determine one another, or expressed in another way, there is a two-way communication between the microscopic and macroscopic domains.[2] With rising flow instability, the situation represented by eq. (9.9) is expected to become more tenable. We will use eq. (9.9) for reasons of mathematical simplicity.

2 The reader is also referred to Section 4.3 of Chapter 4.

By integrating eq. (9.9) and defining the dimensionless quantities

$$S^* = \frac{S}{2k_c J_0^2} \tag{9.11}$$

$$t^* = St \tag{9.12}$$

and

$$J_a^*(t^*) = \frac{J_a(t)}{J_0} \tag{9.13}$$

it may be shown that

$$J_a^*(t^*) = \frac{e^{S^*}}{1 - e^{-t^*}} \sqrt{\pi S^*} \left[\operatorname{erf}(\sqrt{S^* + t^*}) - \operatorname{erf}(\sqrt{S^*})\right] \tag{9.14}$$

As $t^* \to 0$, applying l'Hôpital's rule to eq. (9.14) gives

$$J_a^*(t^* \to 0) = 1 \tag{9.15}$$

while as $t^* \to \infty$, eq. (9.14) becomes

$$J_a^*(t^* \to \infty) = \frac{J_a(t \to \infty)}{J_0} = \frac{J_{\mathrm{lim}}}{J_0} = J_{\mathrm{lim}}^* = e^{S^*} \sqrt{\pi S^*} \left[1 - \operatorname{erf}(\sqrt{S^*})\right] \tag{9.16}$$

where J_{lim} is the value of the limiting or steady-state permeate flux (i.e., as $t \to \infty$).

The three parameters of the surface renewal model can be estimated as follows: (1) Determine the membrane resistance R_m from the value of the initial flux J_0 and eq. (9.2); (2) calculate the dimensionless surface renewal rate S^* from the experimental value of J_{lim} and eq. (9.16); (3) estimate the surface renewal rate S by fitting eq. (9.14) to experimental data of transient permeate flux so as to minimize the root-mean-square (RMS) deviation between predicted and experimental values of the flux; and (4) finally, calculate k_c and α using eqs. 9.11 and 9.3, respectively.

As seen from eq. (9.14), the permeate flux declines with process time due to a progressive buildup of a cake of solids on the membrane surface. At any time t after the start of filtration, the age-averaged mass of cake $m_{c,a}(t)$ accumulated per unit area of the membrane wall is given by

$$m_{c,a}(t) = \int_0^t m_c(\tau) \frac{Se^{-S\tau}}{1 - e^{-St}} \mathrm{d}\tau \tag{9.17}$$

Substituting eq. (9.5) into eq. (9.17) and performing the integration yield

$$m_{c,a}^*(t^*) = \frac{e^{S^*}}{(1 - e^{-t^*})\sqrt{S^*}} \left[\Gamma(3/2, S^*) - \Gamma(3/2, S^* + t^*)\right] - 1 \tag{9.18}$$

where

$$m_{c,a}^*(t^*) = \frac{m_{c,a}(t) J_0 k_c}{c_b} \tag{9.19}$$

is the nondimensional cake mass and $\Gamma(x, y)$ the incomplete gamma function, encountered in Chapter 2, which is defined by

$$\Gamma(x,y) = \int_y^{\infty} \lambda^{x-1} e^{-\lambda} \mathrm{d}\lambda \tag{9.20}$$

As $t^* \to 0$, using l'Hôpital's rule along with eq. (9.18) gives

$$m^*_{\mathrm{c,a}}(t^* \to 0) = 0 \tag{9.21}$$

whereas as $t^* \to \infty$, eq. (9.18) reduces to

$$m^*_{\mathrm{c,a}}(t^* \to \infty) = m^*_{\mathrm{c,lim}} = \frac{m_{\mathrm{c,lim}} J_0 k_{\mathrm{c}}}{c_{\mathrm{b}}} = \frac{e^{S^*}}{\sqrt{S^*}} \Gamma\left(3/2, S^*\right) - 1 \tag{9.22}$$

where $m_{\mathrm{c,lim}}$ is the steady-state value of the cake mass.

According to eqs. 9.14 and 9.18, in dimensionless coordinates, the permeate flux and cake mass are functions of process time, with the surface renewal frequency being the sole governing parameter. It can also be deduced from 9.16 and 9.22 that as $S^* \to 0$ (unstirred batch filtration), $J^*_{\mathrm{lim}} \to 0$ and $m^*_{\mathrm{c,lim}} \to \infty$ while as $S^* \to \infty$ (high intensity of turbulence), $J^*_{\mathrm{lim}} \to 1$ and $m^*_{\mathrm{c,lim}} \to 0$. Equations 9.16 and 9.22 are universal functions.

9.2.2 Approximate Model

If the value of J_0 is high (e.g., a clean membrane), letting $J_0 \to \infty$ in eqs. 9.14 and 9.18 yield

$$J_{\mathrm{a}}(t) = J_{\mathrm{lim}} \frac{\mathrm{erf}(\sqrt{St})}{1 - e^{-St}} \tag{9.23}$$

and

$$m_{\mathrm{c,a}}(t) = m_{\mathrm{c,lim}} \frac{\left[1 - \left(2/\sqrt{\pi}\right) \Gamma\left(3/2, St\right)\right]}{1 - e^{-St}} \tag{9.24}$$

where

$$J_{\mathrm{lim}} = \sqrt{\frac{\pi S}{2k_{\mathrm{c}}}} = \sqrt{\frac{\pi S \Delta p}{2\mu c_{\mathrm{b}} \alpha}} \tag{9.25}$$

and

$$m_{\mathrm{c,lim}} = c_{\mathrm{b}} \sqrt{\frac{\pi}{2Sk_{\mathrm{c}}}} = \sqrt{\frac{\pi c_{\mathrm{b}} \Delta p}{2S\mu\alpha}} = J_{\mathrm{lim}} \frac{c_{\mathrm{b}}}{S} \tag{9.26}$$

Equation (9.23) has the same mathematical form as the transient permeate flux in crossflow ultrafiltration (see Chapter 8). The value of S can be determined from the experimental value of J_{lim} and by fitting eq. (9.23) to transient permeate flux data after which α can be calculated from eq. (9.25).

According to eq. (9.25), an increase in the surface renewal frequency S, say, due to a greater level of flow instability, an increase in the transmembrane pressure drop Δp or a reduction in feed concentration c_b will raise the value of J_{lim}, which is in accord with physical intuition. Further, eq. (9.26) predicts that increases in c_b and Δp will increase $m_{c,lim}$, which, however, can be compensated for by an increase in S. At a fixed value of c_b, the ratio $m_{c,lim}/J_{lim}$ is inversely proportional to S; thus, the introduction of flow instabilities like vortices into the main flow will increase S and decrease this ratio.

Equations 9.25 and 9.26 also show the great influence of both the surface renewal rate S and specific cake resistance α in determining the values of J_{lim} and $m_{c,lim}$. The introduction of flow instabilities like Dean vortices will increase S and decrease α (i.e., a looser, less compact cake), and the limiting permeate flux, which is proportional to $\sqrt{S/\alpha}$, may increase significantly, while there will be a smaller influence on the limiting cake mass since the latter is proportional to $\sqrt{1/(S\alpha)}$.

As $St \to 0$, i.e., as $S \to 0$ (i.e., low flow instability) or $t \to 0$ (i.e., the initial moments of filtration), eqs. 9.23 and 9.24 reduce to:

$$J_a(St \to 0) = \sqrt{\frac{2}{k_c t}} = \sqrt{\frac{2\Delta p}{\mu c_b \alpha t}} \tag{9.27}$$

and

$$m_{c,a}(St \to 0) = \frac{2}{3} c_b \sqrt{\frac{2t}{k_c}} = \frac{2}{3} \sqrt{\frac{2c_b \Delta p t}{\mu \alpha}} \tag{9.28}$$

At $t = 0$, eq. (9.27) predicts an unbounded permeate flux – a consequence of assuming that the initial flux $J_0 \to \infty$ in the derivation of the approximate model. Moreover, during the initial moments of filtration or for low levels of flow instability, eqs. 9.27 and 9.28 indicate that the permeate flux is inversely proportional to the square root of process time t, whereas the cake mass grows as the square root of t, beginning from a value of zero at $t = 0$. This is analogous to the behavior of the rates of absorption and dissolved-gas transfer to the bulk liquid in physical gas absorption in a liquid as can be seen from eqs. 4.11a and 4.24a in Chapter 4. Equations 9.27 and 9.28 also show that for $St \to 0$, the permeate flux and cake mass are inversely and directly proportional, respectively, to the square root of the feed concentration c_b.

The absence of the surface renewal rate S in eqs. 9.27 and 9.28 is a consequence of assuming the Danckwerts age-distribution function (see eq. (9.7)) in the derivation of eqs. 9.14 and 9.18; this distribution approaches a uniform distribution as $St \to 0$. It was also indicated earlier that the introduction of Dean vortices into the flow field is expected to decrease the specific cake resistance α, resulting in a higher permeate flux (see eq. 9.27) compared to when such vortices are absent. This can therefore lead to greater cake deposition on the membrane surface during the first moments of filtration (see eq. (9.28)) because of the higher convective flow of solids towards the membrane wall as a result of a higher permeate flux. We also note that for small values of S, the initial period of filtration (i.e., when eqs. 9.27 and 9.28 hold) will be greater compared to the situation when S is large, which is the meaning of the condition $St \to 0$.

Although the chief purpose of introducing flow instabilities like Dean vortices into the main flow field is to obtain a higher permeate flux, the relative magnitude of the two parameters, α and S, without and with vortices can generate complex behavior of the permeate flux and cake mass curves as a function of process time, and, in some situations, the introduction of such vortices can be detrimental. This is chiefly due to greater cake compaction because of an increased permeate flow due to vortices immediately after the start of filtration, i.e., during the initial moments of filtration. The appendix of the paper of Idan and Chatterjee (2015) explores this in more detail and presents a qualitative analysis of the behavior of the permeate flux and cake mass curves with process time.

In cake filtration, the specific cake resistance α is often expressed by the empirical relation (McCabe *et al.*, 1993):

$$\alpha = \alpha_0 \Delta p^n \tag{9.29}$$

where α_0 and n are empirical constants, and n is the compressibility coefficient of the cake, which is zero for incompressible sludges and lies between 0.2 and 0.8 for compressible ones (McCabe *et al.*, 1993).

Substituting eq. (9.29) into eqs. 9.25 and 9.26 yield

$$J_{\text{lim}} = \sqrt{\frac{\pi S}{2\mu c_{\text{b}}\alpha_0}}\Delta p^{\frac{1-n}{2}} \tag{9.30}$$

and

$$m_{\text{c,lim}} = \sqrt{\frac{\pi c_{\text{b}}}{2S\mu\alpha_0}}\Delta p^{\frac{1-n}{2}} \tag{9.31}$$

The following observations can be made from eqs. 9.30 and 9.31: (1) J_{lim} is directly proportional to $\sqrt{S}$, whereas $m_{\text{c,lim}}$ is inversely proportional to it; (2) J_{lim} is inversely proportional to $\sqrt{c_{\text{b}}}$, whereas $m_{\text{c,lim}}$ is directly proportional to it; (3) both J_{lim} and $m_{\text{c,lim}}$ are inversely proportional to $\sqrt{\mu}$ and $\sqrt{\alpha_0}$; and (4) both are directly proportional to $\Delta p^{\frac{1-n}{2}}$. All of these conclusions are in accord with physical intuition. Although the inverse dependency of $m_{\text{c,lim}}$ on $\sqrt{\mu\alpha_0}$ may seem counter to intuition, smaller values of the permeate viscosity μ and the compressibility coefficient α_0 will result in a higher permeate flux, thereby leading to greater cake deposition by the convective flow of liquid towards the membrane surface.

For a perfectly incompressible cake ($n = 0$), eq. (9.30) predicts a 0.5 power dependence of the limiting permeate flux J_{lim} on the transmembrane pressure drop Δp, with the curve of J_{lim} versus Δp becoming increasingly flatter with increasing n, which is, once again, in accord with physical intuition since for highly compressible cakes, J_{lim} will be much less sensitive to changes in Δp. In comparison, for a perfectly incompressible cake (i.e., $n = 0$), the resistance model used by Mallubhotla and Belfort (1997), which does not include the effect of (axial) liquid flow, predicts the limiting flux to be linearly proportional to Δp. Thus, assuming all other factors to remain invariant, for a perfectly incompressible cake, a doubling of the transmembrane pressure drop will double the limiting permeate flux according to the resistance model but only increase it by a factor of 1.414 according to the surface renewal model.

9.3 Correlation of the Surface Renewal Model with Experimental Permeate Flux Data

9.3.1 Data of Hasan *et al.* (2013)

The bioprocess industry often uses crossflow (also called tangential flow) filtration to separate microbial cells from a liquid suspension after cell growth in a fermenter or bioreactor. The permeate flow rate is known to depend on the transmembrane pressure drop and velocity of the liquid across the membrane surface.

9.3.1.1 Fermentation

Hasan *et al.* (2013) conducted fermentation and membrane separation at both laboratory and pilot-plant scales. For filtration runs with the laboratory membrane unit (see Figure 4 in their paper), a 15-L fermenter was used that contained aerobically grown (1) *Escherichia coli* (rod-shaped bacteria), (2) *Burkholderia cepacia* (elliptically shaped), and (3) *Pichia stipitis* (a hat-shaped yeast), which are used to produce ethanol, polyhydroxyalkanoates, and ethanol, respectively. The glucose minimal salts medium in which cell growth occurred was supplemented with complex nutrients. For runs with the pilot-scale crossflow unit (see Figure 5 in their paper), the broth was generated by (1) anaerobic growth of *Candida pseudotropicalis* using cheese whey as the nutrient medium, and (2) aerobic growth of the same organism using a glycerol growth medium in a 400-L fermenter. An environment of controlled pH, temperature, agitation, and dissolved oxygen (in the case of aerobic runs) was maintained during fermentation.

9.3.1.2 Membrane Characteristics

Laboratory-scale and pilot-scale ceramic microfiltration units manufactured by Hilliard Corporation, Elmira, New York, were used in the work of

Hasan *et al.* (2013). The small-scale unit consisted of three tubular ceramic membrane modules having a pH resistance of 2–13 and thermal resistance of 0–149 °C. The cylindrical membrane modules (2.54 cm diameter) were approximately 30.48 cm long and contained numerous channels (~ 60) of square cross section, having a side of approximately 0.18 cm. The modules had pore sizes of 0.01, 0.2, and 0.45 μm with a total membrane surface area of 0.13 m^2. This unit (0.01 and 0.2 μm pore sizes) was previously used by Hasan *et al.* (2011) to separate hydrolyzate components in the hot-water extract of sugar maple wood.

The pilot-scale membrane unit had 1800 channels with an average pore size of 0.2 μm; each channel was 86.36 cm long and had a square cross section with a side of 0.18 cm. Some channels were converted to permeate conduits, thus allowing the entire filter diameter to be utilized effectively. The total membrane filtration area was 11.15 m^2 with the membrane having the following operating characteristics: Maximum $\Delta p = 413.7$ kPa, maximum inlet pressure = 586.1 kPa, maximum differential pressure = 206.8 kPa, and maximum back pulse pressure = 689.5 kPa.

9.3.1.3 Experimental Runs

Crossflow filtration runs were performed under complete and continuous recycling of permeate and retentate back to the feed vessel. In the small unit, the filtration was carried out over a 2-h period, and at the end of a run, the filter was (sometimes) back pulsed and then removed from its housing for soaking in a 0.25-N alkali solution. After that, it was reinstalled back into its housing, run through with an alkali solution, and rinsed several times with water. During filtration, the temperature was controlled by running cold water through a jacket around the vessel so as to prevent a change in liquid viscosity because of a rise in temperature caused by fermentation. In the pilot-scale unit, a filtration run lasted for 1 hour on average, and the filter module was cleaned by back pulsing and rinsing with water and alkali solution.

The characteristics of the fermentation broths, experimental conditions, and details of the membrane modules used in the experiments of Hasan *et al.* (2013) are shown in Table 1 of their paper.

9.3.1.4 Model Validation

Hasan *et al.* (2013) tested the ability of the surface renewal model to fit experimental data of permeate flow rate in their CFMF of the fermentation broths mentioned in Section 9.3.1.1. The value of the viscosity μ of the filtrate required to calculate R_m from eq. (9.2) was assumed to be the same as that of water at the experimental temperature (McCabe *et al.*, 1993; Perry *et al.*, 1984), i.e., the effects of substrate and salts on the viscosity were neglected.

Figure 6 in their paper exhibits the behavior of the clean water flux at 20 °C as a function of transmembrane pressure drop Δp for the small-scale unit for two different membrane pore sizes. The data are fitted fairly well with straight lines, and from the slopes of the plots, eq. (9.2), and using a value of 0.0011 kg m^{-1} s^{-1} for the viscosity of water at 20 °C (Perry *et al.*, 1984), the values of R_m for the clean membrane are estimated to be 1.14×10^{12} and 1.01×10^{12} m^{-1} for membrane pore sizes of 0.2 and 0.45 μm, respectively. The data for the 0.2 μm membrane pore size shown in Figure 6 of their paper are from an earlier work (Hasan *et al.*, 2011).

Figure 9.3 compares the predictions of the surface renewal model (complete version) with data of permeate flow rate from the small- and pilot-scale units. The method used to determine the optimum values of the three parameters (R_m, k_c, and S) was explained in Section 9.2.1[3] – these are reported in Table 2 in the work of Hasan *et al.* (2013), which also shows the RMS deviations between the theoretical and experimental permeate flow rates. The higher the values of R_m and k_c, the lower the permeate flux curve as can be seen from eq. (9.1). Of the three types of cells used in the filtration runs carried out with the small-scale unit, *P. stipitis* had the lowest value of k_c, followed by *B. cepacia* and *E. coli*, respectively, the value for *P. stipitis* being about eleven times smaller than that for *E. coli*. This points to a great difference in the nature of the cake formed from these three different types of cells. For the large-scale unit, the values of k_c were of comparable magnitude since the same type of cell (*C. pseudotropicalis*) was involved, although grown under two different conditions. The value of the surface renewal frequency S lay in the range $3.0–5.1 \times 10^{-4}$ s^{-1} (average: 4.4×10^{-4} s^{-1}) for the small-scale unit and $7.5–10.1 \times 10^{-4}$ s^{-1} (average: 8.8×10^{-4} s^{-1}) for the large-scale unit, which can be compared to the range of $8.4–17.4 \times 10^{-4}$ s^{-1} found by Chatterjee (2010) for the experimental permeate flux data of Koltuniewicz and Noworyta (1994) who studied the ultrafiltration of skim milk in a hollow-fiber membrane module (Romicon HF-15-43-PM 50) at 30 °C. These authors reported that S varied from 14 to 25.4×10^{-4} s^{-1} for the same data in their version of the surface renewal model.

The following observations can be drawn from the work of Hasan *et al.* (2013):

1) There is good agreement between the theoretical and experimental permeate flow rates, with the average RMS deviation between them being 4.6%.
2) The experimental flux declines with process time and attains a constant or limiting value as predicted by the model (eq. (9.14)).
3) The experimental value of the initial permeate flux J_0, which is conventionally thought to be the pure solvent flux of the membrane (Koltuniewicz and Noworyta, 1994), is not the same for the different experimental runs – this is reflected in the varying values of R_m shown in Table 2 of their paper, which, for

---

3 The value of the specific cake resistance α cannot be determined from eq. (9.3) because the mass concentration of cells c_b in the liquid feed to the filter was not measured by Hasan *et al.* (2013).

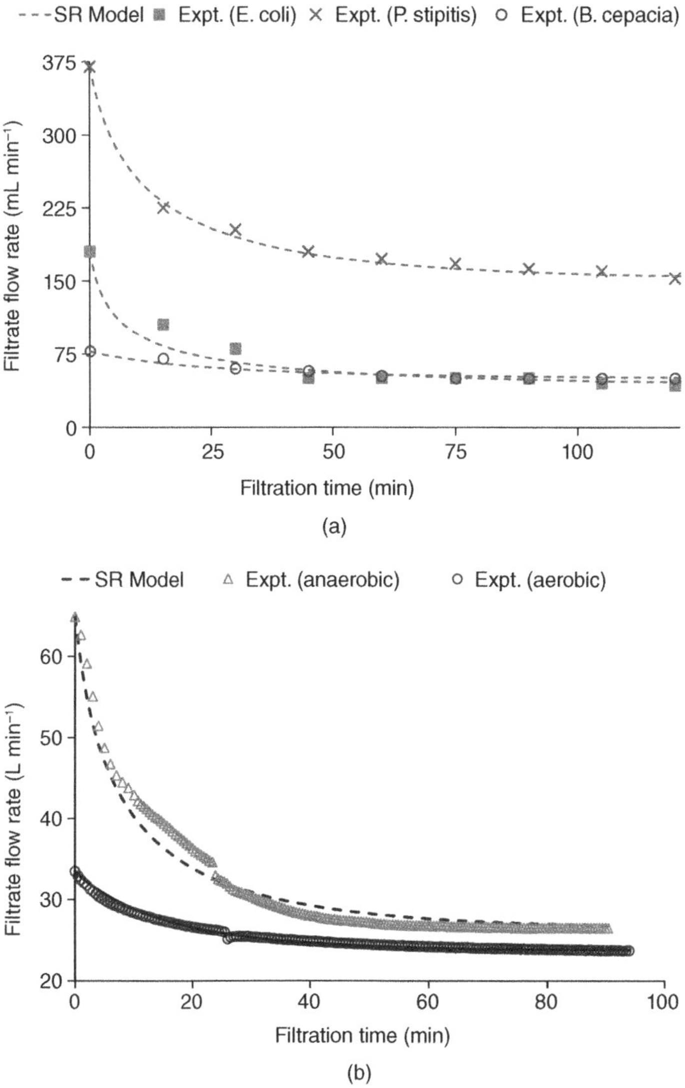

Figure 9.3 Comparison of the predictions of the surface renewal model (complete version, eq. (9.14)) and experimental permeate flow rates in the microfiltration of fermentation broths in the (a) laboratory (*E. coli*, *P. stipitis*, and *B. cepacia*) and (b) pilot (*C. pseudotropicalis*) units. Experimental data are from Hasan *et al.* (2013). Source: Modified from Hasan *et al.* (2013).

the small-scale unit, are approximately 4 to 19 times greater compared to the value of R_m of the clean membrane. Further, R_m varies by a factor of 5.2 for the small-scale unit, whereas for the pilot-scale unit, the factor is 1.2. The following hypotheses are advanced to explain this behavior: (1) The procedure used to clean the membrane module between experimental runs was either not adequate or pore blocking had occurred during the initial moments of filtration. For the two runs in the pilot-scale unit involving *C. pseudotropicalis*, the value of R_m is about 22% higher for the run under anaerobic conditions, in which Δp was 40% greater compared to the run under aerobic conditions. This may be partially due to the additional forcing of solids in the broth into the membrane due to the higher permeate flux, which is caused by the greater transmembrane pressure drop Δp (McCabe *et al.*, 1993). The values of R_m are, thus, for a fouled membrane, and not a clean one. In conventional cake filtration theory, no complete analysis of the buildup of the (initial) resistance of the filter cloth is possible because this resistance depends upon how the pressure is developed and the support geometry; it is hence usual to combine the resistance of the cloth with that of the first few layers of deposited particles, which have a propensity to block the pores of the cloth (Richardson *et al.*, 2002). (2) The viscosity of water (at the experimental temperature) was used by Hasan *et al.* (2013) to calculate R_m. However, the actual viscosity of the filtrate may have been different due to biological activity, which is a function of the cell type and fermentation conditions, and also due to the presence of substrate and salts in the growth medium. (3) Fouling of the membrane during the initial moments of filtration was dependent on the type of cells in the broth. This explains the large variation in R_m for the small-scale unit, in which three different types of cells were used, as compared to the pilot-scale unit, in which only one cell type was used. (4) Table 1 in the paper of Hasan *et al.* (2013) shows that the initial optical density (OD) of the different broths varied to a considerable degree.

In the microfiltration of sugar maple wood extract, Hasan *et al.* (2011) found that the values of the initial permeate flux J_0 and the clean water flux differed significantly, which they attributed to a difference in viscosity. Huang and Morrissey (1998) observed that J_0, which they referred to as the apparent initial permeate flux, depends on the feed concentration – the higher this concentration, the lower is the initial flux. According to them, it is difficult to measure the true initial permeate flux because the system pressure cannot be stabilized instantaneously as the feed is pumped into the filter. The first measurement point can only be obtained after some "lag" time that depends upon the measurement technique, according to Kołtuniewicz (1992), who recommended that pure solvent permeability measurements, after correction for the osmotic pressure effect, be used to estimate J_0 for ultrafiltration. Because of the uncertainty associated with experimental data of permeate flux near the beginning of filtration, it may be best to treat R_m as an empirical parameter, which would also include any resistance to flow in the pipes to and from the filter (McCabe *et al.*, 1993). In the modeling of microfiltration, Arnot *et al.* (2000) also considered J_0 as an adjustable parameter in order to minimize experimental errors in the initial flow rate measurement, although they found that in most cases it was not significantly different from the clean water flux under equivalent operating conditions.

Since only the OD at 540 nm of the feed suspension was measured by Hasan *et al.* (2013) instead of the actual cell concentration in the feed (c_b), it is not possible to determine values of the cake mass $m_{c,a}$ as a function of process time. However, if one assumes, as a very rough measure, that 1 OD unit corresponds to a cell concentration of 0.5 kg m^{-3}, the limiting or steady-state values of the cake mass per unit area of membrane surface, as calculated by eqs. 9.19 and 9.22 are found to be 0.58, 1.26, and 2.12 kg m^{-2} for *E. coli*, *B. cepacia*, and *P. stipitis*, respectively, for the small-scale unit. It is to be reiterated that these are approximate estimates and may not be precise.

Figures 9 and 10 in Hasan *et al.* (2013) illustrate the growth of the cake mass as a function of process time in dimensionless coordinates in the laboratory- and pilot-scale units, respectively, for the different broths as calculated from eq. (9.18). All the curves begin at a value of zero and progress toward a steady state as t^* increases. The values of the dimensionless steady-state permeate flux J^*_{lim}, dimensionless steady-state cake mass $m^*_{\mathrm{c,lim}}$, and dimensionless surface renewal rate S^* are provided in Table 3 of their paper for all the experimental runs. J^*_{lim} approaches a value of 1 with an increase in S^* due to a reduction in $m^*_{\mathrm{c,lim}}$.

9.3.2 Dependence of the Surface Renewal Frequency on Feed Velocity

The surface renewal model presented in this chapter is self-contained, i.e., the three model parameters can be completely determined in a sequential manner by fitting the model to experimental permeate flux data. Two of the parameters (membrane and cake resistances) come from conventional cake filtration theory, while the third (surface renewal rate) depends upon the hydrodynamics in the membrane channel. Koltuniewicz and Noworyta (1995) used the surface renewal model to determine the variation of the surface renewal frequency S with velocity u of the main flow for milk ultrafiltration, BSA concentration, and microfiltration of dodecane-water emulsions in hollow-fiber and flat-plate modules. Table 9.1 provides their experimental details and correlations for S as a function of u. It is seen that S is a power-law function of u, with the exponent of u being greater than 1 for microfiltration and less than 1 in the case of ultrafiltration. Koltuniewicz and Noworyta (1995) attributed this to the dominance of the lateral migration effect in the case of the former and the existence of transient flow regimes in the case of

Table 9.1 Experimental parameters and correlations for the surface renewal frequency *S*.

Experimental parameters	UFMOD	ROMICON	PLEIADE
Membrane type	PVC	Polysulfone	CERAMESH 0.1
Module type	Flat plate	Hollow fiber	Flat plate
Crossflow area (m^2)	10^{-4}	5×10^{-4}	0.62×10^{-4}
Membrane area (m^2)	1×10^{-2}	1.4	0.54×10^{-2}
Hydraulic diameter (m)	2×10^{-3}	1.2×10^{-3}	1×10^{-3}
Solute	BSA	Skimmed milk	Dodecane emulsion
Feed concentration ($g\,L^{-1}$ for skimmed milk and BSA, and ppm for dodecane)	0–50	0–85	0–1000
Feed velocity u ($m\,s^{-1}$)	0–2	0–0.65	0–1.2
Transmembrane pressure drop Δp (kPa)	0–180	0–180	0–120
Temperature (°C)	25	25	25
Surface renewal frequency S (Hz)	$S = 0.748 \times 10^{-3}\ u^{0.75}$	$S = 3.48 \times 10^{-3}\ u^{0.661}$	$S = 0.459 \times 10^{-3}\ u^{4.464}$

Source: Modified from Koltuniewicz and Noworyta (1995).

the latter in the modules that they used in their investigation. In a prior study of ultrafiltration of a BSA solution in a flat-plate module, Kołtuniewicz (1992) presented data that showed an almost linear relationship of S with u over a range of $u = 0.1–1.6\ m\,s^{-1}$ (corresponding to a Reynolds number range of 100–1600) and found that S was independent of Δp. From standard mass-transfer coefficient correlations for flow in a tube (Rautenbach and Albrecht, 1989; Koltuniewicz and Noworyta, 1994), it can be deduced that S varies with u raised to a power that ranges between 0.66 and 1.75 as the flow transitions from laminar to turbulent. Equation (9.25) predicts that the limiting or steady-state flux in CFMF is proportional to $S^{0.5}$, which also holds for crossflow ultrafiltration (Chatterjee, 2010). Thus, the limiting flux should be proportional to $u^{0.33}$ for laminar flow and $u^{0.875}$ in the case of turbulent flow. This conclusion agrees closely with the observation of Rudolph and MacDonald (1994) that for modules that operate in the laminar flow regime, the flux increases as the one-third power of the tangential flow rate (or shear), while for devices operating in the turbulent flow regime, the flux increases in proportion to the tangential flow rate. If one assumes the main flow velocity in the small-scale unit, which was not measured by Hasan *et al.* (2013), to be $1.3\ m\,s^{-1}$

(from Table 1 of Hasan *et al.*, 2011), the correlation shown in Table 9.1 for the flat-plate module UFMOD, whose hydraulic diameter D of 2 mm is comparable to the value of 1.8 mm in the experimental system of Hasan *et al.* (2013), predicts a value of S equal to $9.1 \times 10^{-4}\ \text{s}^{-1}$, which can be compared to the average value of $4.4 \times 10^{-4}\ \text{s}^{-1}$ mentioned earlier for the small-scale unit.

From dimensional considerations, it may be postulated that

$$\frac{SD}{u} = a\left(\frac{Du\rho_s}{\mu_s}\right)^b \tag{9.32}$$

where ρ_s and μ_s are the density and the viscosity of the feed suspension, respectively, while a and b are parameters, with a being dependent on the module configuration (Koltuniewicz and Noworyta, 1995).

The values of b are −0.339, −0.25, and 3.464 for the ROMICON, UFMOD, and PLEIADE modules in Table 9.1. Equation (9.32) may be generalized to include the effect of membrane roughness δ as follows:

$$\frac{SD}{u} = g\,(Re, {}^{\delta}/_{D}) \tag{9.33}$$

where $Re\ (= Du\rho_s/\mu_s)$ is the Reynolds number and g denotes a function that has to be determined experimentally.

9.3.3 Data of Mallubhotla and Belfort (1997)

As mentioned previously, Belfort and co-workers have studied the use of Dean vortices in crossflow membrane filtration. Such vortices reduce concentration polarization and solids buildup on the membrane wall, thereby greatly enhancing the permeate flux compared to operation without vortices. Idan and Chatterjee (2015), whose work is recapitulated in this section, correlated both the complete and approximate surface renewal models presented earlier with the transient permeate flux data reported in Figure 3 of the paper by Mallubhotla and Belfort (1997). These data were obtained from CFMF runs using a polysulfone membrane having a nominal pore size of 0.2 μm and a feed suspension containing polydispersed particles (average nominal diameter $d_p = 25\ \mu\text{m}$, density $\rho_p = 1{,}052\ \text{kg m}^{-3}$) that was made by suspension polymerization of styrene-divinyl-benzene (S/DVB). The particle volume fraction ϕ in the suspension was 0.0015 (base case), which, upon multiplying with the particle density ρ_p, yields a value of $1.578\ \text{kg m}^{-3}$ for the bulk concentration c_b of particles in the suspension.

The vortex intensity is expressed by the Dean number ratio D equal to De/De_c, where the Dean number De can be interpreted as the ratio of centrifugal to viscous forces (Mallubhotla and Belfort, 1997). $De = (2du/\vartheta_s)\sqrt{2d/r_c}$ where d is the half-channel height ($2d = R_2 - R_1$), u is the mean axial liquid velocity, ϑ_s is the kinematic viscosity of the feed suspension, r_c is the centerline radius of the curved channel (see below), and R_1 and R_2 are the radii of curvature of the inner and outer walls of the curved channel, respectively. De_c is the critical value of the Dean number (equal to 35.71 in the experimental system of Mallubhotla and Belfort, 1997) above which the fluid becomes unstable, and vortices are produced. Their test cell, which had an active surface area A of 48.4 cm^2, consisted of a thin channel with width W, height H, and length L of 63.5, 1.6, and 76.2 mm, respectively. There was a 180° U-bend (R_1 = 61.9 mm and R_2 = 63.5 mm) through which the feed passed before it entered the test cell, which arrangement generated Dean vortices in the flow field. By reversing the flow direction, they could conduct their CFMF operation in the absence of such vortices. The experiments were performed at a liquid velocity u = 0.57 m s^{-1}, transmembrane pressure drop Δp = 69.8 kPa, and Dean number ratio D = 4.77 (base case).

As mentioned earlier, the feed suspension employed by Mallubhotla and Belfort (1997) contained polydispersed particles of S/DVB whose average nominal diameter was 25 μm. Measurement of the number and volume size distributions by a Coulter counter revealed the highest frequency of particles to occur between 2 and 10 μm. Since these values are much greater than the nominal pore size of 0.2 μm of the polysulfone membrane used by them, it can be conjectured that there was minimal or negligible pore blocking of the membrane in their microfiltration experiments. Therefore, the value of the initial permeate flux J_0 is assumed to be the same as the pure water flux, which was equal to 221 L m^{-2} h^{-1}. Since Mallubhotla and Belfort (1997) did not report the experimental temperature, a temperature of 20 °C is assumed at which the viscosity of water is estimated to be 0.0011 kg m^{-1} s^{-1} (Perry *et al.*, 1984). This value, the pure water flux value, and the value of the transmembrane pressure drop Δp(= 69.8 kPa), along with eq. (9.2), gives a value of the membrane resistance R_m as 9.6×10^{11} m^{-1}. In the calculations, the viscosity of the filtrate is also assumed to be 0.0011 kg m^{-1} s^{-1}.

A brief summary of the experimental parameters of Mallubhotla and Belfort (1997) is provided in Table 9.2. Their paper contains a schematic of the experimental apparatus, a detailed discussion of the experimental conditions and procedures, and an interpretation of the results obtained.

9.3.3.1 Model Validation

9.3.3.1.1 Complete Model Figure 9.4 shows the fit of the complete model to the transient permeate-flux data obtained by Mallubhotla and Belfort (1997) for CFMF runs both without and with Dean vortices. The following observations can be made: (1) there is a decline in the experimental permeate flux with process time, which eventually attains a steady-state value as predicted by theory, (2) the experimental permeate flux with Dean vortices is significantly greater and approaches the limiting or steady-state value faster compared to that without vortices, and

Table 9.2 Experimental details in the study of Mallubhotla and Belfort (1997).

Experimental detail	Value or name
Temperature	20 °C (assumed)
Type of particles in the feed suspension	Styrene-divinyl-benzene (S/DVB)
Average nominal diameter of particles (d_p)	25 μm
Volume fraction of particles (ϕ)	0.0015 (base case)
Density of particles (ρ_p)	1,052 kg m^{-3}
Width of test cell (W)	63.5 mm
Length of test cell (L)	76.2 mm
Height of test cell (H)	1.6 mm
Active area of test cell (A)	48.4 cm^2
Critical Dean number (De_c)	35.71
Dean number ratio ($D = De/De_c$)	4.77 (base case)
Liquid velocity in membrane channel (u)	0.57 m s^{-1}
Transmembrane pressure drop (Δp)	69.8 kPa
Pure water flux	221 L m^{-2} h^{-1}
Type of membrane	0.2 μm polysulfone

Source: Modified from Mallubhotla and Belfort (1997).

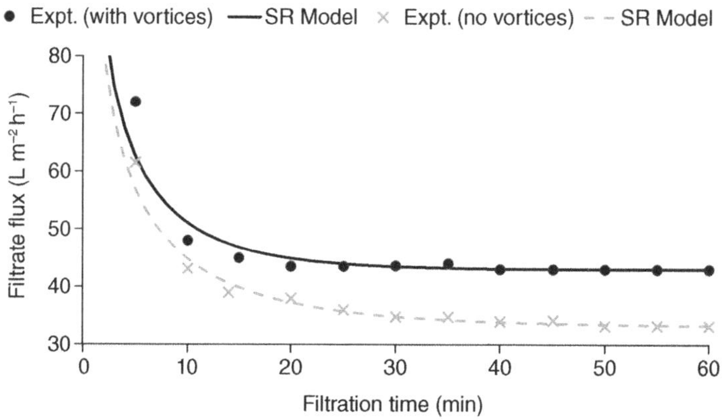

Figure 9.4 Comparison of the predictions of the surface renewal complete model (eq. (9.14)) with the experimental permeate flux data shown in Figure 3 of Mallubhotla and Belfort (1997) for the microfiltration of an S/DVB particulate suspension. Experimental and model parameters are reported in Tables 9.2 and 9.3, respectively. SR denotes surface renewal. Source: Modified from Idan and Chatterjee (2015).

(3) there is a reasonably good fit of the model (eq. (9.14)) to the experimental data, both in the absence and presence of vortices.

Without vortices, the average RMS deviation between the theoretical and experimental permeate fluxes is 2.6%, with the surface renewal rate S having a value of 1.5×10^{-3} s^{-1}. The specific cake resistance α, which measures the resistance offered by the accumulated material on the membrane surface to the flow of permeate, and which depends on the packing density and nature of the cake, is estimated to be 8.42×10^{14} m kg^{-1}, whereas the limiting or steady-state cake mass $m_{c,lim}$, as calculated by eq. (9.22), is found to be 9.70×10^{-3} kg m^{-2}. If this value is multiplied by the value of the active area A of the test cell (Table 9.2), the total mass of accumulated solids on the membrane surface is found to be 47 mg at steady state.

The introduction of flow instabilities through Dean vortices increases the surface renewal frequency by 53% ($S = 2.3 \times 10^{-3}$ s^{-1}), decreases the specific cake resistance by 14.5% ($\alpha = 7.20 \times 10^{14}$ m kg^{-1}), with a resultant increase in the limiting permeate flux J_{lim} by 30% and a decrease in the limiting cake mass by 15.5% ($m_{c,lim} = 8.19 \times 10^{-3}$ kg m^{-2}, total accumulation of solids on the membrane surface at steady state equals 39.7 mg). The RMS deviation between theory and experiment is 4.3%, which is higher than the value of 2.6% when vortices are absent.

Figure 9.5 plots the predicted age averaged cake mass $m_{c,a}$ (calculated by eq. (9.18)) as a function of process time t in the absence and presence of vortices. Both curves begin at a value of zero and approach the corresponding steady-state values mentioned earlier with the progress of filtration. During the first ten minutes, the mass of cake deposited on the membrane wall is greater in the presence of vortices compared to that in their absence. Thereafter, the situation is

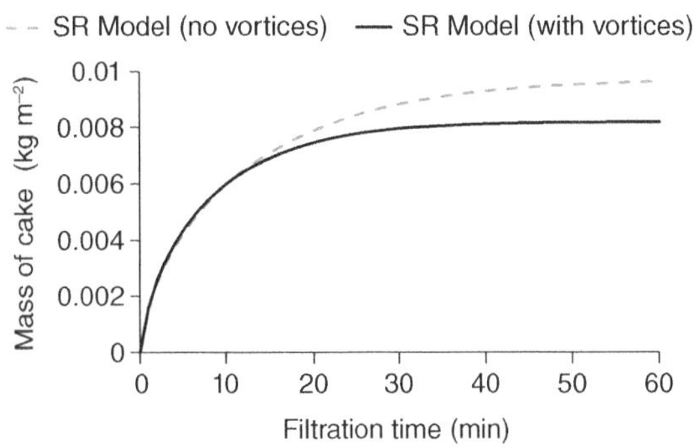

Figure 9.5 Cake buildup with filtration time as predicted by the surface renewal complete model (eq. (9.18)) for the microfiltration of an S/DVB particulate suspension used in the experiments of Mallubhotla and Belfort (1997). Experimental and model parameters are reported in Tables 9.2 and 9.3, respectively. SR denotes surface renewal. Source: Modified from Idan and Chatterjee (2015).

reversed, with the growth of cake in the absence of vortices gradually outpacing that in the presence of vortices. This behavior was theoretically anticipated earlier in Section 9.2.2, although the discussion was presented in the context of the approximate model.

9.3.3.1.2 Approximate Model Calculations performed with the approximate model share the same qualitative features as the complete model, with somewhat different values for the estimated model parameters. Figure 9.6 shows the fit of the approximate model (eq. (9.23)) to the experimental permeate flux data of Mallubhotla and Belfort (1997) without and with Dean vortices. With no vortices, the RMS deviation between the theoretical and experimental permeate fluxes is 2.3%, with the surface renewal frequency S estimated at 1.6×10^{-3} s^{-1}. The specific cake resistance α is estimated to be 1.10×10^{15} $m\,kg^{-1}$, and the limiting cake mass $m_{c,lim}$ (calculated from eq. (9.26)) is found to be 9.1×10^{-3} $kg\,m^{-2}$ (total accumulation of solids on the membrane surface at steady state equals 44 mg).

The introduction of Dean vortices increases the surface renewal frequency S by 50% to 2.4×10^{-3} s^{-1}, decreases the specific cake resistance α by 10.5% to 9.87×10^{14} $m\,kg^{-1}$, enhances the limiting permeate flux J_{lim} by 30%, and reduces the limiting cake mass $m_{c,lim}$ by 13.7% to 7.85×10^{-3} $kg\,m^{-2}$ (total accumulation of solids on the membrane surface at steady state equals 38 mg). The RMS error of fit between theory and experiment is 3.9% compared to 2.3% for the case without vortices.

Figure 9.7 shows the dynamic variation of the theoretical age-averaged cake mass $m_{c,a}$ (calculated from eq. (9.24)) with process time in the absence and presence of vortices. The observed behavior is similar to that for the complete model.

Using subscripts 1 and 2 to denote the situations without and with vortices, respectively, Table 9.3 gives $S_2/S_1 = 1.5$ and $\alpha_1/\alpha_2 = 1.11$ for the approximate model. Since $\alpha_1/\alpha_2 < S_2/S_1$, the behavior of the permeate flux and cake buildup on the membrane surface as a function of process time shown in Figures 9.6 and 9.7 corresponds to Figures A1[a] and A1[b(ii)], respectively, in the appendix of the manuscript of Idan and Chatterjee (2015).

The model parameters, along with RMS deviations between theory and experiment, are gathered together in Table 9.3 for both the complete and approximate models.

Flux Enhancement due to Vortices In the following, an approximate analysis is developed for indicating the relationship between the change in the quantity $\omega = S/\alpha_0$ (due to vortices) and the Dean number ratio D, which, as mentioned earlier, is a measure of vortex intensity.

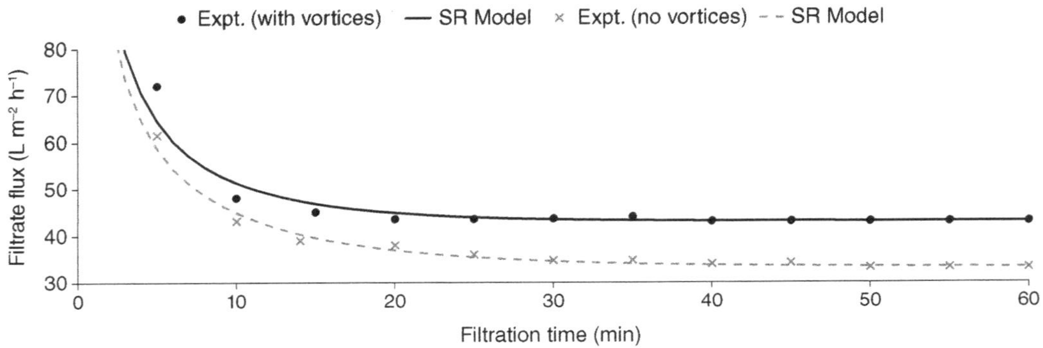

Figure 9.6 Comparison of the predictions of the surface renewal approximate model (eq. (9.23)) with the experimental permeate flux data shown in Figure 3 of Mallubhotla and Belfort (1997) for the microfiltration of an S/DVB particulate suspension. Experimental and model parameters are provided in Tables 9.2 and 9.3, respectively. SR denotes surface renewal. Source: Modified from Idan and Chatterjee (2015).

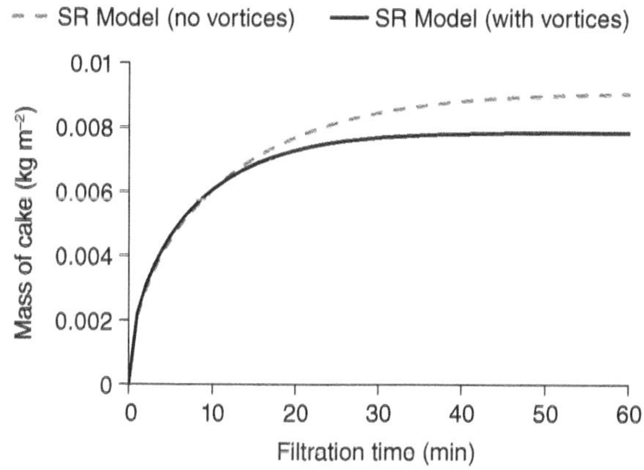

Figure 9.7 Cake buildup with filtration time as predicted by the surface renewal approximate model (eq. (9.24)) for the microfiltration of an S/DVB particulate suspension used in the experiments of Mallubhotla and Belfort (1997). Experimental and model parameters are reported in Tables 9.2 and 9.3, respectively. SR denotes surface renewal. Source: Modified from Idan and Chatterjee (2015).

Table 9.3 Values of the parameters of the surface renewal model and root-mean-square (RMS) deviations between experiment and theory (Figures 9.4 and 9.6).

	No vortices		With vortices	
Parameter	Approximate model	Complete model	Approximate model	Complete model
R_m (m^{-1})	9.6×10^{11}	9.6×10^{11}	9.6×10^{11}	9.6×10^{11}
α (m kg^{-1})	1.10×10^{15}	8.42×10^{14}	9.87×10^{14}	7.20×10^{14}
S (Hz)	1.6×10^{-3}	1.5×10^{-3}	2.4×10^{-3}	2.3×10^{-3}
RMS deviation (%)	2.3	2.6	3.9	4.3

The limiting flux enhancement FE as a function of transmembrane pressure drop Δp is obtained from eq. (9.30) as

$$FE = \frac{J_{\mathrm{lim},2} - J_{\mathrm{lim},1}}{J_{\mathrm{lim},1}} = \sqrt{\frac{S_2/\alpha_2}{S_1/\alpha_1}} - 1 = \sqrt{\frac{S_2/\alpha_{02}}{S_1/\alpha_{01}} \Delta p^{\frac{n_1-n_2}{2}}} - 1 = \sqrt{\frac{\omega_2}{\omega_1} \Delta p^{\frac{n_1-n_2}{2}}} - 1 \tag{9.34}$$

Here, $\omega_1 = S_1/\alpha_{01}$ and $\omega_2 = S_2/\alpha_{02}$, whereas $J_{\mathrm{lim},1}$ and $J_{\mathrm{lim},2}$ are the limiting permeate fluxes without and with vortices, respectively. According to eq. (9.34), the flux enhancement will be far less sensitive to the transmembrane pressure drop than the individual limiting permeate fluxes (see eq. (9.30)). Figure 6 in the paper of Mallubhotla and Belfort (1997) shows the behavior of *FE* as a function of Δp (in the range 36–74 kPa) at four different levels of *D* (obtained by varying the liquid velocity) for an S/DVB particle volume fraction $\phi = 0.00075$ and a 0.2 μm polypropylene membrane. Beyond a value of 36 kPa (the lowest Δp at which the *FE* data are available), *FE* increases only weakly with Δp; i.e., the curve of *FE* versus Δp flattens out. This implies that $n_1 \approx n_2$, i.e., vortices do not affect the compressibility coefficient of the cake, which reduces eq. (9.34) to

$$\frac{\omega_2}{\omega_1} = (1 + FE)^2 \tag{9.35}$$

where *FE* is interpreted as the average flux enhancement over a range of Δp (i.e., 36–74 kPa) at a given value of *D*. Average values of *FE* over the Δp range can be estimated from Figure 6 of Mallubhotla and Belfort (1997) at different values of *D*; such estimates are reported in Table 9.4. The average absolute deviation of the individual *FE* values from the average *FE* value for a particular *D* value is also shown in this table, which gives an idea of the magnitude of the error involved in making the assumption that led to eq. (9.35). As may be seen, the maximum average deviation is 14.5% at $D = 5.4$. The factor $\omega_2/\omega_1 - 1$ can be calculated from the average *FE* value using eq. (9.35); values of this factor are shown in the last column of Table 9.4 and plotted as a function of $D - 1$ in Figure 9.8. The data are represented fairly well by a straight line and can be expressed as

$$\frac{\omega_2}{\omega_1} - 1 = K(D - 1) \tag{9.36}$$

Table 9.4 Data of average *FE* (limiting flux enhancement) as a function of *D* (Dean number ratio) obtained from Figure 6 of Mallubhotla and Belfort (1997) for an S/DVB particle volume fraction (ϕ) of 0.00075 and a 0.2-μm polypropylene membrane in a transmembrane pressure drop (Δp) range of 36–74 kPa (see Table 9.2 for other experimental parameters).

Dean number ratio *D*	Liquid velocity *u* (m s^{-1})	Average limiting *FE*	Average absolute deviation of individual *FE*s from average *FE* (%)	$\omega_2/\omega_1 - 1$ (calculated from eq. (9.35))
1.00	0.119 (calculated)	0 (assumed)	—	0
4.02	0.480	0.23	10	0.50
4.71	0.562	0.28	8.4	0.64
5.40	0.644	0.35	14.5	0.83
6.10	0.728	0.40	7.2	0.96

Source: Modified from Idan and Chatterjee (2015).

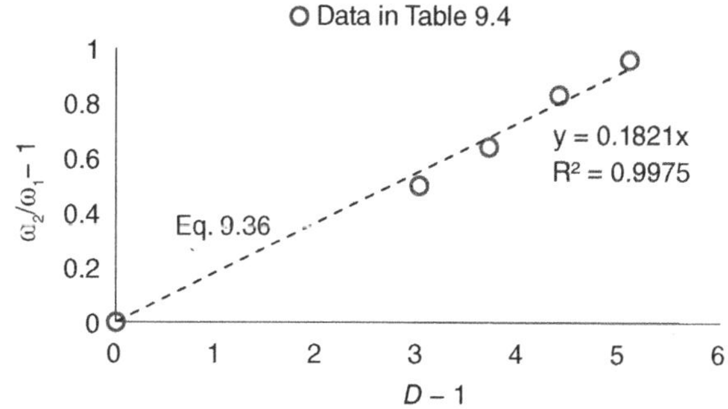

Figure 9.8 Fractional increase of the ratio S/α_0 due to vortices as a function of $D-1$ where D is the Dean number ratio. The dashed line is a fit of eq. (9.36) to the data shown in Table 9.4. Source: Modified from Idan and Chatterjee (2015).

where $K = 0.182$. It should be noted that vortices are only generated when D is larger than 1; thus, the line in Figure 9.8 passes through the origin. Equation (9.36) clearly shows that the fractional increase in the ratio S/α_0 because of vortices is directly related to the vortex intensity, which is represented by D.

Mallubhotla and Belfort (1997) also studied the effect of particle concentration on the steady-state or limiting flux enhancement FE. The data presented in Table 9.5, extracted from Figure 8 of their paper, show that FE increases as the particle volume fraction ϕ in the feed is raised from a value of zero, reaches a maximum of 27% at $\phi = 0.0015$, and then drops down again to zero at $\phi = 0.002$.[4]

Table 9.5 Variation of FE (limiting flux enhancement) with ϕ (particle volume fraction) in the microfiltration of an S/DVB particulate suspension at a value of D (Dean number ratio) equal to 4.77 and transmembrane pressure drop (Δp) of 69.8 kPa for a 0.2-μm polysulfone membrane. Liquid velocity $u = 0.57\ \text{m s}^{-1}$ and critical Dean number $De_c = 35.71$. Data were obtained from Figure 8 of Mallubhotla and Belfort (1997).

ϕ (particle volume fraction)	FE (limiting flux enhancement)	α_1/α_2 (calculated from eq. (9.34))	Figure designation in the appendix of Idan and Chatterjee (2015)
0.0	0	—	—
0.0005	0.22	$0.99 \approx 1$	A2 [a and b]
0.0015	0.27	1.08	A1 [a and b(ii)]
0.0020	0	0.67	A3 [a and b(iii)]
0.0025	0	—	—

Source: Modified from Idan and Chatterjee (2015).

4 The 3% discrepancy between the FE value of 27% shown in Table 9.5 at $\phi = 0.0015$ and the value of 30% at the same value of ϕ mentioned in the discussion of Figures 9.4 and 9.6 is most likely due to human subjectivity in extracting the data of the limiting flux from Figure 3 of their paper.

The authors attributed the flux loss to compression of the deposited cake, increased viscosity, and particle–particle interactions near the membrane surface as ϕ increases.

The following hypothesis is advanced to explain the experimentally observed initial increase and subsequent decrease of FE as ϕ is changed from zero to 0.0015 (Table 9.5). As ϕ is increased, the nature of the cake will change, and the specific cake resistance α is expected to increase, both in the absence and presence of vortices. Although the influence of particle concentration c_b $(= \phi\rho_p)$ does not appear explicitly in eq. (9.34), it is manifested indirectly through the magnitudes of the specific cake resistances α_1 and α_2, with the latter also depending on the strength of the vortex intensity (reflected by the value of S_2). At a fixed value of the Dean number ratio D, the value of S_2/S_1 (>1) would be fixed, assuming that the surface renewal frequency is independent of particle concentration since it is a hydrodynamic parameter. As $\phi \rightarrow 0$ (i.e., $c_b \rightarrow 0$ or $k_c \rightarrow 0$), it follows from eqs. 9.3 and 9.9 that the permeate flux becomes independent of process time and the surface renewal frequency, i.e., it becomes equal to the initial flux J_0.

Hence, $FE \rightarrow 0$, which also implies that cake buildup with or without vortices is the same and approaches zero as may be seen from eqs. 9.5 and 9.17. Table 9.5 also reports numerical values of the ratio α_1/α_2 as a function of FE as determined from eq. (9.34) using a value of $S_2/S_1 = 1.5$ that was obtained from Table 9.3 (approximate model). As ϕ increases from a value of zero, the ratio α_1/α_2 reaches a maximum value of 1.08 at $\phi = 0.0015$ ($FE = 0.27$) and then falls to 0.67 at $\phi = 0.002$ ($FE = 0$). The increase in FE implies that, although both α_1 and α_2 increase with ϕ, α_1 increases at a greater rate than α_2, and therefore in this region, FE increases with ϕ. But this increase occurs only up to a certain value of ϕ beyond which the trend is reversed, and, as ϕ is increased further, α_2 increases at a greater rate than α_1. In this region, FE decreases as ϕ increases, i.e., vortices become less and less effective, with FE becoming zero when a threshold value of ϕ is crossed (0.002 in Table 9.5). On the ascending side of the FE versus ϕ plot (see Figure 8 in Mallubhotla and Belfort, 1997), $\alpha_1/\alpha_2 \approx 1$ at $\phi = 0.0005$, where $FE = 0.22$ (Table 9.5), whereas on the descending side, these same values of α_1/α_2 and FE occur when ϕ is approximately equal to 0.00165. Thus, within the range $0.0005 < \phi < 0.00165$, $\alpha_1/\alpha_2 > 1$, and vortices are effective. Outside this range, $\alpha_1/\alpha_2 < 1$, i.e., vortices increase the specific cake resistance, the reason for which is greater cake compaction (thereby reducing cake porosity) because of a heightened flux of permeate in the presence of vortices immediately after the beginning of filtration, with the compaction effect becoming much more severe at higher levels of ϕ. Thus, in the ascending part of the FE versus ϕ plot, a change in ϕ of 0.0015 leads to an increase of 27% in FE, whereas, in the descending part, a change in ϕ of only 0.0005 causes the same amount of decrease.

It is seen from eq. (9.34) that another critical factor that governs the behavior of *FE* is the ratio, S_2/S_1. Since this factor will be greater than 1 when vortices are introduced into the flow field, there will be some flux enhancement even when $\alpha_1 < \alpha_2$, with the enhancement being especially significant when $\alpha_1 > \alpha_2$, which will occur in a specific particle volume fraction range ($0.0005 < \phi < 0.00165$ in the experiments of Mallubhotla and Belfort, 1997). Thus, there is a range of ϕ within which the use of vortices is advantageous and outside of which the need for them disappears (Mallubhotla and Belfort, 1997). Table 9.5 also lists the designation of the figure in the appendix of the work of Idan and Chatterjee (2015) that corresponds to each individual case.

It was assumed in the preceding discussion that the specific cake resistance α is a function of the feed concentration c_b. For a cake consisting of rigid individual particles (i.e., $n = 0$), α is expected to be constant (equal to α_0, as eq. (9.29) shows), and hence both α_1 and α_2 (and thus their ratio) will be independent of the feed concentration. Also, for this case, $\alpha_1/\alpha_2 > 0$ (i.e., no cake compaction) and the flux enhancement *FE* will therefore be greater than zero (eq. (9.34)) and independent of the particle concentration in the feed, which corresponds to Case 1 in the appendix of Idan and Chatterjee (2015). However, for a compressible cake formed from a flocculated suspension, the resistance of the cake will be governed by floc properties; the method used in preparing the feed suspension; the age and temperature of the material constituting it; floc distortion and breakage due to forces within the cake; and variation of cake properties within the cake layer (McCabe *et al.*, 1993). Moreover, due to consolidation and creep effects, the local (and average) value of α can vary with process time, and the introduction of fluid shear (e.g., vortices) will result in the floc suspension having different material properties compared to those in the absence of shear (Kovalsky *et al.*, 2009). These effects will vary for suspensions of differing feed concentrations, which will be manifested as a dependence of α on c_b. Finally, in the derivation of eq. (9.18) from eqs. 9.5 and 9.17, it was tacitly assumed that the dominant flux of solids toward the surface of the membrane was due to permeate flow (driven by the transmembrane pressure difference) compared to the transport of solids to and from the membrane as a result of the surface renewal mechanism.

It thus becomes evident that the surface renewal model proposed in this chapter, which tacitly assumes a complete cake layer to form on the membrane surface during the initial moments of filtration, and whose resistance (represented by α) thereafter does not change with time, is a highly idealized picture of a very complex process. This is a usual feature of a scientific model, which is an interpretative and partial description of a phenomenon, and which facilitates access (perceptual and intellectual) to that phenomenon (Bailer-Jones, 2009).

Notation

a	parameter in eq. (9.32)
A	active filtration area of test cell (cm^2)
b	parameter in eq. (9.32)
c_b	mass of solids deposited in the filter per unit volume of filtrate (approximately equal to the concentration of solids in the feed or bulk liquid; $kg\,m^{-3}$)
d	half channel height; equal to $(R_2 - R_1)/2$ (mm)
d_p	average nominal diameter of particles in feed suspension (μm)
D	hydraulic diameter of membrane channel (mm or m) or Dean number ratio (De/De_c)
De	Dean number; equal to $(2du/\vartheta_s)\sqrt{2d/r_c}$
De_c	critical Dean number
FE	limiting or steady-state flux enhancement due to vortices; defined by eq. (9.34)
H	height of test cell (mm)
$f(\tau, t)$	transient Danckwerts age-distribution function; given by eq. (9.7) (Hz or s^{-1})
g	function of Re and δ/D in eq. (9.33)
$J(\tau)$	instantaneous permeate flux in a surface element having an age of τ ($m\,s^{-1}$)
$J_a(t)$	age-averaged permeate flux at process time t ($m\,s^{-1}$)
$J_a^*(t^*)$	dimensionless age-averaged permeate flux at dimensionless process time t^*
J_{lim}	limiting or steady-state permeate flux ($m\,s^{-1}$)
J_{lim}^*	dimensionless limiting or steady-state permeate flux; equal to J_{lim}/J_0
J_0	initial permeate flux in a surface element; defined by eq. (9.2) ($m\,s^{-1}$)
k_c	defined by eq. (9.3) ($s\,m^{-2}$)
L	length of test cell (mm)
$m_c(\tau)$	mass of solids accumulated per unit area of membrane wall in a surface element of age τ ($kg\,m^{-2}$)
$m_{c,a}(t)$	age-averaged mass of cake accumulated per unit area of membrane surface at process time t ($kg\,m^{-2}$)
$m_{c,a}^*(t^*)$	dimensionless age-averaged mass of cake at process time t^*
$m_{c,lim}$	limiting or steady-state cake mass per unit area of membrane surface ($kg\,m^{-2}$)
$m_{c,lim}^*$	limiting or steady-state dimensionless cake mass; equal to $m_{c,lim}J_0k_c/c_b$

n	cake compressibility coefficient
r_c	centerline radius of curved channel (mm)
R^2	correlation coefficient
R_m	hydraulic resistance of membrane (m^{-1})
R_1	radius of curvature of inner wall of curved channel (mm)
R_2	radius of curvature of outer wall of curved channel (mm)
Re	Reynolds number; equal to $Du\rho_s/\mu_s$
S	surface renewal frequency (Hz or s^{-1})
S_1	surface renewal frequency when no vortices are present (Hz or s^{-1})
S_2	surface renewal frequency when vortices are present (Hz or s^{-1})
S^*	dimensionless surface renewal frequency; defined by eq. (9.11)
t	process time (s)
t^*	dimensionless process time; defined by eq. (9.12)
u	mean axial liquid or feed velocity in the main flow direction in membrane channel ($m\,s^{-1}$)
W	width of test cell (mm)
x	parameter of $\Gamma(x, y)$
y	parameter of $\Gamma(x, y)$

Greek Letters

α	specific cake resistance ($m\,kg^{-1}$)
α_0	constant in eq. (9.29) ($m\,kg^{-1}$)
α_{01}	value of α_0 when no vortices are present ($m\,kg^{-1}$)
α_{02}	value of α_0 when vortices are present ($m\,kg^{-1}$)
α_1	specific cake resistance when no vortices are present ($m\,kg^{-1}$)
α_2	specific cake resistance when vortices are present ($m\,kg^{-1}$)
$\Gamma(x, y)$	incomplete gamma function; defined by eq. (9.20)
δ	membrane roughness (m)s
Δp	transmembrane pressure drop (Pa or kPa)
λ	variable of integration in eq. (9.4) and eq. (9.20)
μ	viscosity of permeate ($kg\,m^{-1}\,s^{-1}$)
μ_s	viscosity of feed suspension ($kg\,m^{-1}\,s^{-1}$)
ϑ_s	kinematic viscosity of feed suspension ($mm^2\,s^{-1}$)
ρ_p	density of particles in feed suspension ($kg\,m^{-3}$)
ρ_s	density of feed suspension ($kg\,m^{-3}$)
τ	age or residence time of a liquid element at membrane–liquid interface (s)
ϕ	volume fraction of particles in feed suspension

ω	S/α_0 (kg m^{-1} s^{-1})
ω_1	S_1/α_{01} (kg m^{-1} s^{-1})
ω_2	S_2/α_{02} (kg m^{-1} s^{-1})

Acronyms

BOD	biological oxygen demand
CFMF	crossflow microfiltration
COD	chemical oxygen demand
GFD	gallons per square feet per day
MBR	membrane bioreactor
MGD	million gallons per day
OD	optical density
RMS	root mean square
RO	reverse osmosis
S/DVB	styrene-divinyl-benzene
TSS	total suspended solids

Homework Exercises

i) Derive eqs. 9.14 and 9.18 of the complete surface renewal model.

ii) Derive eqs. 9.23 and 9.24 of the approximate surface renewal model.

iii) Check the validity of eqs. 9.27 and 9.28 of the approximate surface renewal model.

iv) Develop a relationship between the flux enhancement *FE* and the Dean number ratio *D* from eqs. 9.35 and 9.36. Plot the relation for $K = 0.182$ and D varying between 1 and 6. Would it be appropriate to use a straight line to describe this relation?

v) A crossflow microfiltration membrane system has been proposed to generate a permeate flow rate of 1,000 L h^{-1} continuously from an aqueous suspension containing polymer particles. A transmembrane pressure drop Δp of 200 kPa will be employed, and the system will operate at a temperature of 20 °C. Develop plots showing the required membrane filtration area (m^2), permeate flux (L m^{-2} h^{-1}), and mass of cake accumulated on the membrane surface (g m^{-2}) as a function of liquid velocity u in a membrane channel for u varying between 0.1 and 0.6 m s^{-1}. Use the correlation for the surface renewal frequency S as a function of u for the ROMICON module shown in Table 9.1 and the parameter values in Table 9.3 for your design. Perform calculations for feed concentrations of 2 and 20 g L^{-1}. Provide a brief discussion of your results.

References

Arnot, T.C., Field, R.W., and Koltuniewicz, A.B. (2000). Cross-flow and dead-end microfiltration of oily-water emulsions. *Journal of Membrane Science* 169 (1): 1–15. Available at: http://dx.doi.org/10.1016/S0376-7388(99)00321-X.

Bailer-Jones, D.M. (2009). *Scientific models in philosophy of science.* Pittsburgh, Pennsylvania: University of Pittsburgh Press.

Bailey, S.M. and Meagher, M.M. (1997a). Crossflow microfiltration of recombinant *Escherichia coli* lysates after high pressure homogenization. *Biotechnology and Bioengineering* 56 (3): 304–310. Available at: http://dx.doi.org/10.1002/(SICI)1097-0290(19971105)56:3%3c304::AID-BIT8%3e3.0.CO;2-N.

Bailey, S.M. and Meagher, M.M. (1997b). The effect of denaturants on the crossflow membrane filtration of Escherichia coli lysates containing inclusion bodies. *Journal of Membrane Science* 131 (1–2): 29–38. Available at: http://dx.doi.org/10.1016/S0376-7388(97)00005-7.

Chatterjee, S.G. (2010). On the use of the surface-renewal concept to describe cross-flow ultrafiltration. *Indian Chemical Engineer* 52 (3): 179–193. Available at: http://dx.doi.org/10.1080/00194506.2010.547927.

Cheryan, M. (1998). *Ultrafiltration and microfiltration handbook*, Seconde. Boca Raton, Florida: CRC Press.

Chung, K.Y., Brewster, M.E., and Belfort, G. (1996). Dean vortices with wall flux in a curved channel membrane system: 2. The velocity field. *AIChE Journal* 42 (2): 347–358. Available at: http://dx.doi.org/10.1002/aic.690420205.

Chung, K.Y., Brewster, M.E., and Belfort, G. (1998). Dean vortices with wall flux in a curved channel membrane system: 3. Concentration polarization in a spiral reverse osmosis slit. *Journal of Chemical Engineering of Japan* 31 (5): 683–693. Available at: http://dx.doi.org/10.1252/jcej.31.683.

Constenla, D.T. and Lozano, J.E. (1996). Predicting stationary permeate flux in the ultrafiltration of apple juice. *Lebensmittel-Wissenschaft und -Technologie* 29: 587–592.

Dąbkowska-Susfał, K., Półgrabska, A., Sobieszuk, P., and Kołtuniewicz, A.B. (2024). Mathematical modeling of the enzymatic hydrolysis of lignocellulosic waste in membrane bioreactor considering transport phenomena. *Chemical Engineering Research and Design* 208: 656–665. Available at: http://dx.doi.org/10.1016/j.cherd.2024.07.034.

Dhagumudi, V. and Yan, D. (2012). Pilot study of pulp & paper mill effluent treatment with MBR-RO system. *Proceedings of the Water Environment Federation* 15: 1721–1733.

Gehlert, G., Luque, S., and Belfort, G. (1998). Comparison of ultra- and microfiltration in the presence and absence of secondary flow with polysaccharides, proteins, and yeast suspensions. *Biotechnology Progress* 14 (6): 931–942. Available at: http://dx.doi.org/10.1021/bp980101p.

Gurian, P.L. (2010). 'Microfiltration cost benchmarking for large facilities', *Drexel University. Department of Civil, Architectural, and Environmental Engineering. Faculty Research and Publications* [Preprint]. Available at: https://researchdiscovery.drexel.edu/esploro/outputs/journalArticle/Microfiltration-Cost-Benchmarking-for-Large-Facilities/991014632438404721 (Accessed: 17 November 2024).

Hasan, A., Peluso, C.R., Hull, T.S. et al. (2013). A surface-renewal model of cross-flow microfiltration. *Brazilian Journal of Chemical Engineering* 30 (1): 167–186. Available at: http://dx.doi.org/10.1590/S0104-66322013000100019.

Hasan, A., Yasarla, R., Ramarao, B.V., and Amidon, T.E. (2011). Separation of lignocellulosic hydrolyzate components using ceramic microfilters. *Journal of Wood Chemistry and Technology* 31 (4): 357–383. Available at: http://dx.doi.org/10.1080/02773813.2011.578232.

Huang, L. and Morrissey, M.T. (1998). Fouling of membranes during microfiltration of surimi wash water: Roles of pore blocking and surface cake formation. *Journal of Membrane Science* 144 (1–2): 113–123. Available at: http://dx.doi.org/10.1016/S0376-7388(98)00038-6.

Idan, G. and Chatterjee, S.G. (2015). Application of a surface-renewal model to permeate-flux data of constant-pressure cross-flow microfiltration with Dean vortices. *Brazilian Journal of Chemical Engineering* 32 (2): 609–627. Available at: http://dx.doi.org/10.1590/0104-6632.20150322s00003417.

Jiang, S. and Chatterjee, S.G. (2015). A surface-renewal model for constant flux cross-flow microfiltration. *Journal of Applied Polymer Science* 132 (21). Available at: http://dx.doi.org/10.1002/app.41778.

Kołtuniewicz, A. (1992). Predicting permeate flux in ultrafiltration on the basis of surface renewal concept. *Journal of Membrane Science* 68 (1–2): 107–118. Available at: http://dx.doi.org/10.1016/0376-7388(92)80153-B.

Koltuniewicz, A. and Noworyta, A. (1994). Dynamic properties of ultrafiltration systems in light of the surface renewal theory. *Industrial & Engineering Chemistry Research* 33 (7): 1771–1779. Available at: http://dx.doi.org/10.1021/ie00031a016.

Koltuniewicz, A. and Noworyta, A. (1995). Method of yield evaluation for pressure-driven membrane processes. *Chemical Engineering Journal* 58: 175–182.

Kovalsky, P., Bushell, G., and Waite, T.D. (2009). Prediction of transmembrane pressure build-up in constant flux microfiltration of compressible materials in the absence and presence of shear. *Journal of Membrane Science* 344 (1–2): 204–210. Available at: http://dx.doi.org/10.1016/j.memsci.2009.08.005.

Kroner, K.H., Schütte, H., Hustedt, H., and Kula, M. (1984). Cross-flow filtration in the downstream processing of enzymes. *Process Biochemistry* 19: 67–74.

Le, M.S., Spark, L.B., Ward, P.S., and Ladwa, N. (1984). Microbial asparaginase recovery by membrane processes. *Journal of Membrane Science* 21 (3): 307–319. Available at: http://dx.doi.org/10.1016/S0376-7388(00)80221-5.

Lesage, F., Midoux, N., Latifi, A., and M. (2002). Momentum transfer in a fixed-bed reactor described by the surface renewal model. *Chemical Engineering Science* 57 (24): 5115–5122. Available at: http://dx.doi.org/10.1016/S0009-2509(02)00426-8.

Luque, S., Mallubhotla, H., Gehlert, G. et al. (1999). A new coiled hollow-fiber module design for enhanced microfiltration performance in biotechnology. *Biotechnology and Bioengineering* 65 (3): 247–257. Available at: http://dx.doi.org/10.1002/(SICI)1097-0290(19991105)65:3%3c247::AID-BIT1%3e3.0.CO;2-H.

Mallubhotla, H. and Belfort, G. (1997). Flux enhancement during Dean vortex microfiltration. 8. Further diagnostics. *Journal of Membrane Science* 125 (1): 75–91. Available at: http://dx.doi.org/10.1016/S0376-7388(96)00125-1.

Mallubhotla, H., Belfort, G., Edelstein, W.A., and Early, T.A. (2001). Dean vortex stability using magnetic resonance flow imaging and numerical analysis. *AIChE Journal* 47 (5): 1126–1140. Available at: http://dx.doi.org/10.1002/aic.690470519.

Mallubhotla, H., Hoffmann, S., Schmidt, M. et al. (1998). Flux enhancement during dean vortex tubular membrane nanofiltration. 10. Design, construction, and system characterization. *Journal of Membrane Science* 141 (2): 183–195. Available at: http://dx.doi.org/10.1016/S0376-7388(97)00302-5.

Mallubhotla, H., Schmidt, M., Hyun Lee, K., and Belfort, G. (1999). Flux enhancement during Dean vortex tubular membrane nanofiltration: 13. Effects of concentration and solute type. *Journal of Membrane Science* 153 (2): 259–269. Available at: http://dx.doi.org/10.1016/S0376-7388(98)00255-5.

Marston, F.A.O., Lowe, P.A., Doel, M.T. et al. (1984). Purification of calf prochymosin (prorennin) synthesized in Escherichia coli. *Nature Biotechnology* 2 (9): 800–804. Available at: http://dx.doi.org/10.1038/nbt0984-800.

McCabe, W.L., Smith, J.C., and Harriott, P. (1993). *Unit operations of chemical engineering*, Fifthe. New York: McGraw-Hill.

Perry, R.H., Green, D.W., and Maloney, J.O. (ed.) (1984). *Perry's chemical engineers' handbook*, Sixthe. New York: McGraw-Hill.

Quezada, C., Estay, H., Cassano, A. et al. (2021). Prediction of permeate flux in ultrafiltration processes: a review of modeling approaches. *Membranes* 11 (5): 368. Available at: http://dx.doi.org/10.3390/membranes11050368.

Rautenbach, R. and Albrecht, R. (1989). *Membrane processes*. New York: John Wiley & Sons.

Richardson, J.F., Harker, J.H., and Backhurst, J.R. (2002). *Coulson and Richardson's chemical engineering. Volume 2*, Fifthe. Oxford: Butterworth-Heinemann.

Rudolph, E.A. and MacDonald, J.H. (1994). Tangential flow filtration systems for clarification and concentration. In: *Bioprocess engineering: Systems, equipment and facilities* (ed. B.K. Lydersen, N.A. D'Elia, and K.L. Nelson), 119–157. New York: John Wiley & Sons.

Sablani, S., Goosen, M., Al-Belushi, R., and Wilf, M. (2001). Concentration polarization in ultrafiltration and reverse osmosis: A critical review. *Desalination* 141 (3): 269–289. Available at: http://dx.doi.org/10.1016/S0011-9164(01)85005-0.

Sarkar, D., Datta, D., Sen, D., and Bhattacharjee, C. (2011). Simulation of continuous stirred rotating disk-membrane module: An approach based on surface renewal theory. *Chemical Engineering Science* 66 (12): 2554–2567. Available at: http://dx.doi.org/10.1016/j.ces.2011.02.056.

Schoner, R.G., Ellis, L.F., and Schoner, B.E. (1985). Isolation and purification of protein granules from Escherichia coli cells overproducing bovine growth hormone. *Nature Biotechnology* 3 (2): 151–154. Available at: http://dx.doi.org/10.1038/nbt0285-151.

Song, L. (1998). Flux decline in crossflow microfiltration and ultrafiltration: mechanisms and modeling of membrane fouling. *Journal of Membrane Science* 139 (2): 183–200. Available at: http://dx.doi.org/10.1016/S0376-7388(97)00263-9.

Titchener-Hooker, N.J., Gritisis, D., Mannweiler, K. et al. (1991). Integrated process design for producing and recovering proteins from inclusion bodies. *BioPharm* 4: 34–38.

Zhang, W. and Chatterjee, S.G. (2015). Influence of residence time distribution on a surface-renewal model of constant-pressure cross-flow microfiltration. *Brazilian Journal of Chemical Engineering* 32 (1): 139–154. Available at: http://dx.doi.org/10.1590/0104-6632.20150321s00003129.

10

Constant Flux Crossflow Microfiltration*

10.1 Introduction

As mentioned in Chapter 9, crossflow membrane filtration (CFMF) has come into wide use in the chemical and biotech industries in many countries and also in wastewater treatment. In this type of filtration, an incoming feed solution or suspension passes over the surface of a membrane, and the permeate flow is that portion of the liquid that passes through the membrane in a direction perpendicular to that of the main flow. The magnitude of the permeate flux depends on the liquid velocity over the membrane surface, permeate viscosity, dissolved/suspended solids concentration, applied transmembrane pressure drop (TMP), temperature, membrane characteristics, and membrane fouling. It was also seen in that chapter that for a constant TMP, the permeation flux declines with the advance of process time due to fouling of the membrane by pore blocking, concentration polarization, and cake buildup on its surface.

The majority of studies reported in the literature on CFMF is for constant TMP operation, with only a few studies being available for operation under constant flux conditions. Some examples of the latter are the works of Ho and Zydney (2002), Stressmann (2008), Kovalsky *et al.* (2009), Miller *et al.* (2013), Miller *et al.* (2014), Kirschner *et al.* (2019), and Peles *et al.* (2023). In constant TMP operation, the permeate flux decline with process time causes a changing environment near the membrane surface as regards the hydrodynamics and accumulation of solids. Thus, maintaining the flux at a constant level by raising the TMP with time will enable a better control of this environment (Miller *et al.*, 2013) and reduce the high initial fouling often seen in constant TMP operation (Stressmann, 2008). The numerical models for CFMF under constant flux conditions presented by Ho and Zydney (2002) and Kovalsky *et al.* (2009), whose data will be examined

*The material in this chapter has been adapted from Jiang and Chatterjee (2015).

Applications of the Surface Renewal Model of Mass Transfer, First Edition.
Hengshuo Huang and Siddharth G. Chatterjee.

Companion website: www.wiley.com/go/huang/surfacerenewalmodel

later, do not explicitly include the effect of flow instability or shear generated by stirring or axial flow of liquid over the membrane surface on membrane performance, although the combined pore-blockage cake filtration model of Ho and Zydney (2002) contains a parameter f, which according to the authors, "can also account for the reduction in protein deposition due to any back-flux phenomena, including the effects of crossflow and any long-range electrostatic interactions."

Stressmann (2008), whose work includes a comprehensive review of the major factors important in the microfiltration of biological solutions, studied constant flux crossflow filtration of bovine serum albumin (BSA) solutions and Chinese hamster ovary (CHO) cell culture broths in a hollow fiber membrane module (0.15 μm polysulfone, 30 cm fiber length, 13 fibers, 1 mm inner fiber diameter) at a permeate flux of 30 $L\,m^{-2}\,h^{-1}$ (LMH) at crossflow velocities of 0.5 and 1 $m\,s^{-1}$ (corresponding to shear rates of 4000 and 8000 s^{-1}, respectively; laminar flow regime) at feed concentrations of 2 and 4 $kg\,m^{-3}$ (BSA), and 9.8 and 10.1 $kg\,m^{-3}$ (CHO cell culture broth). Extending the model of Ho and Zydney (2002) to account for crossflow effects and assuming the cake on the membrane surface to be incompressible due to the very small permeate flux, Stressmann (2008) found an excellent fit of the extended model to the experimentally observed convex-type TMP rise with process time for both BSA and CHO cell culture broth solutions.

Miller *et al.* (2013) developed a complex laboratory-scale system for studying fouling in constant flux CFMF, whose design and construction are described in detail in their paper. They used two different types of membranes: (1) polyvinylidene fluoride (PVDF) hydrophilized microfiltration membrane with a nominal pore size of 0.22 μm and (2) polysulfone ultrafiltration membrane having a nominal molecular weight cutoff of 20 kDa (PS-20). Fouling runs were conducted with the PS-20 membrane and a soyabean-oil emulsion (1500 ppm) using a feed velocity of 0.18 $m\,s^{-1}$ (Reynolds number: 1000), feed pressure of 2.1 barg, and constant fluxes of 40 and 85 LMH. A dramatic difference in the TMP evolution between these two limits was observed, suggesting severe fouling at 85 LMH. By flux-stepping experiments (flux step: 10 LMH; time step: 10 min), they estimated the threshold flux (explained below) of the PS-20 membrane to be 68 LMH. With their system and the PVDF microfiltration membrane, they were also able to replicate a previous literature study on membrane fouling using a polystyrene latex bead (3.0 μm) suspension at a feed pressure of 0.20 barg, and crossflow velocity of 0.2 $m\,s^{-1}$ during fouling and 0.5 $m\,s^{-1}$ during 5 min cleaning stages between the flux steps (operating fluxes: 66, 99, 165, 230, and 273 LMH).

Kirschner *et al.* (2019) conducted constant flux crossflow fouling experiments with latex bead suspensions and soybean oil emulsions using commercial polyether sulfone flat sheet ultrafiltration membranes. Starting from the conventional membrane fouling models, which have been used in the literature for constant TMP operation, they developed a combined intermediate pore blocking and cake filtration model for predicting TMP behavior with time in constant flux crossflow filtration. Their model includes a crossflow term to account for the removal of deposited particles by liquid shear over the membrane surface. They arrived at the conclusion that below the threshold flux, the chief fouling mechanism is intermediate pore blocking, while above it, cake filtration dominates.[1] The value of the threshold flux, usually estimated by flux-stepping experiments, can also be determined from their model.

Peles *et al.* (2023) derived a combined (intermediate) pole blocking and cake filtration model for describing protein fouling in constant flux filtration, which was a further development of the previous model of Ho and Zydney (2002). Their new model, whose parameters were determined by previous constant TMP experiments and which did not include the effect of flow instability or liquid shear, showed good agreement with their experimental measurements. They concluded that at low fluxes, the transient TMP profile is determined primarily by the rate of cake growth since the membrane pores are blocked rapidly, whereas at high fluxes, when the cake resistance becomes very large, the TMP is determined by the slow rate of pore blockage.

As seen in Chapters 1, 8, and 9, the surface renewal concept has been used in a number of investigations to theoretically model constant TMP crossflow microfiltration and ultrafiltration. The only work that has used this concept to model the constant flux mode of operation, which is becoming increasingly more common, for example, in industrial water purification processes (Miller *et al.*, 2013) and continuous bioprocess operations (Peles *et al.*, 2023), is that of Jiang and Chatterjee (2015). This work, recapitulated in this chapter, concerns a mathematical model for constant flux CFMF, which is a synthesis of classical cake filtration theory and the surface renewal concept through which the effect of flow instability, originating from random hydrodynamic impulses generated at the membrane–liquid interface due to membrane roughness or by the use of spacers or turbulence promoters, on membrane performance is taken into account. In contrast to the models of Ho and Zydney (2002) and Kovalsky *et al.* (2009), the surface renewal model provides explicit analytical expressions for the TMP and cake mass buildup on the membrane surface as a function of filtration or process time. It has two versions – the complete model, which includes the effects of cake compressibility, and a subsidiary model, which can be derived from the complete model that is valid for incompressible cakes. The complete model is correlated against the experimental TMP data of Ho and Zydney (2002) for constant flux microfiltration of a BSA solution in a stirred cell using polycarbonate track-etched

1 Kirschner *et al.* (2019) proposed the following physical interpretation of the threshold flux: "… the threshold flux is the flux below which cake buildup is negligible and above which cake filtration becomes the dominant fouling mechanism."

membranes, and also with the TMP data of Kovalsky *et al.* (2009) for constant flux yeast filtration in a stirred cell. The subsidiary model is correlated with the experimental TMP data of Miller *et al.* (2014) for constant flux filtration of a soybean-oil emulsion in a crossflow cell having unmodified and surface-modified fouling-resistant membranes.

10.2 Surface Renewal Model of Crossflow Microfiltration

In the surface renewal model, it is postulated that the dominant fouling mechanism responsible for the decline in permeate flux (constant TMP operation) or increase in TMP (constant flux operation) over time is cake formation, with pore blocking occurring only during the initial moments of filtration, which effect, if important, can be incorporated into the membrane resistance R_m. Flow instabilities perennially bring fresh liquid elements from the bulk liquid to the membrane surface, where they reside for a certain amount of time after which they depart and remix with the bulk liquid. Above the surface elements, the liquid is assumed to be well mixed, and its solid concentration is held constant due to a high rate of transport due to flow instability from this location to the bulk liquid. A cake layer gradually builds up on the membrane wall, thereby causing an increase in the TMP with process time under constant flux conditions. It is assumed that during the residence time τ of a liquid element at the membrane surface, TMP buildup within it can be described by classical cake filtration theory (McCabe *et al.*, 1993). From this theory, the pressure drop Δp_c across the cake in a surface element with a residence time of τ can be expressed as

$$(\Delta p_c)^{1-n} = K_r \tau \tag{10.1}$$

where

$$K_r = \mu J^2 c_b \alpha_0 \tag{10.2}$$

and J is the constant permeate flux, μ the viscosity of the filtrate, and c_b the mass of solids accumulated on the membrane wall per unit volume of filtrate passing through it (approximately equal to the bulk or feed concentration of solids), which is assumed to be constant. The two compressibility parameters of the cake α_0 and n are empirical constants, where n is the compressibility coefficient of the cake, which is zero for incompressible sludges and lies between 0.2 and 0.8 for compressible ones (McCabe *et al.*, 1993). These quantities are related to the specific cake resistance α via the empirical equation

$$\alpha = \alpha_0 \Delta p^n \tag{10.3}$$

where Δp is the (total) TMP in a surface element that is given by

$$\Delta p = \Delta p_c + \Delta p_m \tag{10.4}$$

Here, Δp_m is the pressure drop across the membrane, which can be expressed as

$$\Delta p_m = \mu J R_m \tag{10.5}$$

with R_m being the membrane resistance, as mentioned earlier.

The mass m_c of solids accumulated in a liquid element per unit area of membrane surface during its residence time of τ is given by

$$m_c(\tau) = J c_b \tau \tag{10.6}$$

The surface of the membrane at a given process time t during the filtration is visualized as being populated by a mosaic of liquid elements having ages in the time interval zero to t. If we denote the age distribution function of the surface elements by $f_D(\tau, t)$, the age-averaged cake pressure drop $\Delta p_{c,a}$ and cake mass $m_{c,a}$ accumulated per unit area of the membrane surface at time t are then given by

$$\Delta p_{c,a}(t) = \int_0^t \Delta p_c(\tau) f_D(\tau, t) d\tau \tag{10.7}$$

and

$$m_{c,a}(t) = \int_0^t m_c(\tau) f_D(\tau, t) d\tau \tag{10.8}$$

In eq. 10.7, the cake pressure drop Δp_c in an individual surface element is looked upon as an "information content" or "stress level" of the element. Thus, for specified values of the process time t and imposed constant permeate flux J, an older element will require a greater Δp_c due to more cake buildup within it compared to a younger element (see eq. 10.1).

As shown by Zhang and Chatterjee (2015), different speculative hypotheses about the behavior of liquid elements on the membrane surface, corresponding to different startup conditions, lead to different surface age distribution functions. These, when used in eqs. 10.7 and 10.8, will give different expressions for the age-averaged pressure drop across the cake and cake mass buildup at process time t. As done in previous chapters, the transient Danckwerts age distribution function given by eq. 2.10 (Chapter 2) will be used to represent the ages of surface elements, that is,

$$f_D(\tau, t) = \frac{S e^{-S\tau}}{1 - e^{-St}} \tag{10.9}$$

where S, assumed to be constant, is the rate or frequency of renewal of liquid elements at the membrane wall – a hydrodynamic parameter, which depends upon the level of flow instability or turbulence. As seen in Chapter 9, it increases with the velocity of the main flow.

10.2.1 Complete Model: Compressible Cake ($n \neq 0$)

From eqs. 9.1 to 9.9, it can be shown that

$$\Delta p_{\mathrm{a}}(t) = \Delta p_{\mathrm{c,a}}(t) + \Delta p_{\mathrm{m}} = \frac{K_{\mathrm{r}}^{p-1} S^{1-p}}{1 - e^{-St}} [\Gamma(p) - \Gamma(p, St)] + \mu J R_{\mathrm{m}} \tag{10.10}$$

and

$$m_{\mathrm{c,a}}(t) = J c_{\mathrm{b}} \left(\frac{1}{S} - \frac{t e^{-St}}{1 - e^{-St}} \right) \tag{10.11}$$

Here, Δp_{a} is the total TMP at process time t, and $\Gamma(x, y)$ is the incomplete gamma function defined by

$$\Gamma(x, y) = \int_{y}^{\infty} \lambda^{x-1} e^{-\lambda} \mathrm{d}\lambda \tag{10.12}$$

with p given by

$$p = \frac{2 - n}{1 - n} \tag{10.13}$$

As mentioned previously, n usually lies between 0.2 and 0.8, and hence p is expected to be a positive quantity. The first term on the right-hand side of eq. 10.10, which increases with process time, is the contribution of cake buildup to the TMP, whereas the second term, which stays invariant, is the contribution of the membrane. Equation 10.10 contains the possibility of an inflection point occurring in the theoretical TMP profile, that is, at a certain value of t, $\mathrm{d}^2\Delta p_{\mathrm{a}}/\mathrm{d}t^2 = 0$. According to eq. 10.11, the transient cake mass $m_{\mathrm{c,a}}$ is independent of the compressibility parameter α_0 and viscosity μ of the permeate and is solely governed by the feed concentration c_{b}, permeate flux J, and surface renewal frequency S.

Let us examine the behavior of eqs. 10.10 and 10.11 as $St \to 0$, that is, as $S \to 0$ (low flow instability) or as $t \to 0$ (the initial moments of filtration). Differentiating eq. 10.10 with respect to t results in

$$\frac{\mathrm{d}\Delta p_{\mathrm{a}}}{\mathrm{d}t} = K_{\mathrm{r}}^{p-1} S^{2-p} e^{-St} \left[\frac{(St)^{p-1}(1 - e^{-St}) + \Gamma(p, St) - \Gamma(p)}{(1 - e^{-St})^2} \right] \tag{10.14}$$

As $St \to 0$, eq. 10.14 reduces to

$$\frac{\mathrm{d}\Delta p_{\mathrm{a}}}{\mathrm{d}t}(St \to 0) = K_{\mathrm{r}}^{p-1} t^{p-2} \left(\frac{p-1}{p} \right) \tag{10.15}$$

which upon integration yields[2]

$$\Delta p_{\mathrm{a}}(St \to 0) = \frac{K_{\mathrm{r}}^{p-1}}{p} t^{p-1} + \Delta p_{\mathrm{m}} \tag{10.16}$$

For an incompressible cake, $n = 0$ and eq. 10.13 gives $p = 2$. As n changes from 0 to 0.8 (i.e., as cake compressibility increases), p increases from 2 to 6.

10.2.1.1 Variation of TMP with Process Time

According to eq. 10.15, at a fixed value of K_{r}, the rate of change of the TMP with process or filtration time is proportional to t^{p-2} for $t \to 0$ or $S \to 0$. Also, eq. 10.16 predicts that during the initial moments of filtration or when the level of flow instability is low, the TMP will vary with process time raised to a power of $p - 1$. Thus for $p = 2$ (incompressible cake), the TMP will increase linearly with process time as $St \to 0$, whereas for $p = 6$ (a highly compressible cake), it will increase as the fifth power of process time during the initial moments of filtration or for low levels of flow instability, thus displaying very sharp concavity with respect to the process time (i.e., horizontal) axis.

10.2.1.2 Variation of TMP with Permeate Flux

According to eq. 10.2, K_{r} is proportional to the square of the permeate flux, J. Furthermore, as can be inferred from eq. 10.15, the rate of increase of TMP with process time is proportional to $J^{2(p-1)}$ for $t \to 0$ or $S \to 0$. Thus, as p changes from 2 to 6 (i.e., as the cake becomes more compressible), this rate will change proportionately as J^2 to J^{10} for $St \to 0$, severely increasing the concavity of the TMP versus process time curve.

Thus, the shape of the TMP versus process time profile is very sensitive to the cake compressibility parameter n and permeate flux J during the initial moments of filtration or for low levels of flow instability. Such concave experimental TMP curves and the effect of permeate flux on the shape of the TMP profile can be seen in the experimental data reported by Ho and Zydney (2002) and Kovalsky *et al.* (2009) for constant flux microfiltration, which will be discussed later.

Differentiating eq. 10.11 with respect to t gives

$$\frac{\mathrm{d}m_{\mathrm{c,a}}}{\mathrm{d}t} = Jc_{\mathrm{b}} \left[\frac{e^{St}(St - 1) + 1}{(e^{St} - 1)^2} \right] \tag{10.17}$$

2 In the development presented by Jiang and Chatterjee (2015), the factor $(p - 1)/p$ in eq. 10.15 is absent, and the denominator of the first term on the right-hand side of eq. 10.16 is $p - 1$. This will only affect the magnitude of α_0 and not the nature of the TMP versus time relationship. The maximum difference in α_0 between these two formulations for the data examined in this chapter is about 15%. Tables 10.1–10.3 show the values of α_0 obtained by using eq. 10.16, which can be compared with those reported by Jiang and Chatterjee (2015) for the same data.

It therefore follows that as $St \to 0$

$$\frac{dm_{c,a}}{dt}(St \to 0) = Jc_b \tag{10.18}$$

i.e.,

$$m_{c,a}(St \to 0) = Jc_b t \tag{10.19}$$

According to eq. 10.19, the mass of cake on the membrane surface increases linearly with process time during the initial moments of filtration or for low levels of flow instability. It is also seen from eqs. 10.16 and 10.19 that under these same conditions, the TMP and mass of cake accumulated on the membrane surface are independent of the surface renewal frequency S – a consequence of the Danckwerts age distribution (eq. 10.9) used in the analysis, which approaches a uniform distribution as $St \to 0$.

As $t \to \infty$, eqs. 10.10 and 10.11 simplify to

$$\Delta p_a(t \to \infty) = \Delta p_{\lim} = \left(\frac{K_r}{S}\right)^{p-1} \Gamma(p) + \mu J R_m \tag{10.20}$$

and

$$m_{c,a}(t \to \infty) = m_{c,\lim} = \frac{Jc_b}{S} \tag{10.21}$$

where $\Delta p_{\lim}$ and $m_{c,\lim}$ are the limiting TMP and mass of cake accumulated per unit area of membrane surface, respectively, when steady state is reached. Equations 10.20 and 10.21 show that the limiting TMP and the accumulated mass of cake decrease as the level of flow instability or turbulence, expressed by the magnitude of S, rises.

As $St \to \infty$, it can be shown that eqs. 10.14 and 10.17 reduce to

$$\frac{d\Delta p_a}{dt}(St \to \infty) = 0 \tag{10.22}$$

and

$$\frac{dm_{c,a}}{dt}(St \to \infty) = 0 \tag{10.23}$$

that is, as $St \to \infty$, the TMP and cake mass profiles level out. This flattening of the profiles will occur earlier in the process when the surface renewal frequency S is high, that is, steady state will be approached more quickly.

The four parameters of the complete model (R_m, α_0, n, and S) can be estimated in principle by the following procedure: (1) The membrane resistance R_m can be calculated from the equation

$$R_m = \frac{1}{\mu P} \tag{10.24}$$

where P is the pure-water permeance of the membrane, or it can be estimated from the experimental TMP value at the beginning of filtration. These two values of R_m should be close to one another if pore blocking of the membrane during the initial moments of filtration is negligible. (2) For a finite level of flow instability, eq. 10.16 is plotted on logarithmic coordinates using initial experimental TMP versus process time data, which will allow the parameter p, and thus n, to be estimated (see eq. 10.13). (3) For different guess values of S, $\ln[\Delta p_a(t) - \mu J R_m]$ is plotted versus $\ln[\{\Gamma(p) - \Gamma(p, St)\}/(1 - e^{-St})]$ using experimental TMP versus process time data (see eq. 10.10). The value of S which yields a magnitude of 1 for the average slope of the plot is the correct value of S. (4) K_r can be calculated from the intercept of this plot (see eq. 10.10) after which α_0 is determined from eq. 10.2. If the level of flow instability is low, eq. 10.16 will approximately hold for all values of t. For such a case, K_r and p (and hence n) can be estimated from the intercept and slope of the transient experimental TMP data plotted on logarithmic coordinates (see eq. 10.16), after which α_0 can be calculated from eq. 10.2.

10.2.2 Subsidiary Model: Incompressible Cake ($n = 0$)

When $n = 0$ (i.e., $p = 2$), eq. 10.10 reduces to

$$\Delta p_a(t) = \Delta p_{c,a}(t) + \Delta p_m = K_r\left(\frac{1}{S} - \frac{te^{-St}}{1 - e^{-St}}\right) + \mu J R_m \tag{10.25}$$

in the derivation of which the following relations have been used:

$$\Gamma(2) = 1 \tag{10.26}$$

and

$$\Gamma(2, St) = (1 + St)e^{-St} \tag{10.27}$$

Equation 10.27 is a special case of the formula (Abramowitz and Stegun, 1965)

$$\Gamma(m, y) = \Gamma(m)\left[1 + y + \frac{y^2}{2!} + \ldots \frac{y^{m-1}}{(m-1)!}\right] e^{-y} \tag{10.28}$$

where m is a positive integer.

Substituting $n = 0$ (i.e., $p = 2$) into eqs. 10.16 and 10.20 yields

$$\Delta p_a(St \to 0) = \frac{K_r}{2} t + \Delta p_m \tag{10.29}$$

and

$$\Delta p_a(t \to \infty) = \Delta p_{\lim} = \frac{K_r}{S} + \mu J R_m \tag{10.30}$$

According to eq. 10.29, the TMP will increase linearly with process time during the early stages of filtration ($t \to 0$) or for low levels of flow instability ($S \to 0$), as mentioned previously. This equation also shows that $K_r/2$ is the slope of the TMP versus process time curve during the initial moments of filtration or for small levels of flow instability. The greater the values of μ, c_b, J, and α_0, the greater will be the slope. Since K_r is proportional to J^2 (eq. 10.2), the TMP will increase as the square of the permeate flux as $t \rightarrow 0$ or $S \rightarrow 0$. For example, an increase by a factor of two in J will increase the slope by a factor of four.

The rate of buildup of the mass of cake with process time on the membrane wall is given by eq. 10.11 whereas its magnitude as $St \rightarrow 0$ or $t \rightarrow \infty$ is given by eqs. 10.19 or 10.21.

The three parameters of the subsidiary model (R_m, α_0, and S) can be determined as follows: The membrane resistance R_m is calculated as for the complete model. A value for the surface renewal frequency S is assumed, and K_r is calculated from eq. 10.30 and the experimental value of Δp_{lim}. These values of S and K_r are then substituted into eq. 10.25, whose fit to experimental transient TMP data is examined. This procedure is repeated for different (assumed) values of S until the RMS deviation between the theoretical and experimental TMP is minimized. At the conclusion of this process, the "best" value of S will be known after which α_0 can be calculated from eq. 10.2. If a value of Δp_{lim} is not available, for example, if the experiment did not reach steady state or was terminated before it was attained, K_r can be estimated by fitting eq. 10.29 to initial experimental transient TMP data, after which optimum values of S and α_0 can be determined as indicated earlier. However, this method of extrapolation for extracting K_r from initial TMP data may not accurately represent the data at large process times.

The shape of the experimental TMP profile with process time will determine the choice between the complete and subsidiary models. For convex TMP profiles that approach a plateau with the progress of process time, for example, in those situations where the effects of crossflow mitigate cake buildup on the membrane surface, as in the work of Stressmann (2008), the subsidiary model should be adequate, provided cake compressibility is not important. If the TMP profile is initially concave and then becomes convex with the advance of process time, and subsequently levels out, the complete model is recommended, which should also be used if the TMP profile is concave for all values of process time, as commonly observed in dead-end filtration.

10.3 Correlation of the Surface Renewal Model with Experimental TMP Data

10.3.1 Data of Ho and Zydney (2002)

The microfiltration of BSA solutions in a 25-mm diameter stirred ultrafiltration cell using polycarbonate track-etched (PCTE) membranes having two different porosities (3 and 10%) was investigated by Ho and Zydney (2002). The concentration of BSA in the feed solution was 2 kg m^{-3}. In the following, gentle stirring (i.e., $S \rightarrow 0$) and an experimental temperature, which was not reported in their work, of 25 °C are assumed.

Figure 10.1 compares the complete model (eq. 10.16) with the experimental TMP profiles of Ho and Zydney (2002) for a PCTE membrane having a porosity of 10% for three different permeate fluxes. The experimental TMP develops slowly at first, after which there is a rapid increase because of cake buildup on the membrane surface. The model successfully captures the concave shape of the experimental TMP profile, which is markedly sensitive to the level of the imposed permeate flux, as was anticipated earlier in the theoretical section. Values of the three model parameters (R_m, α_0, and n) are given in Table 10.1 along with RMS errors, whose average value is 11.5%. The value of R_m varies to some extent between the membranes, with an average value of 3.4×10^{10} m^{-1}. A separate clean membrane was

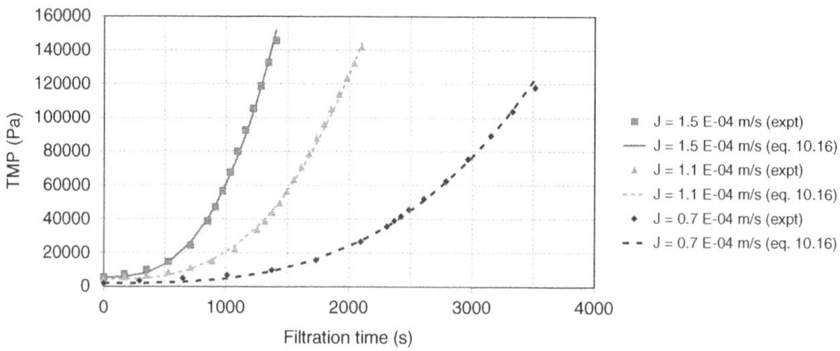

Figure 10.1 Comparison of the complete model (eq. 10.16) and experimental TMP profiles (Figure 1 of Ho and Zydney, 2002) in the microfiltration of a BSA solution with a PCTE membrane. Values of the model parameters are shown in Table 10.1, and the feed concentration is 2 kg m^{-3}. Source: Jiang and Chatterjee (2015) / with permission of John Wiley & Sons.

Table 10.1 Parameter values of the complete model (eq. 10.16) and RMS deviations between theory and experiment for the TMP data in Figure 1 of Ho and Zydney (2002).

Parameter	$J = 0.7 \times 10^{-4}$ (m s^{-1})	$J = 1.1 \times 10^{-4}$ (m s^{-1})	$J = 1.5 \times 10^{-4}$ (m s^{-1})
R_m (m^{-1})	2.84×10^{10}	3.47×10^{10}	3.89×10^{10}
α_0 (m kg^{-1})	2.49×10^{9}	2.11×10^{9}	1.75×10^{9}
n	0.67	0.65	0.65
RMS deviation (%)	16.8	9.6	8.1

Source: Jiang and Chatterjee (2015) / with permission of John Wiley & Sons.

used by Ho and Zydney (2002) in each experiment; hence, the variability in R_m can be attributed to the different surface characteristics of the individual PCTE membrane sheets. The individual values of α_0 are of the same order of magnitude, with an average value of 1.91×10^9 m kg^{-1}, whereas the values of n are quite consistent (average value equals 0.66). These are significantly different from the values of $\alpha_0 = 1.7 \pm 0.02 \times 10^{12}$ m kg^{-1} and $n = 0.78 \pm 0.01$ obtained by Ho and Zydney (2002) by fitting their five-parameter numerical model to these same data. These five parameters are the membrane resistance (whose values they did not report), resistance of a single protein aggregate, pore-blockage parameter, and the two compressibility parameters (α_0 and n). Through independent measurements, as explained later, they estimated values of α_0 and n to be 3×10^{12} m kg^{-1} and 0.82, respectively, which compare favorably with their values mentioned earlier.

The explanation for the discrepancy between the values of the compressibility parameters (α_0 and n) as reported by Ho and Zydney (2002) and those obtained by fitting the surface renewal model to their TMP data is as follows:

Using their resistance-in-series model of constant-pressure microfiltration, Ho and Zydney (2000) calculated the specific cake resistance α from the following equations:

$$J_{\text{lim}} = \frac{\Delta p}{\mu R_{\text{tot}}} \tag{10.31}$$

where

$$R_{\text{tot}} = R_{\text{m}} + \alpha m_{\text{p}} \tag{10.32}$$

Here, m_p is the (steady-state) mass of the protein (i.e., cake) layer deposit per unit area of the membrane surface, and R_{tot} is the total resistance (membrane plus cake). By measuring the steady-state flux of a saline solution through a heavily fouled membrane and the difference in weights of the clean and fouled membrane (for estimating m_p), Ho and Zydney (2000) calculated values of α at different TMP values from eqs. 10.31 and 10.32. These values were plotted according to eq. 10.3 on logarithmic coordinates from which α_0 and n were estimated. Assuming that the resistance of the cake is much greater than that of the membrane in eq. 10.32 and using eqs. 10.3 and 10.31 gives the approximate expression

$$J_{\text{lim}} \approx \frac{\Delta p^{1-n}}{\mu \alpha_0 m_{\text{p}}} \tag{10.33}$$

which shows that J_{lim} is proportional to Δp^{1-n}, assuming that all other quantities in eq. 10.33 are constant.

However, according to the constant pressure surface renewal model of microfiltration (see eq. 9.30 in Chapter 9 or eq. 21 in Idan and Chatterjee, 2015)

$$J_{\text{lim}} = \sqrt{\frac{\pi S}{2\mu c_{\text{b}} \alpha_0}} \Delta p^{\frac{1-n}{2}} \tag{10.34}$$

that is, J_{lim} is proportional to $\Delta p^{\frac{1-n}{2}}$, assuming that all other quantities in eq. 10.34 remain invariant.

Thus, the theoretical frameworks of the resistance-in-series and surface renewal models of membrane filtration are radically different from one another,[3] which is analogous to the difference between the film and penetration or surface renewal models of gas absorption in a liquid, as discussed in Chapter 1. This is a possible cause for the discrepancy between the values of α_0 and n reported here and those obtained by Ho and Zydney (2002). Therefore, unless the specific cake resistance α can be directly measured as a function of the applied pressure drop Δp, say by nonintrusive measurements of the cake structure, under different compressive loads, it is best to consider α_0 and n as relative (i.e., as opposed to absolute) quantities since their values will depend upon the theoretical lens used to analyze the experimental data.

The predicted cake-mass buildup as a function of process or filtration time (calculated from eq. 10.19), corresponding to the three permeate fluxes of Figure 10.1, is shown in Figure 10.2. The buildup occurs in a linear manner with time, and the higher the permeate flux, the greater its value is at a given value of the process time.

Figure 10.3 shows comparisons between the complete model and the experimental TMP profiles of Ho and Zydney (2002) for the PCTE and PCTE-L (porosity of 3%) membranes at an imposed permeate flux of 1.3×10^{-4} m s^{-1}, and Table 10.2 reports values of the model parameters and RMS errors (average RMS error = 10.9%). The values of α_0 and n for the PCTE membrane are 4.2×10^8 m kg^{-1} and 0.76, respectively, which are considerably different from the (average) values

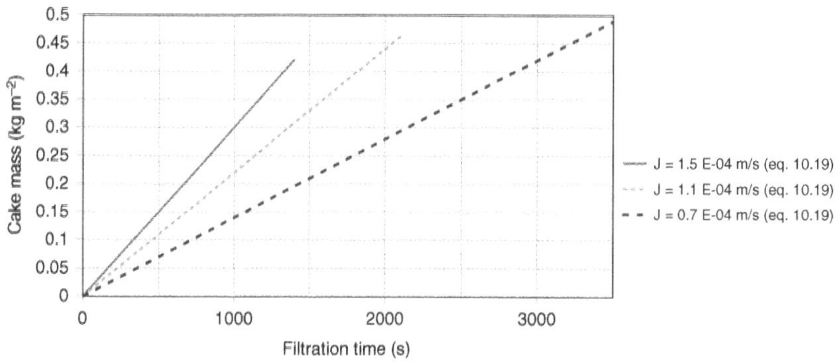

Figure 10.2 Predicted cake buildup (eq. 10.19) for the experiments of Ho and Zydney (2002) in the microfiltration of a BSA Solution with a PCTE membrane. Values of the model parameters are shown in Table 10.1, and the feed concentration is 2 kg m^{-3}. Source: Jiang and Chatterjee (2015) / with permission of John Wiley & Sons.

3 See also Section 9.2.2 of Chapter 9.

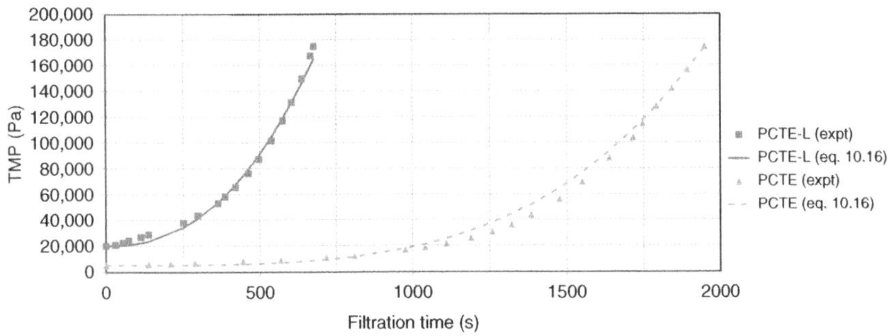

Figure 10.3 Comparison of the complete model (eq. 10.16) and experimental TMP profiles (Figure 2 of Ho and Zydney, 2002) in the microfiltration of a BSA solution with PCTE-L and PCTE membranes. Values of the model parameters are shown in Table 10.2. The feed concentration and permeate flux are 2 kg m^{-3} and 1.3×10^{-4} m s^{-1}, respectively. Source: Jiang and Chatterjee (2015) / with permission of John Wiley & Sons.

Table 10.2 Parameter values of the complete model (eq. 10.16) and RMS deviations between theory and experiment for the TMP data in Figure 2 of Ho and Zydney (2002).

Parameter	PCTE-L	PCTE
R_m (m^{-1})	1.71×10^{11}	4.37×10^{10}
α_0 (m kg^{-1})	1.30×10^{10}	4.42×10^{8}
n	0.57	0.76
RMS deviation (%)	7.6	14.1

Source: Jiang and Chatterjee (2015) / with permission of John Wiley & Sons.

of 1.91×10^9 m kg^{-1} and 0.66 estimated earlier for this type of membrane – see Jiang and Chatterjee (2015) for an explanation.

The predicted growth of the mass of cake with filtration time on the surface of the membrane (PCTE and PCTE-L) is plotted in Figure 10.4 for a permeate flux of 1.3×10^{-4} m s^{-1}.

10.3.2 Data of Kovalsky *et al.* (2009)

The experimental TMP data of Kovalsky *et al.* (2009) for filtration of a 10 kg m^{-3} yeast suspension in a stirred cell (filtration area = 19.6 cm^2; stirring speed = 5 RPM) are compared against the complete model (eq. 10.16) in Figure 10.5 for three different levels of the imposed permeate flux. A temperature

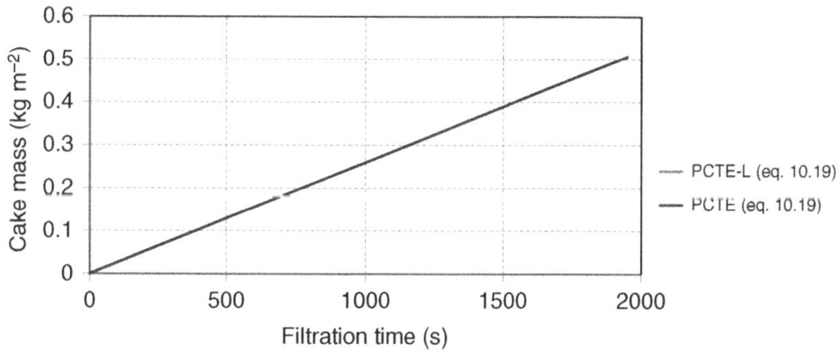

Figure 10.4 Predicted cake buildup (eq. 10.19) for the experiments of Ho and Zydney (2002) in the microfiltration of a BSA solution with PCTE-L and PCTE membranes. Values of the model parameters are shown in Table 10.2. The feed concentration and permeate flux are 2 kg m^{-3} and 1.3×10^{-4} m s^{-1}, respectively. Source: Jiang and Chatterjee (2015) / with permission of John Wiley & Sons.

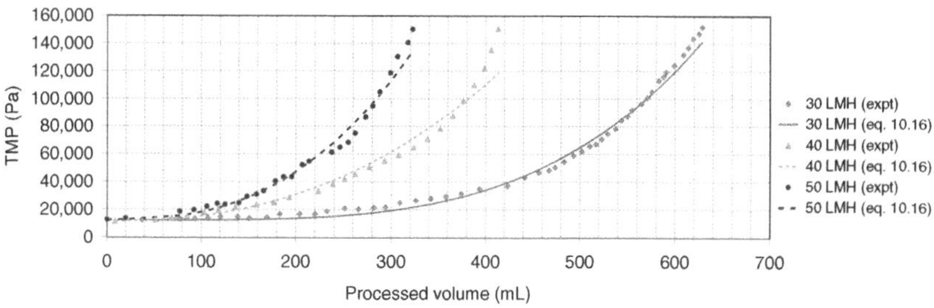

Figure 10.5 Comparison of the complete model (eq. 10.16) and experimental TMP profiles (Figure 5 of Kovalsky *et al.*, 2009) in the microfiltration of a yeast suspension at pH 2.7. Values of the model parameters are provided in Table 10.3. Feed concentration = 10 kg m^{-3} and LMH = L m^{-2} h^{-1}. Source: Jiang and Chatterjee (2015) / with permission of John Wiley & Sons.

of 25 °C is assumed in the theoretical calculations. It is observed once again that the model, whose parameters are given in Table 10.3, is able to resolve the concave shape of the experimental TMP profile fairly well (average RMS error = 9.2%; see Table 10.3). The discrepancy between theory and experiment near the end of an experimental run is likely due to creep and consolidation effects, which were considered in the numerical model of Kovalsky *et al.* (2009).

The linear growth in the mass of cake with filtration time is shown in Figure 10.6.

10.3.3 Data of Miller *et al.* (2014)

Miller *et al.* (2014) investigated the use of unmodified and surface-modified membranes for constant flux CFMF. Their work contains a detailed description of the

Table 10.3 Parameter values of the complete model (eq. 10.16) and RMS deviations between theory and experiment for the TMP data shown in Figure 5 of Kovalsky *et al.* (2009).

Parameter	30 LMH	40 LMH	50 LMH
R_m (m^{-1})	1.63×10^{12}	1.17×10^{12}	1.02×10^{12}
α_0 ($m\,kg^{-1}$)	1.25×10^{9}	1.06×10^{10}	6.68×10^{9}
n	0.75	0.58	0.62
RMS deviation (%)	10.5	8.8	8.4

LMH = $L\,m^{-2}\,h^{-1}$.
Source: Jiang and Chatterjee (2015) / with permission of John Wiley & Sons.

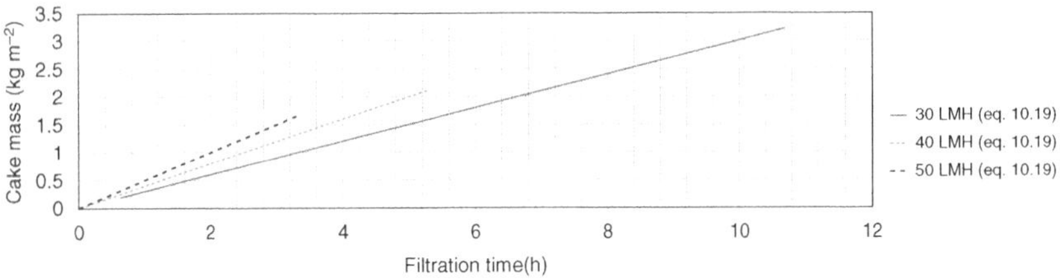

Figure 10.6 Predicted cake buildup (eq. 10.19) for the experiments of Kovalsky *et al.* (2009) in the microfiltration of a yeast suspension at pH 2.7. Values of the model parameters are provided in Table 10.3. Feed concentration = 10 kg m^{-3} and LMH = $L\,m^{-2}\,h^{-1}$. Source: Jiang and Chatterjee (2015) / with permission of John Wiley & Sons.

materials used and experimental conditions and procedures and also a schematic of their constant flux CFMF system. A brief overview of their experimental work is provided next.

The base (i.e., unmodified) ultrafiltration (UF) membrane material used by Miller *et al.* (2014) was (hydrophobic) polysulfone with molecular weight cutoffs of 10 and 20 kDa, which were designated as PS-10 and PS-20, respectively. Two additional hydrophilic surface-modified membranes, called PDA-modified and PDA75-modified, were produced from PS-20. Polyethylene glycol (PEG) was grafted onto the surface of some of the PDA-modified sheets; such membranes were referred to as PDA-g-PEG- modified. The chief objective of this surface modification was to produce a hydrophilic surface on the base hydrophobic membrane for attracting water molecules, which would function as a buffer between hydrophobic foulants and the membrane surface, thereby restricting their adsorption on it and also within the pores of the membrane. The steric hindrance offered

Table 10.4 Characteristics of the membranes used by Miller *et al.* (2014).

Membrane	Pure-water permeance (LMH bar^{-1})	Effective pore radius (nm)
unmodified (PS-20)	900 ± 200	4.2
PDA-modified (PS-20)	700 ± 100	3.7
PDA-g-PEG-modified (PS-20)	570 ± 70	3.3
PDA75-modified (PS-20)	570	—
unmodified (PS-10)	570	—

Source: Jiang and Chatterjee (2015) / with permission of John Wiley & Sons.

by long PEG chains extending from the membrane surface is also thought to further reduce the interaction between the surface and potential foulants. Table 10.4 shows the pure water permeance P (measured by dead-end filtration) and estimated pore radius of the unmodified, PDA-modified, and PDA-g-PEG-modified PS-20 membranes and also the P-value of the unmodified PS-10 membrane. It is seen from this table that surface modification decreased the pure water permeance of the PDA-modified and PDA-g-PEG-modified membranes by 22 and 37%, respectively and also reduced the effective pore radius compared to the unmodified (PS-20) membrane.

The feed solution used in the experiments of Miller *et al.* (2014) was a 1500 ppm (1.5 kg m^{-3}) emulsion of soybean oil in water that had an average oil droplet size of 1.4 μm, with almost all droplets lying in the size range of 0.8–3.0 μm. Thus, the average droplet size was two orders of magnitude greater than the effective pore radius of the unmodified (PS-20) membrane. The feed temperature and axial velocity were maintained at 25 °C and 0.18 m s^{-1} (Reynolds number = 1000), respectively, in all the fouling experiments. The feed pressure was 2.1 barg (30 psig) was kept constant, and the permeate and retentate were recycled back to the feed tanks. The membrane filtration area was 19.4 cm^2. A constant permeate flux was maintained by means of feedback control of a peristaltic pump installed on the permeate line. As the membrane fouled during an experimental run, the pressure on the permeate side of the membrane decreased, causing the TMP to rise. In cases of severe fouling, the pressure on the permeate side dropped to atmospheric pressure, and the experiment was terminated. Membrane rejection was estimated by measuring the TOC (total organic carbon) content of the feed and permeate solutions. The membrane rejection values were quite high and were in the range of 96.5–99.1%, with most of them lying above 98%. Five constant permeate flux levels were used in the experiments – 25, 40, 55, 70, 85, and 100 LMH. The experimental parameters of Miller *et al.* (2014) are given in Table 10.5.

Table 10.5 Experimental parameters of Miller *et al.* (2014).

Parameter	Description or value
base membrane type	UF polysulfone with molecular weight cutoffs of 10 and 20 kDa
type of feed suspension	soybean-oil emulsion in water
droplet average size in feed suspension	1.4 μm
feed concentration	1.5 kg m^{-3}
feed pressure	2.1 barg (30 psig)
membrane filtration area	19.4 cm^2
feed axial velocity	0.18 m s^{-1} (Reynolds number = 1000)
experimental temperature	25 °C
constant flux levels	25, 40, 55, 70, 85, and 100 LMH

Source: Jiang and Chatterjee (2015) / with permission of John Wiley & Sons.

It was mentioned earlier that the feed suspension of Miller *et al.* (2014) contained soybean oil droplets whose average size (1.4 μm) was much larger than the effective pore radius of the unmodified (PS-20) membrane (4.2 nm), which was reduced further due to surface modification (see Table 10.4). It can therefore be speculated that there was minimal or negligible pore blockage of the membranes used in their experiments unless there was significant droplet breakage into much finer sizes or droplet deformation due to shearing forces and the applied TMP and subsequent penetration into the pores of the membrane, especially at high permeate fluxes.[4] Thus, it can be postulated that the primary cause for the increase in TMP with process time observed in the experiments of Miller *et al.* (2014) is because of cake buildup on the membrane surface. It is therefore assumed that the membrane resistance R_m can be calculated by eq. 10.24 using values of the (average) membrane pure-water permeance P shown in Table 10.4 and a value of 8.98×10^{-4} kg m^{-1} s^{-1} for the viscosity of water at 25 °C (McCabe *et al.*, 1993). This value is also used for the viscosity of the filtrate in the calculations, whose results are presented in what follows. Values of R_m for the three membranes are reported in Table 10.6. Miller *et al.* (2014) state that "…all of the fouling experiments started, to good approximation, essentially instantaneously at a V/A value of zero." V/A is the ratio of the total or cumulative volume of permeate at process time t to the filtration area and is a measure of process time, which can be obtained by dividing this ratio (reported by Miller *et al.*, 2014 in cm) by the permeate flux.

4 Kirschner *et al.* (2019) cite prior literature according to which emulsified oil droplets, despite having a diameter larger than that of the membrane pores, can nevertheless deform and squeeze through them at high permeate fluxes.

Table 10.6 Parameter values of the subsidiary model (eq. 10.25) and RMS deviations between theory and experiment for the TMP data shown in Figure 2a of Miller *et al.* (2014).

Parameter	Unmodified (PS-20)	PDA-modified (PS-20)	PDA-*g*-PEG-modified (PS-20)
R_m (m^{-1})	4.45×10^{11}	5.73×10^{11}	7.03×10^{11}
S (s^{-1})	4.2×10^{-3}	8.5×10^{-3}	5.7×10^{-3}
α_0 (m kg^{-1})	9.07×10^{13}	2.53×10^{14}	4.23×10^{14}
n	0	0	0
RMS deviation (%)	2.9	5.8	6.8

Source: Jiang and Chatterjee (2015) / with permission of John Wiley & Sons.

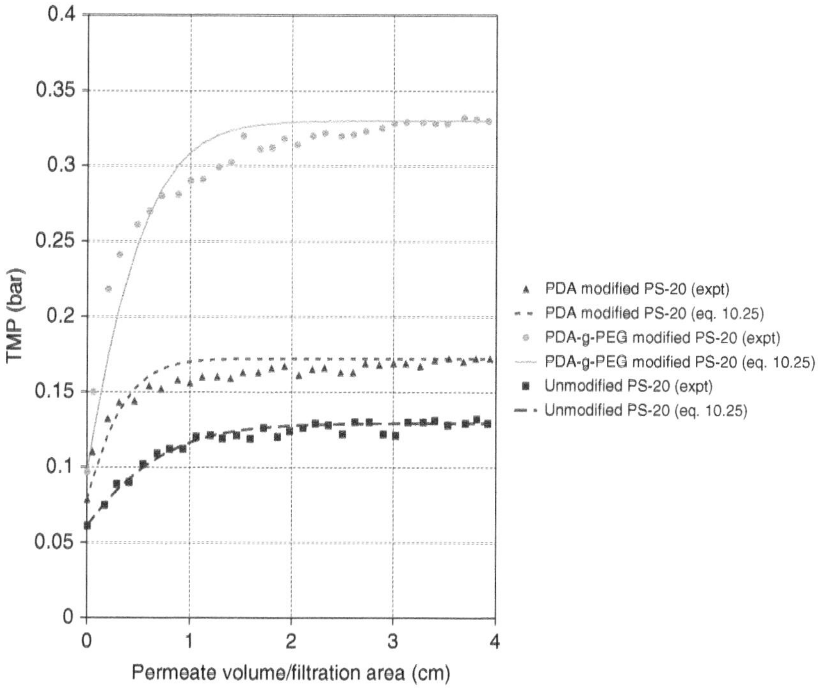

Figure 10.7 Comparison of the theoretical (subsidiary model; eq. 10.25) and experimental (Figure 2a of Miller *et al.*, 2014) TMP profiles in the microfiltration of a soybean-oil suspension. Values of the experimental and model parameters are given in Tables 10.4–10.6. Permeate flux = 55 LMH. Source: Jiang and Chatterjee (2015) / with permission of John Wiley & Sons.

Figure 10.7 compares predictions of the subsidiary surface renewal model (eq. 10.25) with experimental TMP data of Miller *et al.* (2014) extracted from Figure 2a of their paper for constant flux CFMF runs of soybean oil emulsion at a permeate flux of 55 LMH ($J = 1.528 \times 10^{-5}$ m s^{-1}) using the three membranes mentioned earlier. The following observations can be gleaned from this figure: (1) For all membranes, the experimental TMP curve increases rapidly with process time at the beginning and then at an extremely slow rate, eventually approaching a plateau (i.e., steady state), as mentioned by Miller *et al.* (2014), that is, it is convex-shaped.[5] (2) The theoretical TMP curves start at different levels of Δp_{m} (pressure drop across the membrane; determined from eq. 10.5) due to different values of the pure-water permeance P of the membranes (Table 10.4). (3) The unmodified (PS-20) and PDA-g-PEG-modified membranes show the lowest and highest TMPs, respectively, while those of the PDA-modified membrane lie in between. (4) The unmodified (PS-20) and PDA-modified membranes have flatter TMP profiles compared to the PDA-g-PEG-modified membrane, exhibiting a faster approach to steady state. (5) The fit of the model (eq. 10.25) to the experimental TMP data is quite good in the case of the unmodified membrane and inferior for the modified membranes.

For the unmodified (PS-20) membrane, the average RMS deviation between theoretical and experimental TMPs is 2.9%, with the surface renewal frequency S being estimated at 4.2×10^{-3} s^{-1}, which is comparable to the values of S reported in the literature by other researchers (Koltuniewicz and Noworyta, 1994; Chatterjee, 2010; Hasan *et al.*, 2013). The compressibility parameter α_0 (specific cake resistance for a unit value of TMP), which measures the resistance offered by the accumulated material on the membrane surface to the flow of permeate and depends upon the packing density and nature of the cake, is estimated to be 9.07×10^{13} m kg^{-1}. The mass of cake accumulated per unit area of the membrane surface at steady state, $m_{\text{c,lim}}$, is calculated by eq. 10.21 to be 5.46×10^{-3} kg m^{-2}. Multiplying this value with the membrane filtration area of 19.4 cm^2 yields the total mass of solids accumulated on the membrane surface at steady state of 10.6 mg. For the PDA-modified membrane, the values are as follows: $S = 8.5 \times 10^{-3}$ s^{-1}, $\alpha_0 = 2.53 \times 10^{14}$ m kg^{-1}, and $m_{\text{c,lim}} = 2.7 \times 10^{-3}$ kg m^{-2}. The mass of solids on the membrane surface at steady state equals 5.2 mg, whereas the RMS deviation between theory and experiment is 5.8%, which is higher than the value of 2.9% for the unmodified membrane. For the PDA-g-PEG-modified membrane, $S = 5.7 \times 10^{-3}$ s^{-1}, $\alpha_0 = 4.23 \times 10^{14}$ m kg^{-1} and $m_{\text{c,lim}} = 4 \times 10^{-3}$ kg m^{-2}. The limiting mass of solids on the membrane surface and RMS error are 7.8 mg and 6.8%, respectively. Values of the above model parameters, along with RMS

5 In fitting the model to the data, it is assumed that the last experimental TMP value shown in the figure (at $V/A \approx 3.9$ cm) is equal to Δp_{lim}, the limiting or steady-state TMP in the model.

deviations between theory and experiment, are collected together in Table 10.6 for all three membranes.[6]

According to Miller *et al.* (2014), the higher TMP in the case of the modified membranes, compared to that for the unmodified membrane, is a result of increased mass-transfer resistance due to surface treatment. However, this can only explain the greater initial resistance of the modified membranes (see Table 10.6), which is manifested in an increased Δp_m at the start of filtration (Figure 10.7). The subsequent increase in the TMP with process time for all three membranes is attributed to cake formation in the surface renewal model (eq. 10.25), and the different rates of TMP increase for the three membranes are due to different α_0 values. As Table 10.6 shows, the values of R_m for the PDA-modified and PDA-g-PEG-modified membranes are 29 and 58% larger, while the α_0 values of these membranes are 179 and 367% higher, respectively, compared to the corresponding values for the unmodified (PS-20) membrane. This explains, to some extent, the high TMP of the PDA-g-PEG-modified membrane, the intermediate TMP of the PDA-modified membrane, and the low TMP of the unmodified (PS-20) membrane.

Figure 10.8 exhibits the predicted age-averaged cake-mass profile (calculated from eq. 10.11) for the three membranes. The curves in this figure start from a value of zero and, as filtration progresses, approach the corresponding steady-state values mentioned earlier.[7] Although the axial velocity of the feed suspension was kept at the same level (0.18 m s^{-1}) in the experimental runs of Miller *et al.* (2014), Table 10.6 shows that the values of the surface renewal frequency S, which depend upon the prevailing hydrodynamic conditions at the membrane surface, are different for the three membranes. It can be speculated that these differences are due to differences in membrane surface roughness among the three membranes, since roughness will have an effect on microscale hydrodynamics. However, no firm conclusion can be drawn because of the large variance of the membrane permeance about its mean value for the three membranes, as can be observed in Table 10.4. It can also be concluded from this table and Figure 10.8 that the higher the value of S, the lower is the cake buildup per unit membrane surface area.

The results discussed thus far were concerned with the performance of membranes having different pure-water permeances. In order to compare membrane performance on an equivalent basis, Miller *et al.* (2014) carried out experimental runs with the PS-10 and PDA75-modified (PS-20) membranes that had a pure-water permeance of 570 LMH bar^{-1}, that is, the same as that of the

6 A value of $n = 0$ has been assumed since, in a single pass, the permeate flux is much smaller compared to the crossflow velocity.

7 Such convex-type predicted cake buildup curves during the microfiltration of BSA and CHO cell culture broth solutions with 0.45 μm membranes at a constant permeate flux of 30 LMH can be seen in Figure 3.4 of Stressmann (2008).

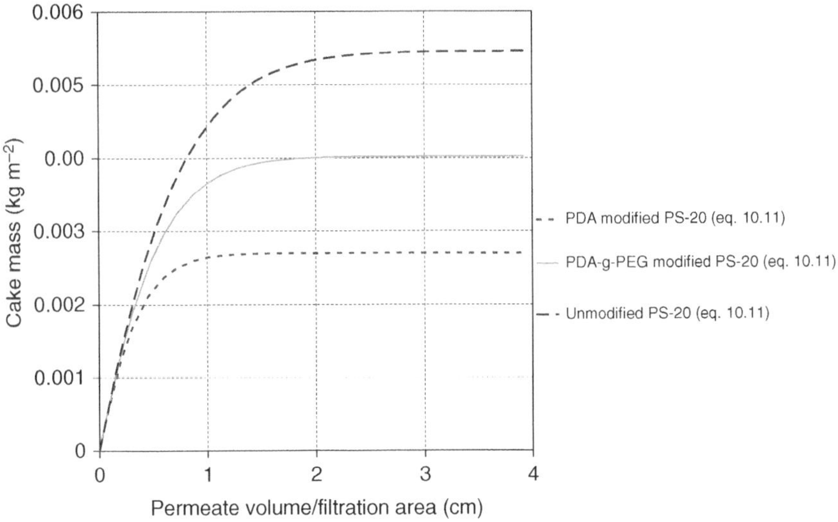

Figure 10.8 Predicted cake buildup (eq. 10.11) in the experiments of Miller *et al.* (2014) during the microfiltration of a soybean-oil suspension. Values of the experimental and model parameters are provided in Tables 10.4–10.6. Permeate flux = 55 LMH. Source: Jiang and Chatterjee (2015) / with permission of John Wiley & Sons.

PDA-g-PEG-modified (PS-20) membrane. Jiang and Chatterjee (2015) compared the subsidiary surface renewal model (eq. 10.25) with the experimental TMP data of Miller *et al.* (2014), extracted from Figure 6 of their paper, and found a fairly good fit of the model to the convex-type experimental TMP profile for the unmodified membrane but there were substantial errors for the modified membranes. The subsidiary model was unable to fully capture the initially convex shape and subsequent slow rise of the experimental TMP for the modified membranes. The inferior fit of the subsidiary model to the experimental TMP data for the modified membranes is likely due to hydrophilicity, steric hindrance, and other surface (e.g., electrochemical) effects that resulted from modifying the membrane surface, which are not explicitly accounted for in the surface renewal model. Also, in contrast to Figure 10.7, the unmodified PS-10 membrane had a significantly higher TMP compared to the surface-modified membranes – their TMP profiles were close to one another and much flatter. The superior performance of the modified membranes was attributed by Miller *et al.* (2014) to the beneficial effects of hydrophilicity and steric hindrance, which retarded and countered the accumulation of solids on the membrane surface caused by the transverse flow of liquid toward it.

Miller *et al.* (2014) also reported TMP data at higher permeate fluxes of 70 and 85 LMH (see Figures 2b and 2c in their paper) when cake-compressibility effects are expected to become significant. Although the experimental TMP curves for the PDA-modified and PDA-g-PEG modified membranes still exhibited convex-type behavior for a permeate flux of 70 LMH, that for the unmodified (PS-20) membrane was initially convex-shaped until a certain value of the V/A ratio was reached, after which it became concave and rose sharply. At the flux of 85 LMH, the TMP curves for all three membranes exhibited this latter type of behavior with the curves crisscrossing one another. According to Miller *et al.* (2014), this behavior occurs when the threshold flux of the membrane is crossed, which brings about the onset of intense fouling.

The chief feature of the surface renewal model developed in this chapter, and also its constant TMP counterparts presented in Chapters 8 and 9, is its ability to explicitly account for flow instabilities generated at the membrane surface (due to membrane roughness, presence of spacers, stirring, etc.) through the hydrodynamic parameter, S. The model has the ability to correlate both convex- and concave-shaped experimental TMP profiles observed experimentally but may not be suitable for representing more complex TMP behavior. The assumption that pore blocking of the membrane is either negligible or occurs only during the initial moments of filtration is another limitation of the model for there is evidence in the literature, which suggests that fouling can transition from one mechanism to another with the progress of filtration. Membrane pore blocking models (e.g., complete, intermediate, and standard) have been used in the literature, either singly or in combination with the mechanism of cake filtration, to represent membrane fouling. The incorporation of such models into the surface renewal model and also effects like hydrophilicity and steric hindrance, which arise from surface modification of the membrane, would make the model more widely applicable.

Notation

A	filtration area of membrane (cm^2 or m^2)
c_b	mass of solids deposited in the filter per unit volume of filtrate (approximately equal to the concentration of solids in the feed or bulk liquid; $kg\,m^{-3}$)
$f(\tau, t)$	transient Danckwerts age distribution function; given by eq. 10.9 (Hz or s^{-1})
J	constant permeate flux ($m\,s^{-1}$)
$J_a(t)$	age-averaged permeate flux at process time t ($m\,s^{-1}$)
K_r	defined by eq. 10.2
m	positive integer (1, 2, 3, …)
m_c	mass of solids accumulated per unit area of membrane wall in a surface element of age τ ($kg\,m^{-2}$)

$m_{c,a}$	age-averaged mass of cake accumulated per unit area of membrane surface at process time t (kg m^{-2})
$m_{c,lim}$	limiting or steady-state cake mass per unit area of membrane surface (kg m^{-2})
m_p	steady-state mass of protein layer deposit per unit area of membrane surface (kg m^{-2})
n	cake compressibility coefficient
p	defined by eq. 10.13
P	pure-water permeance of membrane (L m^{-2} h^{-1} bar^{-1} or m^2 s kg^{-1})
R_m	hydraulic resistance of membrane (m^{-1})
R_{tot}	total resistance (membrane plus cake) defined by eq. 10.32 (m^{-1})
S	surface renewal frequency (Hz or s^{-1})
t	filtration or process time (s)
V	cumulative or total volume of permeate at process time t (m^3)
x	parameter of $\Gamma(x, y)$
y	parameter of $\Gamma(x, y)$

Greek Letters

α	specific cake resistance (m kg^{-1})
α_0	cake compressibility parameter in eq. 10.3 (m kg^{-1})
$\Gamma(x, y)$	incomplete gamma function; defined by eq. 10.12
Δp	total transmembrane pressure drop in a surface element of age τ (Pa)
Δp_a	age-averaged total transmembrane pressure drop at process time t (Pa)
Δp_c	pressure drop across cake in a surface element of age τ (Pa)
$\Delta p_{c,a}$	age-averaged pressure drop across cake at process time t (Pa)
Δp_{lim}	limiting or steady-state total transmembrane pressure drop (Pa)
Δp_m	pressure drop across membrane (Pa)
λ	variable of integration in eq. 10.12
μ	viscosity of permeate (kg m^{-1} s^{-1})
τ	age or residence time of a liquid element at membrane–liquid interface (s)

Acronyms

BSA	bovine serum albumin
CHO	Chinese hamster ovary
CFMF	crossflow membrane filtration
LMH	liters per square meter per hour

PCTE	polycarbonate track etched
PCTE-L	polycarbonate track-etched low pore density
PDA	polydopamine
PEG	polyethylene glycol
PS	polysulfone
PVDF	polyvinylidene fluoride
RMS	root mean square
RO	reverse osmosis
RPM	revolutions per minute
TMP	transmembrane pressure drop
TOC	total organic carbon
UF	ultrafiltration

Homework Exercises

i) Derive eqs. 10.10 and 10.11 of the complete surface renewal model.
ii) Show that eq. 10.14 reduces to eq. 10.15 as $St \rightarrow 0$.
iii) Show that eq. 10.17 reduces to eq. 10.18 as $St \rightarrow 0$.
iv) Show that as $t \rightarrow \infty$, eqs. 10.10 and 10.11 approach the limiting forms given by eqs. 10.20 and 10.21, respectively.
v) Derive eq. 10.25 of the subsidiary surface renewal model.
vi) A crossflow microfiltration cell has been proposed for generating a constant permeate flow rate of $1000\,\mathrm{L\,h^{-1}}$ from an aqueous yeast suspension of concentration $2\,\mathrm{g\,L^{-1}}$. A maximum total transmembrane pressure drop of 200 kPa is to be reached at the end of a filtration run of 1 h, and the operating temperature will be 20 °C. Assuming low flow instability and values of α_0 (cake compressibility parameter) and R_m (membrane resistance) of $5\times10^{14}\ \mathrm{m\,kg^{-1}}$ and $1\times10^{12}\ \mathrm{m^{-1}}$, respectively, develop plots showing the required membrane filtration area ($\mathrm{m^2}$), permeate flux ($\mathrm{L\,m^{-2}\,h^{-1}}$), and mass of solids accumulated on the membrane surface ($\mathrm{g\,m^{-2}}$) as functions of the compressibility coefficient n varying from 0 to 0.8. Provide a brief discussion of your results.
vii) A batch (i.e., no feed inflow or retentate outflow) stirred microfiltration cell, operating at constant flux, is to be employed for concentrating a polymer suspension. Following the lines of reasoning given at the beginning of Section 10.2, develop the governing equations that describe the evolution of the total transmembrane pressure drop and mass of cake accumulated per unit area of the membrane surface as a function of process time. State the chief assumptions used in your derivation.

viii) A gently stirred microfiltration module containing a flat-sheet porous membrane and having feed, permeate, and retentate ports is to be used for removing cells and cell debris from a fermentation broth containing a valuable soluble product. The feed and permeate flows to and from the module will be 1 and $0.8\,\mathrm{L\,h^{-1}}$, respectively, and the concentration of cells in the feed stream is expected to be $1\,\mathrm{g\,L^{-1}}$. The maximum total transmembrane pressure drop (TMP) to be reached at the end of a filtration run is 100 kPa, and the system will operate at a temperature of 25 °C. The values of α_0 (cake compressibility parameter) and R_m (membrane resistance) have been estimated to be $5 \times 10^{14}\ \mathrm{m\,kg^{-1}}$ and $1 \times 10^{12}\ \mathrm{m^{-1}}$, respectively. Develop plots showing the required membrane filtration area ($\mathrm{m^2}$), maximum operating filtration time (h), and mass of solids accumulated on the membrane surface ($\mathrm{g\,m^{-2}}$) at the end of a filtration run as functions of the permeate flux for fluxes of 5, 10, 15, and $20\,\mathrm{L\,m^{-2}\,h^{-1}}$ and values of the cake compressibility coefficient n of 0 and 0.1. Also, plot the TMP evolution with process time, state your assumptions, and discuss your results briefly.

References

Abramowitz, M. and Stegun, I.A. (eds.) (1965). *Handbook of mathematical functions—with formulas, graphs, and mathematical tables*. New York: Dover Publications.

Chatterjee, S.G. (2010). "On the use of the surface-renewal concept to describe cross-flow ultrafiltration," *Indian Chemical Engineer*, 52(3), pp. 179–193. Available at: http://dx.doi.org/10.1080/00194506.2010.547927.

Hasan, A., Peluso, C.R., Hull, T.S., Fieschko, J. and Chatterjee, S.G. (2013). "A surface-renewal model of cross-flow microfiltration," *Brazilian Journal of Chemical Engineering*, 30(1), pp. 167–186. Available at: http://dx.doi.org/10.1590/S0104-66322013000100019.

Ho, C.-C. and Zydney, A.L. (2000). "A combined pore blockage and cake filtration model for protein fouling during microfiltration," *Journal of Colloid and Interface Science*, 232(2), pp. 389–399. Available at: http://dx.doi.org/10.1006/jcis.2000.7231.

Ho, C.-C. and Zydney, A.L. (2002). "Transmembrane pressure profiles during constant flux microfiltration of bovine serum albumin," *Journal of Membrane Science*, 209(2), pp. 363–377. Available at: http://dx.doi.org/10.1016/S0376-7388(02)00282-X.

Idan, G. and Chatterjee, S.G. (2015). "Application of a surface-renewal model to permeate-flux data of constant-pressure cross-flow microfiltration with Dean vortices," *Brazilian Journal of Chemical Engineering*, 32(2), pp. 609–627. Available at: http://dx.doi.org/10.1590/0104-6632.20150322s00003417.

Jiang, S. and Chatterjee, S.G. (2015). "A surface-renewal model for constant flux cross-flow microfiltration," *Journal of Applied Polymer Science*, 132(21). Available at: http://dx.doi.org/10.1002/app.41778.

Kirschner, A.Y., Cheng, Y.-H., Paul, D.R., Field, R.W. and Freeman, B.D. (2019). "Fouling mechanisms in constant flux crossflow ultrafiltration," *Journal of Membrane Science*, 574, pp. 65–75. Available at: http://dx.doi.org/10.1016/j.memsci.2018.12.001.

Koltuniewicz, A. and Noworyta, A. (1994). "Dynamic properties of ultrafiltration systems in light of the surface renewal theory," *Industrial & Engineering Chemistry Research*, 33(7), pp. 1771–1779. Available at: http://dx.doi.org/10.1021/ie00031a016.

Kovalsky, P., Bushell, G. and Waite, T.D. (2009). "Prediction of transmembrane pressure build-up in constant flux microfiltration of compressible materials in the absence and presence of shear," *Journal of Membrane Science*, 344(1–2), pp. 204–210. Available at: http://dx.doi.org/10.1016/j.memsci.2009.08.005.

McCabe, W.L., Smith, J.C. and Harriott, P. (1993). *Unit operations of chemical engineering*. Fifth edition. New York: McGraw-Hill.

Miller, D.J., Kasemset, S., Wang, L., Paul, D.R. and Freeman, B.D. (2014). "Constant flux crossflow filtration evaluation of surface-modified fouling-resistant membranes," *Journal of Membrane Science*, 452, pp. 171–183. Available at: http://dx.doi.org/10.1016/j.memsci.2013.10.037.

Miller, D.J., Paul, D.R. and Freeman, B.D. (2013). "A crossflow filtration system for constant permeate flux membrane fouling characterization," *Review of Scientific Instruments*, 84(3). Available at: http://dx.doi.org/10.1063/1.4794909.

Peles, J., Cacace, B., Carbrello, C., Giglia, S. and Zydney, A.L. (2023). "Protein fouling during constant-flux virus filtration: Mechanisms and modeling," *Biotechnology and Bioengineering*, 120(11), pp. 3357–3367. Available at: http://dx.doi.org/10.1002/bit.28511.

Stressmann, M. (2008). *Membrane fouling in constant permeate flux cross-flow microfiltration of biological solutions*. Waterloo, Ontario: PhD thesis: University of Waterloo.

Zhang, W. and Chatterjee, S.G. (2015). "Influence of residence time distribution on a surface-renewal model of constant-pressure cross-flow microfiltration," *Brazilian Journal of Chemical Engineering*, 32(1), pp. 139–154. Available at: http://dx.doi.org/10.1590/0104-6632.20150321s00003129.

Index

Applications of the Surface Renewal Model of Mass Transfer, First Edition.
Hengshuo Huang and Siddharth G. Chatterjee.

Companion website: www.wiley.com/go/huang/surfacerenewalmodel

Printed and bound by CPI Group (UK) Ltd, Croydon, CR0 4YY

17/09/2026

14964232-0004